Petroleum, Industry and Governments: A Study of the Involvement of Industry and Governments in the Production and Use of Petroleum

SECOND EDITION

Kluwer Law International

Petroleum, Industry and Governments: A Study of the Involvement of Industry and Governments in the Production and Use of Petroleum

SECOND EDITION

Bernard G. Taverne

Wolters Kluwer
Law & Business

AUSTIN BOSTON CHICAGO NEW YORK THE NETHERLANDS

Published by:
Kluwer Law International
P.O. Box 316
2400 AH Alphen aan den Rijn
The Netherlands
Website: www.kluwerlaw.com

Sold and distributed in North, Central and South America by:
Aspen Publishers, Inc.
7201 McKinney Circle
Frederick, MD 21704
United States of America
Email: customer.care@aspenpubl.com

Sold and distributed in all other countries by:
Turpin Distribution Services Ltd.
Stratton Business Park
Pegasus Drive, Biggleswade
Bedfordshire SG18 8TQ
United Kingdom
Email: kluwerlaw@turpin-distribution.com

ISBN 978-90-411-2663-4

International Energy and Resources Law and Policy Series

VOLUME 15

The titles published in this series are listed at the end of this volume.

Preface to the First Edition

Petroleum, Industry and Governments describes and discusses the main aspects of the involvement of industry and governments with the production and use of petroleum. Within this setting this book gives an account of the technical and environmental aspects of the production of oil and natural gas, and the latter's position and function in the economy. It reviews government policies and attitudes towards petroleum and national and international regulation of onshore or offshore petroleum operations, including petroleum taxation. The book describes the development and working of the oil and natural gas markets, the cost of supply and the establishment of producer prices. Attention is paid to the internationalization of the extractive industry, the way this industry is organized and the old and new forms and manifestations of political risk, with which this industry has to cope. In addition this book contains a special chapter on production economics and the final chapter is devoted to the main forms of inter-company production agreements (joint ventures and unitization agreements). This book elaborates on the subjects described and discussed in the author's 'Introduction to the Regulation of the Petroleum Industry (Laws, Contracts and Conventions)' (first published 1994) and 'Cooperative Agreements in the Extractive Petroleum Industry' (first published 1996). However, this book is not merely an updated version of the earlier publications. Here, the occasion has been utilized to expand the scope of the discussion by paying attention to the value of petroleum for the economy and to its other economic aspects. As a result of this extension into the field of economy and economics, it is believed that this book is of special importance and interest to petroleum venture managers, as well as for independent consultants and other professionals with a commercial business background who are required to give advice with respect to the economic, regulatory and cooperative aspects of petroleum operations. In writing this book, the author has drawn on a total thirty-seven years' experience with the technical, economic and legal aspects of the international petroleum industry acquired during employment with the Royal Dutch/Shell Group, as well as his appointment as part-time professor at the Delft University of Technology.

The book is based and relies on original material consisting of the officially or otherwise published texts of laws, contracts, etc., official statements and policy documents and further on statistical data as made available to the public by international organizations and institutes, by international oil companies and trade journals.

Leidschendam, The Netherlands
April 1999

Preface to the Second Edition

In the year 2000, just one year after the publication of the first edition (1999), the global oil market reached a turning point. Against all expectations, predictions and growing concerns about climate change, the demand for oil started to rise quite steadily in response to an expanding world economy. From the years 2000 to 2006, world consumption increased from 76.3 million b/d to 84 million b/d. Over this period, China's oil consumption rose from 4.8 million b/d to 7.4 million b/d, even overtaking Japan thereby. International prices reacted more sharply, in particular after 2005 the price reached spectacular levels: in November 2007 Brent Dated was nearing the USD 100/b mark. To put the matter in its proper perspective, exactly the same crude had in 1998 been priced at USD 12.72. Even when correcting for the current weakness of the US dollar, this is an astonishing development. Nonetheless in the first quarter of 2008, these trends continued. In the last part of said quarter, the Brent price regularly exceeded USD 100/b in line with a further weakening of the US dollar. The world's annual consumption of natural gas also increased over the period: from 2,428 billion cubic metres in 2000 to 2,851 billion cubic metres in 2006. In light of these trends, the second edition undertook a new attempt, using the latest available information about future consumption, available reserves etc., to try to estimate the time that is left to the world economy before the oil and natural gas reserves are exhausted and practicable and acceptable alternatives must have been found to replace them. It is further investigated whether such alternatives exist.

Next to worrying about the prices of petroleum and the lifetime of the reserves, governments had to focus their attention on CO_2-emissions caused by the use of oil and natural gas and to make a start with taking measures and formulating policies which should help or which were thought to be necessary to reduce said emissions and so to fulfil the commitments accepted by them in this regard under the Kyoto Protocol (effective 16 February 2005). The often conflicting policy options open to governments and the consequences, if any, for both oil and natural gas and the petroleum industry are reviewed and discussed.

In the intervening years, governments have been active in further developing petroleum legislation although the changes introduced have not led to any significant difference with the systems and provisions described in the first edition. Nonetheless, legislation or contracts referred to and presented in the first edition as examples of 'current' legislation or contracts have been kept still 'current' by including in the second edition any changes and additions that were found to have been made thereto over the last years.

Finally, the text of the second edition has been restructured, subjects have been rearranged in different chapters, the chapter 'on Government Policies' has completely been rewritten and the chapter 'on Production Economics' has been deleted and replaced by a new chapter 'on Charters and Interstate Agreements'. To the latter, a description and discussion of the 1959 Treaty on Antarctica and its Protocol of 1991 have been added.

Leidschendam, The Netherlands
April 2008

Table of Contents

List of Tables

Chapter 1
Introduction to Petroleum

1.1 PHYSICAL ASPECTS

1.1.1 Description

Petroleum (literally meaning 'rock oil') consists of, naturally occurring and through geological processes, generated hydrocarbons and associated non-hydrocarbon substances. These hydrocarbons are hereinafter referred to as 'natural' hydrocarbons in contrast to hydrocarbons manufactured by industry.

Note. Natural hydrocarbons consist of molecules almost exclusively made up of carbon (C) and hydrogen (H) atoms in ratios ranging from four (methane, CH_4) to less than one (asphaltic bitumen). The end member of this range is carbon (C) itself, the main constituent of coals. The number of carbon atoms per molecule determines molecular size and structural complexity. Natural hydrocarbons, including natural carbon, may contain, as an integral part of their molecular structure, small amounts of other elements such as sulphur (S), nitrogen (N) and various metals.

Depending on molecular composition, prevailing reservoir temperature and pressure, natural hydrocarbons may occur in a semi-solid, liquid or gaseous phase, respectively referred to as asphaltic bitumen (including tar and natural asphalt), extra-heavy oil, heavy oil, light oil (the latter also referred to as conventional oil) and natural gas.

1.1.2 Physicochemical Properties

The physicochemical properties of the hydrocarbon molecule depend on the H/C ratio, the structural complexity and the number of carbon atoms comprised in the molecule. The H/C ratio ranges from 4/1 (e.g., gaseous hydrocarbons, starting with methane) to 2/1 (e.g., the diesel oil to gasoline range) to 1/1 (e.g., fuel oils to asphaltic bitumen).

The size of the hydrocarbon molecule is determined by the number of carbon atoms per molecule. This number is related to the H/C ratio as this ratio becomes smaller in the larger molecules (with a corresponding increase of the specific gravity). Methane (CH_4) is the structurally simplest, smallest and lightest hydrocarbon molecule, possessing only a single carbon atom accompanied by four hydrogen atoms. This gives the molecule an H/C ratio of 4, which is the highest H/C ratio possible. At room temperature and atmospheric pressure methane is a gas. By contrast, heavy hydrocarbons with H/C ratios around 1, known as asphaltenes, may have forty or more carbon atoms per molecule. The H/C ratio, the number of carbon atoms per molecule and the structural complexity of the hydrocarbon molecule determine the latter's physicochemical properties, such as specific gravity, viscosity, volatility, heat content and energy density.

Note. Density of oils is expressed in degrees of API gravity, a standard of the American Petroleum Institute which is computed as (141.5/spec. grav.) – 131.5. This means that the lower the specific gravity the higher the API number and vice versa. A specific gravity of 0.92 corresponds with 22 degrees API gravity. The latter number is used for separating light oil, also referred to as conventional oil, from heavy oil, including extra-heavy oil and asphaltic bitumen. The lightest of light oil has a specific gravity of 0.80 which corresponds with 45 degrees API gravity. In stark contrast, extra-heavy oil and asphaltic bitumen have a specific gravity of 1.00 or more which corresponds with an API gravity of 10 degrees or less. Viscosity of oils is measured in centipoise (cP) and ranges at prevailing reservoir temperature from less than 100 cP for light oil to more than 10,000 cP for asphaltic bitumen. It follows that asphaltic bitumen is not only very dense but also very viscous.

From the economic point of view the most important property of hydrocarbons is their heat content (or calorific value) per standard unit of volume (cubic metres at standard conditions) or per unit of mass (metric ton). One metric ton of the average light oil has a heat content of 42 GJ, which compares with 38 GJ as delivered by one thousand cubic metres of natural gas under standard conditions of temperature and pressure. Stating it differently, one thousand cubic metres of natural gas is equivalent in heat content to 0.9 metric tons of crude oil (0.9 toe). It follows that the transportation of natural gas on the basis of energy units delivered is much more expensive than that of crude oil, because the gas if it cannot be used on the spot has to be transported in a pipeline under high pressure or by ocean-going tanker after liquefaction.

Note. Due to their different volatility the respective hydrocarbon components of oil can be separated from one another in the refinery by the process of distillation. By this process heavy to light distillates are produced, in the range heavy residue/fuel oil/gas oil/kerosene/gasoline/naphta, with corresponding specific gravities of 1.00 to 0.70. Each of those distillates has its own specific use and application in the respective sectors of the economy. In this context, hydrocarbons with a H/C ratio of 2:1, which occupy the range of diesel oil to gasoline, are more suitable for use in internal combustion engines and, therefore, very valuable as transport fuel; so valuable in fact that in the refinery, this range of hydrocarbons is manufactured by catalytic conversion processes from hydrocarbons with a lower H/C ratio, i.e., from the heavier distillates. Finally, from methane, the least complex member of the hydrocarbon family, the same range of hydrocarbon products can be manufactured albeit at considerable cost and with the loss of half of the heat content of the original methane.

1.1.3 Generation

Solid, liquid and gaseous natural hydrocarbons and coals were (and still are) generated by the chemical conversion of organic matter that was deposited in sediments. Said organic matter may be of marine origin, such as plankton and algae, or may have a terrestrial origin (consisting mostly of land plants). Organic matter of marine origin form the base material for hydrocarbons; organic matter of terrestrial origin form the base material for coals but also for gaseous hydrocarbons, mostly methane, that are co-generated with the coals. The chemical conversion of all this organic material is set in motion by geological processes, whereby marine or terrestrial sediments, and the organic matter contained therein, become deeply buried and in the process exposed to high temperatures and pressures. In the end the sediment is converted into rock (shale, limestone) and their organic contents into hydrocarbons and coals.

When methane that is co-generated with coals together with carbon dioxide (CO_2) remains behind in the coal seams in which it was generated, it is referred to as coal bed methane. This methane is partly chemically- and partly physically-bound to the carbon atoms. It may be produced from the coal itself. In this situation, the coal beds function as source rock, as well as reservoir rock.

Coals consist for a large part (seventy per cent or more) of carbon (C) and for the remainder of volatile and mineral matter, among which are hydrogen and hydrocarbons, sulphur compounds and other minerals. Depending on its composition, coals are classified as either bituminous coal, including anthracite (referred to as hard coal), or sub-bituminous coal, including lignite and brown coal.

1.1.4 Migration, Accumulation and Occurrence

After their generation in the source rock, natural hydrocarbons as well as methane co-generated with coals are squeezed out of the rock and start their migration upwards through permeable rock formations and their faults and fissures in the direction of the surface. The migration through the rock strata is slow, at least as far as the liquid hydrocarbons are concerned. But the migration for all hydrocarbons will come to an end when they either emerge at the surface or become trapped and accumulate at a place where the rock strata, through which they are migrating, are either cut off or covered by an impervious type of rock, e.g., a salt dome or salt layer.

The rock formation, in which the natural hydrocarbons and associated non-hydrocarbons are eventually entrapped and accumulated, is referred to as the reservoir rock or reservoir. Reservoirs are sedimentary rocks with sufficient pore volume, fissures and permeability allowing the accumulation of natural hydrocarbons as described. Fissured limestone (i.e., rocks mostly consisting of calcium carbonate [$CaCO_3$]) and relatively clean well-sorted sandstone are examples of suitable reservoirs. In a reservoir, the natural hydrocarbons will replace formation water and may separate into a liquid (oil) and a gaseous phase (natural gas) with the latter on top (gas-cap gas). As a result, a gas/oil contact and an oil/water contact may be encountered in a reservoir.

As geological time passes, movements in the earth crust create new faults and/or fissures or re-activate old ones, thereby opening up new channels of communication between the rock layers. A new process of migration may start and the natural hydrocarbons may move to and accumulate in another reservoir or resume their migration to the surface. In some instances, only the gaseous phase plus the lighter components of the liquid phase will escape and move onwards to another reservoir or to the surface. What remains behind in the reservoir after such secondary migration is heavy oil including extra-heavy oil or asphaltic bitumen with a low H/C ratio (1:1 or even less), very viscous and very dense as described above. If the primary or secondary migration is not interrupted by any geological barrier or stopped by the earlier transformation of the liquid hydrocarbons into non-moving heavy oil or bitumen, natural hydrocarbons will eventually emerge at the surface. In this context, 'surface' could be the surface of the dry land or of the bottom of the sea. When surfacing on dry land, gaseous hydrocarbons, whether appearing as free gas or as gas dissolved in the crude oil, will escape into the atmosphere or start burning (if ignited by lightning or other ways). When natural gas surfaces in the deep ocean, a chemical reaction with water (H_2O) may ensue, resulting in the formation of gas hydrates. The latter are crystals of a methane/water compound. Any liquid hydrocarbons emerging at the surface will first lose their volatile components and over time turn into oil or tar sands. In the end, all semi-solid hydrocarbons occurring at the surface, and for as long as remaining in contact with the atmosphere, will be oxidized and converted into CO_2 and water and by this process the natural carbon cycle will be closed.

Where reservoir rock is exposed at the surface, any oil contained therein will slowly flow, ooze or drip out of the rock (oil seepages, oil springs). In this way, tar pits or asphalt lakes are formed.

Natural gas dissolved in the crude oil is in a technical sense called solution gas. It is produced jointly and simultaneously with oil and referred to as associated gas. When the oil in which the gas is dissolved flows upward in the production tubing, the pressure is gradually diminished and solution gas is set free in the production tubing. This phenomenon has a tremendous flow accelerating effect. Natural gas constituting a separate gaseous phase (free gas) may or may not remain in contact with the oil. If the latter is the case, the free natural gas is referred to as 'non-associated' (natural) gas. In a technical sense, natural gas that was co-generated with coal instead of with oil is also referred to as non-associated gas.

Natural hydrocarbons are found mixed with or accompanied by various non-hydrocarbon substances. This association is a matter of physical bonding as opposed to the chemical bonding of the elements described below. The association or bonding process takes place during the subsoil generation processes or during the hydrocarbons' migration through the rock formations. The chemically-bound elements most commonly encountered in heavy oils are sulphur and various metallic compounds. Sulphur has to be removed to prevent the substance from being released into the atmosphere as a by-product of the combustion process. The metallic compounds frequently give problems in chemical conversion processes in which catalysts are used.

Natural gas, whether generated in association with oil or with coals, also contains varying proportions of non-hydrocarbon gases. To a large extent, these non-hydrocarbon gases consist of nitrogen (N_2) and carbon dioxide (CO_2). To a lesser extent, natural gas may contain inert gases, such as argon and helium, or sulphur compounds and also hydrogen sulphide (H_2S). Sulphur compounds have to be removed from the natural gas at the first opportunity, in any case before the gas enters the main transportation facilities to protect the latter from the occurrence of corrosion damage.

A special type of sedimentary rock known as oil shale contains kerogen (to be considered a proto-hydrocarbon) in a finely dispersed form. An oil shale is a source rock that through geological processes has been uplifted and become exposed at the surface. The geological uplifting has interrupted the natural hydrocarbon generation processes as these are dependent on high temperatures and intensive pressures. After mechanically crushing the rock, kerogen is extracted therefrom and converted into regular hydrocarbons.

1.2 THE EARTH'S CARBON CYCLE

The generation of coal (C) and hydrocarbons (CxHy) from organic matter, as described above, and their ultimate conversion into CO_2 and water form part of the Earth's carbon cycle. In this context, this cycle can best be described as an interaction between atmospheric CO_2 and the Earth's biosphere and lithosphere that take place according to physical, chemical and biological processes.

The most important biological/chemical process in which CO_2 is involved is photosynthesis that takes place within terrestrial plants and marine phytoplankton, respiration by organic organism and the latter's decomposition and oxidation, and the forming of calcium carbonate ($CaCO_3$) by marine organism for building their skeletons and shells.

Note 1. Photosynthesis is the most important biological/chemical process occurring in the biosphere. It is the process in which photons energize the reaction whereby, in the presence of water, CO_2 is converted into oxygen and carbohydrates, the building blocks of plants and phytoplankton (the latter being the microscopic plants that are the basis of the food chain in the ocean).

Note 2. The forming of CaCO3 by marine organism is thought to become more difficult if by absorbing more and more CO_2 the oceans get more acidic. The calcareous remains of these marine organism form chalk.

Physical and chemical processes are at work when atmospheric CO_2 is absorbed by the oceans; when coal and hydrocarbons are generated in the deep underground in the way as described before; when rain water washes out CO_2 from the atmosphere and chemically binds the latter to the rocks in the form of carbonates (chemical weathering); or when CO_2 is returned to the atmosphere through volcanic eruptions, or by the natural oxidation of hydrocarbons, or coal deposits occurring at the surface or by the oxidation of biomass (plant material).

Note. The CO_2 that is released into the atmosphere through volcanic eruptions originates from CaCO3-sediments such as chalk and limestone that take part in the geological process of the movement of plates (plate tectonics). Despite the time lag between CO_2-locking up and CO_2-release, volcanic eruptions appear the most important non-anthropogenic source of atmospheric CO_2. In the geological record, evidence has been found of a direct relationship between the intensity of volcanic activity and the average global surface temperature.

Plants, soils and the oceans are referred to as carbon sinks. About half of the CO_2 taken up by plants and soils is returned to the atmosphere through respiration. The same amount is eventually returned through decomposition (natural oxidation). In contrast to carbon sinks, respiration and decomposition form with respect to the atmosphere carbon sources. If the remains of organism, among which most prominently plant material, usually referred to as biomass, rather than being decomposed are deposited in marine or terrestrial sediments they will under the right geological conditions and on a geological timescale, be turned into hydrocarbons or coal respectively as described before.

Unless trapped in a reservoir, transformed into a dense, very viscous residual oil or transformed into gas hydrates, natural hydrocarbons will on a geological timescale find their way to the surface. Gaseous hydrocarbons (mainly methane) will escape into the atmosphere, liquid hydrocarbons will first lose their volatile components, turn into bitumen and thereafter, in a time-consuming process, be oxidized and turned into CO_2 and water.

Coal stays in the underground layer in which it was generated and in this way carbon is permanently removed from the carbon cycle. Should however through geological processes coal layers come into contact with the surface and the atmosphere, the methane contained in the layers will be released and the coal itself (like the surface deposits of semi-solid hydrocarbons) will be oxidized. This time-consuming process is speeded up if, after ignition through natural causes, the coal layers start burning; the subsurface coal fires in China form an important example of this phenomenon. In both instances, CO_2 is generated that subsequently is returned to the atmosphere.

CO_2 is also locked up in limestone and chalk. This CO_2 will only through the geological process of plate tectonics and accompanying volcanic eruptions eventually, on a geological timescale, return to the atmosphere. Despite the time lag between CO_2-lock up and release, volcanic eruptions appear to be the prime natural carbon source in the carbon cycle.

The presence of CO_2 in the atmosphere is of fundamental importance for regulating the average global temperature on the surface of the planet. This is due to the physical molecular property of CO_2 (a property that it shares with water vapour and methane) to absorb infrared light, that is radiated back from the Earth's surface, and of re-emitting the same kind of radiation. That part of the re-emitted radiation that travels back to the Earth's surface will cause an increase of the temperature at that surface. From the study of ice cores, some covering the past 650,000 years, a close parallel connection between atmospheric CO_2 concentrations and surface temperatures was found; in periods of low temperatures, the corresponding atmospheric CO_2 concentrations were equally low and vice versa. This relationship has also been confirmed by geological evidence.

Note. In the extreme case of a 0 CO_2 concentration, i.e., in case no CO_2 (or water vapour) would be present in the atmosphere, the average global temperature would be around -20 degrees Celsius compared to the currently prevailing average temperature of +15 degrees, a difference of 35 degrees. The so-called greenhouse effect was first recognized by Joseph Fourier in 1824. By the turn of the century, Svante Arrhenius in Sweden and Thomas C. Chamberlin in the US made the connection between the burning of fossil fuels and this greenhouse effect, predicting that the use of fossil fuels would lead to global warming.

The industrial production of hydrocarbons, coal and biomass (hereinafter carbon fuels), and using the latter as fuel to generate heat, electricity and mechanical power has added a completely new and anthropogenic source of CO_2 to the Earth's carbon cycle. This anthropogenic source can indeed be described as completely new because firstly the anthropogenic oxidation by way of burning proceeds much faster even explosive compared to natural oxidation; and secondly, the carbon fuels that are being burned have been specially mined or extracted or grown as part of an industrial or agricultural process. Without this anthropogenic activity, the natural hydrocarbons and coals would have remained in situ and part of the biomass would have been buried and neither of them would, on a human timescale at least, have been oxidized and generated CO_2.

The new and anthropogenic source of CO_2 has revealed itself by the fact that since the time of the industrial revolution (1750) and in the absence of significant natural CO_2-sources such as volcanic eruptions in that period, the value of the global atmospheric concentration of CO_2 has markedly increased from a pre-industrial value of about 280 ppm (0.028 per cent by volume) to 379 ppm (0.0379 per cent by volume) in 2005. The consequences of this phenomenon for global warming and climate change and government policies in response thereto are discussed in Chapter 4.

1.3 EXTRACTION AND PRODUCTION TECHNIQUES

In the pre-drilling days, oil was collected at the places where it spontaneously emerged at the surface. Where oil leaked or seeped into pools or creeks it was skimmed from the surface of the water with cloths and rags and subsequently bottled and sold. Tar was collected from tar pits or, where the tar was mixed with sand (tar sands); it was extracted from the sand by washing the same. In some places, e.g., in Pechelbronn, France, shafts and galleries were dug to get better access to the oil or tar and to increase its recovery. These methods to produce oil proved inadequate when more oil was needed to meet a rising demand for kerosene. This new demand for kerosene was triggered by the invention of a more suitable kerosene-burning lamp (around 1850) in which kerosene could burn without smoke or smell, something that had not been possible in the earlier types of lamp. More efficient production methods were called for in order to meet the increased demand. In the US at the time, salt water was produced by boring holes into the salt water reservoir. In these salt water wells oil was regularly pumped up together with the salt water. Entrepreneurs got the idea that oil should be bored for in a similar manner. This idea was put to the test, and following indications of oil at the surface, by an oil well which was bored by Drake in 1859 near Titusville, a small town in Pennsylvania. On 27 August 1859 at a depth of only 69.5 feet, oil was struck. This well is traditionally considered to be the first oil well in history, and as such its boring (or drilling, as boring later was called, after the invention of the rotary drill pipe technique) marked the founding of the extractive petroleum industry.

Note 1. In 1860 already hundreds of 100-foot wells had been drilled in southwestern Ontario, Canada, and that the well drilled by Williams three years earlier (in 1857) might have been the discovery well and thus be entitled to lay claim on being the first oil well in history. Outside North America, the first oil wells were drilled around Baku in the Caucasus in 1871. In Pechelbronn, which may be considered the West European counterpart of Titusville, the first well was drilled in 1879, twenty years after the Drake well.

Note 2. The rotary drill pipe technique had its inception in 1882, when the Baker brothers, drilling contractors of Yankton, Dak, tried a crude bit on a rotating string of pipe, pumping water down the pipe to raise the cuttings in the annular space between pipe and walls of hole. About 1895, they teamed up with three Texas water-well drillers and built the first rotary outfit in Spindletop oil field, Texas, in 1901. Since about 1933 the technique has greatly been improvedupon. Drilling to 3000 metres became possible. (Source: R. Peele, *Mining Engineers' Handbook*, 3rd edition, volume I, John Wiley & Sons, Inc., New York, 1956).

The idea that oil should be looked for in the underground and produced by means of drilling holes had a sound geological basis, since there is no chance of finding natural hydrocarbons other than tar sands and asphalt lakes at the surface. Conventional hydrocarbons i.e., light oil and natural gas and the substances associated with them are only found trapped in sedimentary rocks that are not in communication with the surface. From this geological point of view, drilling holes into a underground reservoir may be considered to be no more than establishing the otherwise failing communication with the surface.

Where previously only vertical or slightly deviated holes were drilled, in recent times holes are drilled that end up in a horizontal trajectory that can more or less trace a rock layer, making in this way a much greater part of the borehole accessible to the petroleum contained in the rock.

Note. After the well has been drilled, tubulars are installed in the borehole. The lower part of the space between the outer casing and the borehole is sealed off by the injection of cement in order to prevent, behind the casing, petroleum leaking or escaping from the reservoir. Thereafter the casing at the level of the reservoir earmarked for production is perforated, unless it is decided to leave the bottom of the hole open.

As from the moment the drilling fluid is removed from the well and the counter-pressure is lifted, petroleum in the vicinity of the well and accompanied by formation water and rock particles, starts moving towards and into the well. Whether or not this initial movement will be followed by a natural flow of petroleum through the pores and fissures of the reservoir in the direction of the well bore and how efficient the extraction of the petroleum is depend on:

(i) reservoir rock characteristics, foremost among which is the permeability;
(ii) drive mechanisms (e.g., water-drive or gas cap drive), by the working of which the movable part of the petroleum is displaced from the pores and fissures of the reservoir;
(iii) well spacing and orientation (vertical to horizontal); and
(iv) the characteristics of the oil to be extracted, foremost among which is its viscosity.

If petroleum emerging at the wellhead consists of a mixture of oil and (associated) natural gas in different proportions, the field concerned is referred to as an oil field. If petroleum emerging at the wellhead consists of either free gas or coal-generated gas, the field concerned is referred to as a gas field and the natural gas is referred to as non-associated gas.

Note. Natural gas is a mixture of gaseous hydrocarbons and non-hydrocarbon gases such as carbon dioxide and nitrogen. The hydrocarbon component of natural gas consists mainly of methane (CH_4), but in varying degrees it may be accompanied by slightly larger or heavier (in the sense of containing more carbon atoms per molecule) but still gaseous hydrocarbons, such as ethane (C_2H_6), propane (C_3H_8), butane (C_4H_{10}) and pentane (C_5H_{12}). Natural gas containing these heavier hydrocarbons is called a wet or rich gas; otherwise, if mainly consisting of methane, it is known as a dry gas. If economically justified and/or if necessary for the liquefaction of the natural gas, the hydrocarbons in the range of ethane to pentane are recovered from the natural gas in the form of natural gas liquids (NGLs). The recovery takes place in a special gas plant, before the gas is transported to the markets (or is flared or recycled, if there happens to be no market for the gas). NGLs behave at room temperature and atmospheric pressure as gases. The main components of NGL, viz. propane (C_3H_8) and butane (C_4H_{10}) are known as liquid petroleum gases (LPG).

In an oil field the recovery of light oil (i.e., oil having an API gravity of at least 22 degrees and a viscosity less than 100 cP) including condensate is by itself efficient, provided the reservoir has a satisfactory permeability. By contrast, oil lacking the lighter components usually has a high viscosity or may even have become a semi-solid bitumen, resisting any displacement in the reservoir. In such a situation, or if the naturally available drive mechanism is too weak, reservoir stimulating measures have to be taken, costs permitting. Reservoir-stimulating measures, also named Enhanced Oil Recovery (EOR) techniques, range from the injection of natural gas or water (to repressurize the reservoir or to strengthen a failing water-drive mechanism), to injection of CO_2, to injection of steam or special chemical products (to mobilize very viscous oil). All these injection methods have this in common; they require the drilling of wells at the appropriate spots. Application of EOR techniques is therefore expensive, but may result in an impressive increase of the oil extraction efficiency. In some instances, recovery factors of up to eighty per cent may be achievable, compared to recovery factors between fifteen per cent and thirty per cent achieved by conventional, non-stimulated production. It should be added that in some oil fields, due to excellent reservoir characteristics and the presence of strong, long-enduring natural drive mechanisms, very high extraction efficiencies are attained without applying any non-conventional techniques.

Most difficult to extract from their reservoirs and requiring non-conventional methods are extra heavy oil and natural bitumen. Currently all attention is focused on the natural bitumen contained in the oil sands located in Alberta, Canada, but the tar sands located in the Russian Federation are also attracting attention. This natural bitumen has an API gravity of less than 10 degrees and a viscosity at prevailing reservoir temperatures greater than 10,000 cP. Its extraction requires special heat treatment provided by hot water and steam. Where these oil sands can be mined in open pits, the bitumen is separated from the sand in an extraction/separation plant near the mining site and further processed in the next plant into petroleum coke and hydrocarbon vapours. These vapours condense into naphta, kerosene and gasoil. The larger part of the oil sand deposits however are too deep below the surface for open pit mining. In this situation the bitumen has to be separated from the sand in situ.

Note. The extraction process that is applied in this case is known as steam-assisted gravity drainage (SAGD). The process requires the use of stacked pairs of horizontal wells. By injecting steam through the upper wells the bitumen becomes mobilized whereupon gravity causes the mobilized fluid to move to the lower wells. From there the bitumen is pumped to the surface and further processed into the end-products mentioned above. (Source: US Geological Survey Fact Sheet FS-070-03, August 2003.)

The extraction of non-associated gas is generally much more efficient than that of the average crude oil. A recovery factor of eighty per cent to ninety per cent may be expected, but minimum permeability requirements should be satisfied. This means that, where natural gas is concerned, enhanced recovery methods can be dispensed with, making the production costs of gas that much less expensive.

After having entered the well, petroleum will flow upwards to the wellhead at the surface through the production tubing installed in the borehole.

Note. If the liquid phase contains a large amount of solution gas, this gas is gradually set free and gas bubbles appear as the borehole pressure decreases in the direction of the wellhead. At the same time the emerging gas bubbles will expand with decreasing pressure, giving a tremendous boost to the flow of the oil.

If the petroleum concerned consists of an undersaturated oil, i.e., oil that contains hardly any natural gas in solution, and in absence of any abnormal reservoir pressure, stimulating, flow-assisting measures, mechanical or otherwise may have to be applied in order to let the oil flow upwards through the production tubing. Such measures involve the installation in the production tubing of the appropriate equipment, such as gaslift valves, pumps and pump rods.

Petroleum emerging at the wellhead is accompanied by rock particles and formation water. After passing the wellhead the first treatment (in any field) consists in removing these particles and water. In an oil field, water has to be separated from oil; this separation is critical because the production water cannot be discarded if it contains more oil than allowed by the applicable regulations. The next treatment (in an oil field) is the separation of oil from (associated) gas. As a matter of fact this oil/gas separation treatment can be described as the first stage of the refining process. After their separation, the oil and gas are each conditioned for transportation by pipeline and/or ocean-going tanker to their own, separate markets, except that associated gas, for which there is no commercial destination, even not in the vicinity of the oil field, will be flared or recycled, but not before natural gas liquids (NGLs) will be recovered from the gas if the latter happens to be a wet or rich gas.

Coal-generated natural gas, that remains behind in the coal layer (and is referred to as coal bed methane), is partly physically- and partly chemically-bound to the carbon atoms. Coal bed methane is produced by drilling holes into the coal layers. When a borehole penetrates the targeted layer, the pressure therein is lowered and at least the physical bounds between the methane and the coal are broken. The methane will start flowing through the pores and the natural fractures present in the coal into the borehole. Usually, the coal layer around the borehole is fractured (a production-enhancing technique borrowed from the oil extraction methods) and the fractures made are filled with sand. Coal bed methane is also set free in a coal mine, when coal layers are getting exposed to lower pressures in the course of the mining operations. When this happens in an underground coal mine, methane becomes mixed with the air circulating in the mine. Air/methane mixtures containing between four-and-a-half per cent and fourteen per cent methane will, when ignited, cause an explosion.

Interestingly, experiments are undertaken to produce methane from coal layers by means of injecting CO_2 in the same. CO_2 should break the in the coal layer existing bonds between methane and the carbon atoms and thereby take the place of the methane. Such substitution results in an enhanced recovery of the latter.

1.4 SAFETY, HEALTH AND ENVIRONMENTAL ASPECTS

1.4.1 Safety, Health and Environmental Risks Associated with Petroleum Operations

From the beginning of the production of petroleum as an industrial undertaking, it was recognized that persons carrying out or involved in petroleum operations as well as the environment (the latter in this context to be understood as the whole physical and biological world in which humans live and in which the petroleum operations are carried out) were put at risk by:

(i) an oil well eruption (blow-out) at the drilling or production site, a grounding or collision of an oil tanker at sea, or the rupture of a pipeline and the subsequent massive release of oil into the environment (oil spillage); or
(ii) the temporary or permanent occupation of a portion of the land or sea surface or the sea bottom (spatial occupation) required for drilling locations, for the installation and building of (offshore) petroleum facilities and installations, or for construction and laying of (submarine) pipelines.

More recent concerns, in which human health and safety are not directly involved, are environmental risk caused by:

(iii) subsidence and earth tremors. The extraction of petroleum causes compaction in varying degrees of the reservoir rock from which the petroleum is extracted, depending on the characteristics of that rock. This compaction may in turn lead to subsidence at the surface and may even cause earth tremors;
(iv) the disposal of waste products generated in the course of the petroleum operations; or
(v) abandonment, removal and disposal of production installations, in particular of those placed at sea.

With reference to (i). An oil blow-out, i.e., the uncontrolled, explosive outflow of oil from a well, that may have been caused by a faulty drilling operation or by an accident at the drilling or production site, and the grounding or collision of an oil tanker at sea are serious safety risks. A massive release of oil may follow. Massive oil spills have generally a limited areal impact and as such never cause permanent damage to the environment. Oil escaping from a well usually catches fire and is burned off hence causing little environmental damage.

Note. This has convincingly been demonstrated in the aftermath of the Gulf War in 1991. Iraqi troops retreating from Kuwait set most oil wells on fire. It took more than a year before all fires were extinguished. All the while the smoke of the burning wells was spread over a large area. The sky over Kuwait was darkened for a long time. Smoke particles have even been found in the Himalaya Mountains. Nevertheless any trace of this man-made calamity has since disappeared from the environment. No lasting damage could ever be found.

By contrast, oil escaping at sea from a damaged or broken up tanker, floats on the water, forming a slick that is moved around by current and wind. If such oil slick stays at sea, it will soon break up by wave action and further disappear through evaporation and oxidation or it will be decomposed through biological processes. Any heavy residue will eventually sink to the sea bottom, and do no further environmental damage. On the other hand, if an oil slick is driven up a beach or into an estuary, a viscous layer of oil will spread out over the shore line, with disastrous effects on various forms of marine coastal life, such as seabirds, seals and crustaceans.

Note. Fortunately, due to improved navigational instruments, massive oil spills at sea are actually a rare occurrence. It appears that major, spectacular events are separated by years and together account for only a small fraction of the yearly amount of oil that is released into the sea originating from different sources, such as ports and deliberate or accidental discharges from tankers and other ships. In 1967 the 'Torrey Canyon' wrecked off the coast of Cornwall, spilling more than 30,000 tons of oil into the sea. This particular disaster prompted the International Maritime Organization (IMO) to convene in 1973 a conference of all maritime nations, in order to mutually agree on measures to be taken to prevent pollution from ships. At this conference the International Convention for the Prevention of Pollution from Ships (MARPOL) was adopted. This very important convention was modified in 1978, and finally entered into force on 1 October 1983. A much bigger disaster occurred on 16 March 1978, when the 'Amoco Cadiz', after its steering gear failed in rough seas, ran aground off the coast of Brest, France, spilling its entire cargo of 226,000 tons of oil. Global concern about the environmental risks associated with petroleum and its products rose to a pitch when on 24 March 1989 the oil tanker 'Exxon Valdez' grounded in Prince William Sound, Alaska, releasing thereby about 36 thousand tons of oil into the sea. The spillage turned into a disaster when the greater part of the spilled oil ended up at the shore and fouled more than 1,200 miles of Alaskan coastline. The 'Valdez' incident prompted environmentalists to accuse oil companies of irresponsible behaviour. Furthermore, a coalition of American ethical investment organizations and environmental groups, CERES, saw in the incident an opportunity to put the matter of environmental risks in a broader context. Already in the same year (September 1989), CERES launched a draft code of environmental conduct, which became known as the Valdez Principles. This Code was meant for corporations, whose operations directly or indirectly influence the Earth's biosphere, and was to be subscribed to on a voluntary basis. Subscribers to the Code pledged as follows:

- to minimize and seek to eliminate the release of pollutants that cause damage to the air, water or Earth or its inhabitants;
- to safeguard habitats in rivers, lakes, wetlands, coastal zones and oceans; to protect wildlife habitat, open spaces and wilderness, while preserving biodiversity;
- to minimize the creation of waste, especially hazardous waste, and dispose of waste through safe and responsible methods; and
- to minimize contributing to the greenhouse effect, the ozone depletion of the ozone layer, acid rain or smog.

The 1989 Valdez Principles represent a list of deeply felt environmental concerns and objectives of that time, but did not contain any concrete rules, recommendations or solutions. Although the preparation and launching of the Code was triggered by the grounding of an ocean-going oil tanker, the environmental concerns expressed comprise much more than damage caused by oil spills at sea. For instance, the reference to the greenhouse effect, depletion of the ozone layer, acid rain and smog directly involve the combustion and burning of hydrocarbons and have no bearing on the environmental risk presented by transport of oil by an ocean-going tanker.

In December 1992 the tanker world was again faced with an accident. This time it concerned the 'Aegean Sea', which tanker carried 80,000 tons of oil. On 5 January 1993 the oil tanker 'Braer' carrying 80,000 tons ran aground on its way from Norway to Canada on the coast of one of the Shetland Islands and three years later, in January 1996 a similar sized oil tanker, the 'Sea Empress', was grounded in Milford Haven. In the same year another IMO Convention was adopted providing for construction standards for oil (product) tankers. The new construction standards were aimed to make such ships safer in the sense of less likely to lose oil in case of a collision or grounding at sea. To put oil spillage at sea in a proper perspective, it should be noted that oil spillage, whether accidental or deliberate, account only for a small fraction of the total pollution of the sea, the major contributor being polluting substances released from land-based sources, i.e., sewage installations, chemical industry and agriculture, and carried to the sea through the river system. It should be added that not only oil but also natural gas may be spilled at sea. Such spillage may occur in the event of an accident involving a LNG tanker. Such an accident may result in a massive release of LNG into the atmosphere. When (accidentally) exposed to the atmosphere, the LNG cargo will evaporate quickly and form a low-lying cloud around the tanker. Such a cloud of methane forms with the air, within a range of concentrations, an explosive mixture (as in underground coal mines). Moreover, as long as the cloud lingers around the tanker it will suffocate any person aboard the ship. If a wind is blowing, the cloud will drift away and become diluted. After that the danger of explosion and suffocation will disappear. In the end any visible trace of methane will have disappeared. It follows that the environmental risk of operating a LNG tanker is, strictly speaking, negligible. The real, far more important risk is that human health and life will be endangered if a massive release of LNG occurs.

The environmental damage caused by oil escaping from a ruptured pipeline shall not extend beyond the place where the breach occurred. Where natural gas is escaping from a ruptured pipeline, there exists the risk of an explosion, depending on the composition of the ensuing air/methane mixture. Therefore, a sudden natural gas release from a pipeline is not likely to cause much environmental damage but is more a matter of a human health and safety.

With reference to (ii). Petroleum equipment, facilities and installations, including pipelines need a place (a site of operation) at the surface or even on the bottom of the sea, where they can be installed and be used and operated. Obviously, every effort will be made to minimize the number and areal extent of the sites required. But there is little room for manoeuvre when it comes to selecting the geographical location of suitable drilling and/or production sites or the routing of pipelines. If the selected site of operation is located within a controlled area, which means an area where the occupation of the desired site requires planning permission, under applicable zoning legislation or an area declared to be an environmentally sensitive area (fragile ecosystem, e.g., all types of wetlands and tropical rain forests), permission under the applicable regulations will be required before the site can be acquired or occupied. If the desired site is located in a non-controlled area, an authorized person is free to try to rent or to buy (from the public or private owner of the land) or simply occupy the selected site, subject only to the applicable conditions of the petroleum authorization concerned. Offshore, the competent authorities have to take into account the interests of shipping and fishing, the laying of submarine pipelines and cables, and the establishment of underwater national parks in coastal waters.

Note. The laying and operating of long-distance pipelines may have a negative impact on the environment by the mere fact of requiring extended strips of land for their routing and by their physical presence (compressor stations along the route), in particular when the pipeline is not buried in the ground. When running above the ground, for instance where soil conditions (e.g., permafrost) do not permit it to be buried, the pipeline forms a physical and visible obstacle in the landscape. If such a pipeline has to run through a wildlife refuge, measures have to be taken to prevent that its presence above the ground has a negative effect on the movements of wildlife, in addition to measures that should prevent any leakage. Such an extraordinary construction is exemplified by the 800-mile long Trans-Alaskan pipeline, running from Prudhoe Bay to Valdez. This pipeline was opened in June 1977 and carries daily 1.45 million barrels of crude oil. The pipeline had to be raised five feet off the ground in order not to disturb the yearly migrations of the caribou herds. Recently, a controversy has arisen in connection with the intended routing of a long-distance gas pipeline (one thousand km) connecting natural gas fields in Northern Argentina with the copper mines in Northern Chile. Environmental groups protested against the plans on the ground that the pipeline would traverse rain forests and the territory of a local group of Indian population while an alternative routing was available. Offshore there is the additional risk that an incorrectly buried segment of a submarine pipeline will become buoyant and rise from the sea bottom. When this happens, there is the risk that fishing gear will become entangled therein. The next challenge will be provided by the construction of the Nordstream gas pipeline projected to run through the Baltic Sea.

With reference to (iii). Subsidence is a physical phenomenon that manifests itself over a relatively large surface area. It affects the water system of the land surface and the water table. In a minor manner, it may change the course of rivers and create lakes. The most famous and spectacular example is the subsidence that for many years takes place in and around Lake Maracaibo in Venezuela. Around the lake, dikes had to be built and regularly be increased in height in order to prevent the flooding of the surrounding area. In low-lying coastal or inland areas, i.e., surface areas below sea-level, subsidence causes a relative rise of the water table, necessitating the installation of extra pump capacity to keep the water levels in such areas down and/or measures to strengthen and heighten the sea dikes. Compaction of the reservoir rock may also activate faults and in this manner be the cause of earth tremors. Subsidence and earth tremors do not constitute a human health risk and generally do not cause direct damage to railways, houses and other buildings.

With reference to (iv). In the course of the operations, waste products are produced which could be harmful to the environment (and, depending on the circumstances, also to human health) if not properly disposed of. Distinction may be made between two types of waste products. On the one hand, there are waste products directly related to and connected with the objective of searching for petroleum and producing oil and natural gas, such as spent drilling fluids, drill cuttings, production water and reservoir rock particles, but there is also associated natural gas that cannot be used commercially or re-injected into the reservoir and has to be flared. On the other hand, there are waste products generated as a result of making use of petroleum facilities, such as spent or obsolete material (used drilling bits, spent lubricating oil, all kinds of synthetic material).

With reference to (v). At the end of the production life of an oil or gas field, the petroleum facilities that had been used in such a field have to be abandoned. Depending on the petroleum facility, concerned abandonment may consist in:

(i) the plugging of production wells, removal of wellheads; and
(ii) the removal and dismantling of any other production facility present in the field, from production equipment to field pipelines to field storage tanks.

On the continental shelf, the removal and disposal of abandoned or disused production installations is regulated by international conventions, in particular by the 1982 Convention on the Law of the Sea; the 1992 Convention for the Protection of the Marine Environment of the North-East Atlantic (OSPARCON) and by the 1989 IMO Guidelines and Standards on Removal (see Chapter 9).

1.4.2 Concluding Remarks

Over the years it has become apparent that, if properly executed, and when taking all the reasonable precautionary or preventive measures that modern technology makes available, the operation of petroleum facilities, onshore as well as offshore, does not present a danger for human health and safety and the environment generally. Environmental concern is only justified with respect to long-distance transportation of oil by ship. If anything goes wrong, if an oil tanker is grounded and in one way or another spills its cargo into the sea, the surrounding environmentally-sensitive areas, such as beaches and coastal zones will suffer, albeit no permanent damage has ever been demonstrated. Great progress has been made in the context of MARPOL to make transportation by ship safer and more reliable.

Note. As far as petroleum facilities are concerned, it appears that the use and operation of said facilities may have an adverse effect on health, safety and the environment generally:

- only in exceptional or unavoidable physical circumstances beyond the control of the operator;
- by the deliberate contravention of rules and instructions;
- in cases of irresponsible human behaviour; or
- when a deliberate attempt is made to cause damage (sabotage).

Under normal circumstances however, the environmental risks involved can, in a technical sense, be reduced to any desired level. It all boils down to finding an answer to the question as to what are acceptable levels of safety, health and environmental disturbance, e.g.:

- how stringent safety rules should be;
- how much noise is acceptable;
- how much oil can be allowed to escape from containers, storage tanks and pipelines;
- to what degree may wildlife be disturbed by the presence and operation of a petroleum facility, etc.

In the past, the possession of a petroleum authorization more or less guaranteed that the licensee or the contractor would always have a site at its disposal to install and build the necessary facilities and that the operation of these facilities would be permitted, provided adequate measures were taken to ensure that their operation would be safe and not endanger human health. In modern petroleum legislation, an attempt is made to ensure that the desired protection of the environment should be taken into account as an integral part of the conditions imposed on the execution of petroleum operations. To this end any authorization, be it a licence or a contract of work, will incorporate a reference to planning and environmental legislation that are relevant in respect to petroleum operations and require that the conditions imposed upon the holder of the authorization by the other legislation are adhered to.

Note. Depending on how the authorized area is situated, it is in theory and practice possible that the holder of the authorization does not succeed in obtaining the necessary planning and environmental permits with respect to the site of a drilling operation or field installation that was found by him to be the most suitable one. In such a case, the said holder will be unable to exercise in an optimal manner his rights under his authorization. It is submitted that in such a situation modern petroleum legislation should provide for financial compensation for the loss of opportunity or for the extra costs the authorized person incurs in having to divert or cancel his operations. Obviously, it suits all parties better if the competent authority abstains from granting authorizations in respect of areas which appear to contain so many environmentally sensitive or excluded parts that it is clear from the outset that there will be no room for undertaking petroleum operations in the sense as authorized.

1.4.3 Health and Environmental Risks Associated with Using Oil Products and Natural Gas as Fuel

When oil products or natural gas are used as transport fuel or as fuel for generating heat and power, the carbon (C) content of these hydrocarbons (CxHy) is transformed into CO_2, or depending on the completeness of the combustion into carbon monoxide (CO), whereas the hydrogen (H_2) content is transformed into water (H_2O).

Note. The transformation into simple oxides marks an enormous loss in chemical complexity. Only when oil is used in the petrochemical industry, advantage can be taken of its complex structure.

The air (N/O) used in the combustion process gives rise to various molecular combinations of nitrogen and oxide (NOx), such as nitric oxide (NO) and nitrogen dioxide (NO_2). If the hydrocarbons contain sulphur (S), sulphur dioxide (SO_2) will be formed. Unless special measures are taken for their sequestration, the gases will after the combustion process be released into the atmosphere and as long as the gases remain in the atmosphere they may take part in chemical reactions or alter the latter's physical properties or conditions. The changes brought about in this environment could either be beneficial, negative or neutral. In any case, each gas acts and reacts in its own distinct manner, the intensity thereof depending on its atmospheric concentration and on the time the gas concerned resides in the atmosphere. More specifically, the emissions of CO_2 are reinforcing the heat trapping properties of the atmosphere, which result in an increase of the average temperature at the surface (referred to as the greenhouse effect or global warming) with adverse consequences for the world climate and the rise of the sea level (discussed elsewhere in this chapter and in Chapter 4), whereas NOx gases contribute to smog and acid rain, similarly as SO_2 will do. Finally, combustion of petroleum products, in particular that of diesel oil in diesel engines, causes minute solid particles, also referred to as aerosols, to be emitted into the atmosphere. The aerosol's residence time is short, but during their stay in the atmosphere, aerosols reflect the incoming sun light and have therefore a cooling effect. Furthermore the presence in the local atmosphere of all these exhaust gases and particles constitute a constant human health risk.

Chapter 2
The Petroleum Industry

2.1 ORGANIZATION AND DEVELOPMENT

2.1.1 Early Activities, Early Organization

The boring of the first hole for the express purpose of finding and producing oil has marked the start of the production of oil as an industrial undertaking, or so to speak the birthday of the extractive petroleum industry. This historic operation took place in the year 1859 near the village of Titusville in northwestern Pennsylvania. On 27 August 1859, oil was struck in this first borehole at a depth of 69.5 feet. On this occasion the 'boring' operation was performed by a steam-powered cable-tool drilling rig under supervision by L.E. Drake. The latter's example was soon followed and by November 1960 already about seventy-five wells were producing, and about fifteen small distillation plants (refineries) were in operation for refining the crude oil into kerosene, a product that at that time was much in demand as illuminating oil to be burned in oil lamps. To meet this demand for kerosene, a completely new industry came into existence that was focused on searching and producing oil from which the desired kerosene could be distilled.

The oil producers were independent, non-organized individuals who worked on small-sized plots of land, on which they tried to install as many drilling derricks as space allowed in order to be able to produce oil at the highest rates possible. They sold their production individually to the buyers from the distillation plants. In this manner, a market for oil developed in which many small volume producers traded with many small volume kerosene refiners. The price offered to the individual producer depended on the competitive situation prevailing among the buyers. On the side of the producers, situations of over-supply alternated with periods of shortage due to the wasteful way in which the oil was being produced. This instability on the supply side caused hefty price fluctuations.

Note. Actually, the producers operating on a small piece of land were in most cases producing from the same reservoir from which their neighbours were producing. Producers were aware of this, but instead of becoming motivated to pool their interests and jointly develop the common reservoir, they were only intent to drill as many wells as could be accommodated within the boundaries of their land. In this manner they hoped to maximize production, and to secure for themselves, in competition with the neighbours, a maximum share of the common pool (reservoir). The result was that too many wells were drilled, wells that were not necessary for the ultimate recovery of oil from the common reservoir, thus causing a waste of capital. Apart therefrom, by producing each well at maximum capacity, the well started to produce water much earlier than would be the case under careful reservoir management working with lower production rates. As a result the well's life expectancy was shortened and ultimate recovery was much lower than could have been achieved.

In contrast to the producers, the refinery owners (known as the oil companies) aimed at organization and concentration of commercial power. Rockefeller's Standard Oil Company, which company had been established in 1870 and was restructured as the Standard Oil Trust in 1885, took the lead. Seeking and organizing a concentration of refining capacity was accompanied by expanding activities into the transport and distribution functions. The oil needed to feed the refineries was bought from the producer at the wellhead and the buyer in question took the responsibility for transporting the oil (by pipeline or railroad) to his refinery and for the distribution of the refined products, which in those early days mainly consisted of kerosene. When most refineries came under the control of Standard Oil, the producer price was from then on dictated by the buyer. When around 1885 oil was discovered outside Pennsylvania and oil production had started to spread over the US continent, Standard Oil, later followed by the other oil companies, changed its policy of restricting itself to monopolizing the refinery sector. Instead the oil companies aimed for control over production, not so much for getting control over prices which they already had, but for ensuring an orderly development of the production operations, something the producers themselves, operating as they did within the boundaries of their small plots, were not able or willing to achieve. To this end, oil companies started to acquire production leases for their own risk and account. The pattern that now established itself on the US continent was that of an oil company following and watching the oil prospector waiting for him to make a discovery. The oil company would then try and buy or lease any land on which the discovery was made, together with any land in the surrounding area. By doing so, the oil company sought to forestall that any prospective acreage would fall into the hands of speculators. By cutting out the speculators and collecting large-sized leases, oil companies took the place of the independent producers. As a result of this policy, the chance was reduced that an oil field would be shared and produced by a multitude of persons, resulting in a waste of capital, over-production and reservoir damage.

Note. At a later stage, when the science behind reservoir behaviour and drive mechanism became better understood, and the benefits of water injection in the margins of the oil accumulation were recognized, the producers, in the meantime replaced by the oil companies, finally accepted the need for cooperation in the matter of producing a shared reservoir (see Chapter 11).

2.1.2 Integration and Internationalization (Western Private Sector)

After having gradually replaced the individual small-scale producer, oil companies were no longer satisfied with watching and following prospectors and buying up land where the latter had made discoveries, but decided to undertake the business of prospecting and exploration themselves. To follow up on this new policy, the organization of an oil company had to be expanded and to include the prospecting and exploration function. This so-called vertical integration gradually extended into all activities connected with petroleum: from the search for petroleum, to the production of oil, to the refining of oil, to the distribution and sale of refined products. All these activities became comprised and combined within a single commercial enterprise under a centralized direction.

Meanwhile the oil industry got an international dimension due to the activities of European oil companies which, prompted by the need to satisfy the demand for kerosene, were forced to search for oil everywhere in the world albeit they had to stay outside the US which was the acknowledged domain of the US oil companies. The European-led outward expansion started already in the last decades of the nineteenth century when oil was discovered in Baku on the Apsheron Peninsula in 1871 and in Atjeh, Sumatra in 1885. The kerosene that was manufactured from this oil was exported to Europe and China and created thereby the beginning of an international market outside the US, albeit a market for an oil product and not yet for oil itself. This market for kerosene started to decline at the turn of the century due to the rapid introduction of electricity and electric light. Nonetheless the demand for oil and the products refined therefrom, including the once so important kerosene, did not slacken. New applications and uses for oil products were found opening up new markets in the industrialized countries. In response to this new demand and in search for new supplies, the petroleum industry ventured out of its nineteenth century US and non-US producing areas and moved to new areas worldwide e.g., first Persia and that part of the Ottoman Empire that would become Iraq, later Mexico, Venezuela and then the Arabian Peninsula but also, and in fact initially of greater importance, to new areas inside the US itself, e.g., Texas.

Note. In the process of this expansion many important discoveries were made in the aforesaid countries or regions: to name the most famous ones: 'Spindletop' in Texas in 1901; 'Masjid-i-Sulaiman' field in Persia in 1908; 'Golden Lane' fields in Mexico in 1910; first oil in Venezuela in 1914; 'Kirkuk' field in Iraq in 1927; 'Seria' field in Brunei in 1929, which field, it should be noted, produced in 1940 17,000 b/d and thus became a strategic target in the Pacific War; 'Black Giant' field, East Texas, 1930; first oil in Bahrain in 1932; first oil in Kuwait and Saudi Arabia in 1938.

As mentioned above, part of the global expansion took place in the Middle East. During these years and for the first time in the history of the petroleum industry, the major US and European oil companies formed alliances and joint ventures to cover the risks of and sharing the investments needed for undertaking exploration and production operations in this huge and largely unknown area.

The move into the Middle East started in Persia. In that country, William Knox D'Arcy, more of an investor than an oil prospector, was persuaded to acquire (28 May 1901) a concession for oil. D'Arcy applied for (and got) an area that covered the whole country with (deliberate) deletion of the five northern provinces, the ones bordering on Tsarist Russia. About seven years later, on 25 May 1908 to be precise, oil was struck in the concession area at Masjid-I-Suleiman. In 1909 D'Arcy's concession was taken over by the Anglo-Persian Oil Company a newly-incorporated company that for raising the necessary finance organized a public offering of its shares in that year. For strategic reasons the UK government acquired in August 1914, a few days after the outbreak of the First World War, a controlling interest of 51.7 per cent in the company. In 1933 the concession was unilaterally cancelled by Reza Shah. A new concession was negotiated, whereby the concession area was reduced to 100,000 square miles located in southwestern Iran, still an extremely large area compared with the leases obtainable at the time in the US. In 1935 Anglo-Persian was renamed Anglo-Iranian.

Shortly before the First World War, British and German interests turned their attention to the Ottoman Empire, to the region that later would become Iraq, encouraged as they were by the success of Anglo-Persian's venture in neighbouring Persia and by the numerous oil occurrences and early discoveries in and around that region. Through the intermediary of the Armenian oil investor Gulbenkian, they established a company in London by the name of the Turkish Petroleum Company (TPC). But before starting their drive to acquire exploration acreage in Iraq, a restructuring of TPC was agreed allowing for the participation of Anglo-Persian in the venture. On 19 March 1914 such agreement was reached whereby the distribution of the shareholding in TPC was fixed as follows: Anglo-Persian Oil Company: fifty per cent; Anglo-Saxon Oil Company (Royal Dutch/Shell): twenty-five per cent; Deutsche Bank: twenty-five per cent; Anglo-Persian and Anglo-Saxon each transferred a 2.5 per cent beneficiary shareholding to Gulbenkian, since then known as Mr Five Percent. On 28 June 1914 the Ottoman Empire granted a concession to TPC, but this concession became invalid after the dissolution of the Empire. In 1920 the shareholding of the Deutsche Bank was transferred to the Compagnie Francaise des Petroles (CFP). On 24 March 1925 TPC received a concession from the post-war Iraqi government.

For some time, US oil groups had expressed their interest to become involved in the exploitation of oil in Iraq and objected to being excluded therefrom by the British and French governments. When on 15 October 1927 oil was discovered in the concession area at Baba Gurgur near Kirkuk, the need for a quick solution grew. After lengthy negotiations involving the said governments, as well as the shareholding companies, an agreement was signed in London on 31 July 1928. Under the terms of this agreement, a group of US companies was permitted to participate in TPC, at the expense of the shareholding owned by Anglo-Persian. The latter shareholder agreed to transfer to the US group one half of its original shareholding, i.e., 23.75 per cent. The US group reduced itself to only two companies, namely Standard Oil Company of New Jersey and Socony-Vacuum Oil Company (Mobil), each with a shareholding of 11.875 per cent and represented by their joint company named the Near East Development Corporation. After the shareholding had been rearranged as agreed, the name of TPC was changed to Iraq Petroleum Company (IPC), to which company the Iraqi government granted a new concession (24 March 1931). The area of this new concession covered the Iraqi territory east of the Tigris River. The territory west of the Tigris River was later given in concession (in two separate concessions) to affiliates of IPC (the Mosul Petroleum Company and the Basrah Petroleum Company respectively). As part of the agreement, the US companies agreed not to seek concessions independently in the territory that before the First World War belonged to the Ottoman Empire, with exception of Kuwait and the Khanaqin district of Iraq. The so-reserved area, which was outlined in red on a map of the Middle East, comprised the whole Arabian Peninsula, with exception of Kuwait. Hence it included the countries which later proved to possess the largest oil reserves in the world. The Red Line Agreement, as the agreement became known, did not stop other major US companies from seeking concessions in the reserved area. For instance, in 1930 Standard Oil Company of California took over in Bahrain a concession from the Eastern and General Syndicate of Major Holmes and on 25 May 1933 the company was granted a concession by the Saudi government, which was further extended in 1939. In both concessions Standard of California was later joined by the Texas Company (later renamed Texaco). After the Second World War, the two concessionaires, realizing the enormous potential of their Saudi concession and the marketing effort and capital investment that the development thereof would require, looked for other US companies for joining them in this venture. The two US shareholders in IPC, Standard Oil of New Jersey and Socony-Vacuum (Mobil) were interested, but were restrained by the terms of the Red Line Agreement. After protracted negotiations with the other IPC shareholders, it was finally agreed in 1948 to terminate this Agreement, which opened the way for the participation of the US shareholders in the Arabian American Oil Company (Aramco). In order to forestall further penetration of US oil companies in the area reserved under the Red Line Agreement, IPC secured through affiliates concessions in Qatar (1934), Oman (1937) and Abu Dhabi (1939). In Kuwait, not included in the reserved area, a concession was granted on 23 December 1934 to Kuwait Oil Company, equally owned by Gulf Oil and Anglo-Persian.

Much later, one more joint venture between the major European and US oil companies was founded in the Middle East. It concerned Anglo-Persian's 1933 concession. On 20 March 1951, this concession was nationalized and taken over by the newly founded National Iranian Oil Company (NIOC). After three-and-a-half years of negotiations a contract of work, covering the area of the former concession, was made between NIOC and Iranian Oil Participants Ltd, a UK company better known as the Consortium. Shareholders and their shareholding in the Consortium were as follows: forty per cent for Anglo-Iranian Oil Company, the former concessionaire and on this occasion renamed British Petroleum Company (BP) (representing a loss of sixty per cent economic interest); fourteen per cent for Royal Dutch/Shell; six per cent for CFP; forty per cent for a group of US oil companies, distributed among them as follows: seven per cent for each of Standard of New Jersey, Standard of California, Socony Mobil, Texaco and Gulf Oil, and five per cent for Iricon Agency, Ltd., the latter owned by nine US oil companies, among which Standard of Ohio, Atlantic Refining and Getty Oil, the latter each with 0.417 per cent (see Chapter 11).

After the Second World War, the geographical expansion of the exploration and production activities of the petroleum industry continued, partly in response to a heightened demand for oil products, partly motivated by the wish of the Western States to lessen their dependence on oil from the Middle East. In the process, major oil discoveries were made in Alberta, Canada ('Leduc' field, 1947); Nigeria and Algeria (1956) and in Libya 'Zelten' field, 1959). About a decade later major oil was discovered in 'Prudhoe Bay', on the North Slope, Alaska (1967/1968), with oil flowing in 1977 after the completion of the 800-hundred mile Trans-Alaska pipeline to Valdez, and on the Norwegian North Sea continental shelf ('Ekofisk' field, 1969). One year thereafter, oil was found on the UK North Sea continental shelf ('Forties' field, 1970). In 1972 an important oil discovery was made in Mexico (the 'Reforma' field). All these discoveries demonstrated the existence of oil provinces. Throughout these years, the oil (and natural gas) potential of the Soviet Union, later the Russian Federation, Azerbaijan and Kazakhstan became established.

Note. Nowadays, new oil (and natural gas) discoveries are still continuously being made albeit in technically challenging areas, e.g., Brazil.

Due to the internationalization and at the same time integration of their activities, the oil companies became gradually, and depending on their size, structured as an international oil group (IOG) consisting of a series of companies located and active in different functions in different places in the world directly or indirectly controlled and owned through shareholding by a parent company positioned at the top of the corporate structure. Said group companies are referred to as affiliated companies or affiliates. In any one country to be referred to as the host country, an IOG may have several affiliates each of them bound to and undertaking a different function. In this manner in one and the same host country, an IOG may have an exploration/production venture, may have a refinery and be active in marketing oil products, each function bound to an affiliate specifically established for the purpose. Each such affiliate will be liquidated upon termination of that particular activity for which it was designated. Moreover within any IOG, certain activities are undertaken which are solely meant for the benefit of the IOG itself, e.g., research, finance and internal services. These activities are also organized on the basis of separate corporate identities. Summing up, when fully developed, an IOG consists of an aggregation of separate companies (group companies/ affiliates), either belonging to the category of operating companies or to that of specialized service companies, bound together by shareholding. The aforementioned parent company itself is a public company which is incorporated and has its head office in one of the Western States (OECD countries) and whose shares are traded at any recognized Stock Exchange. For providing financial and other for its shareholders relevant information as required by the rules of the Stock Exchange where the parent company is listed, the latter is considered and treated as a single commercial entity. The OECD Member State, in which the parent company is incorporated and has established its head office, is referred to as the IOG's home country. On the other hand, any group company/affiliate is by preference incorporated in the host country; but if a proper legal infrastructure is absent, it is of necessity incorporated in the home country of the parent company. Specialized service companies, with exception of finance companies, or operating companies exercising a centralized function, such as trading on the international oil market, are usually established in the home country. Finance companies are by preference established in countries with a favourable fiscal regime, which could be the home country.

2.1.3 Joint Ventures for Exploration and Production

Following the example set by the Middle East joint ventures mentioned above, it has become standard practice among IOGs to form exploration and production joint ventures between their respective operating companies (affiliates) and never to seek a large majority in any one single venture. To this end, the affiliates enter into agreements of cooperation providing rules for jointly applying for a petroleum authorization and, if successful, for jointly conducting exploration and production operations within the limits of the applied for petroleum authorization (see Chapter 11).

2.1.4 State-Owned Commercial Oil Enterprises

In many countries the State concerned owns one or more commercial oil enterprises. This is the case not only in countries with a centrally organized or developing economy but also, at least in the past, in countries with a free market economy. In the latter case, governments expected their state-owned enterprise to compete with the private sector with the ultimate aim to safeguard the supply of oil to the domestic market. A very early, but incomplete example of a state-owned commercial oil enterprise was the Anglo-Persian Oil Company (later British Petroleum) in which the UK government for strategic reasons acquired in August 1914, a few days after the outbreak of the First World War, a controlling interest of 51.7 per cent. Examples of 100 per cent state-owned commercial companies are the companies that were established in 1953 in respectively France and Italy, viz. Entreprise de Recherches et d'Áctivites Pétrolières (ERAP) and its subsidiary Société Nationale Elf Acquitaine and Ente Nationale Idrocarburi (ENI) and its subsidiary Agip Mineraria, and the company that was established in 1975 in the UK, viz. the British National Oil Corporation (BNOC).

2.1.5 Governments and IOGs

IOGs are represented by their affiliates in many countries and fall, partially, under many jurisdictions. Only the government of the home country of the IOG, through its jurisdiction, fiscal and administrative, over the parent company and head office, is in a position to exert influence over the global activities of the IOG. This position allows such government, in extreme cases, to compel the IOG's parent company to cause the suspension or even termination of the IOG's activities (as undertaken by a particular affiliate) in any host country, simply by threatening to withdraw fiscal and/or other privileges. In contrast, the government of a host country has only jurisdiction over the locally established IOG-company or rather over that company's assets in the country. Among such assets may feature petroleum exploration and production assets and the underlying petroleum authorization.

2.2 GLOBAL EXPANSION OF PRODUCTION OPERATIONS

2.2.1 Oil Production

Table 2.1. Geographical Distribution and Volume of Oil Production (thousand b/d) 1885-1975

	1885	*1905*	*1939*	*1965*	*1970*	*1975*
US	60	370	3,460	9,014	11,297	10,008
Canada	-	-	-	920	1,473	1,735
Mexico	-	-	114	362	487	806
Venezuela	-	-	570	3,503	3,754	2,422
Tsarist Russia/FSU	38	151	577	4,858	7,127	9,916
Rumania	-	-	145	266	284	311
Netherlands Indies/ Indonesia	-	21	183	486	854	1,306
Persia/Iran	-	-	211	1,908	3,848	5,387
Saudi Arabia	-	-	12	2,219	3,851	7,216
Iraq	-	-	81	1,313	1,549	2,271
Kuwait	-	-	-	2,371	3,036	2,132
Libya	-	-	-	1,220	3,357	1,514
Nigeria	-	-	-	274	1,084	1,785
UK	-	-	-	2	4	34
Norway	-	-	-	-	-	189
China	-	-	-	227	615	1,545
OECD	-	-	-	10,779	13,920	13,670
OPEC*	-	-	-	14,386	23,509	27,168
World	101	590	5,669	31,803	48,061	55,825

* OPEC was established on 1 September 1960.
Sources used: American Petroleum Institute, L.U. De Sitter, 1950; Petroleum Economist; as from 1965: BP Statistical Review of World Energy 2007; Oil & Gas Journal.

The years shown in Table 2.1 have not arbitrarily been chosen. In the selected years new oil provinces were discovered, new inventions using or needing oil products for their operation came to the market and/or political events took place which made a lasting impact on the demand for oil and oil products.

In respect of the year 1885. This year saw the first appearance of a motor car (a Daimler/Benz) on the road in Germany. The year 1885 was the year, in which for the first time oil was discovered outside Pennsylvania, namely in Northwestern Ohio, and also the year, in which the first well was drilled in Atjeh, Sumatra. Finally, 1885 was the year in which Thomas Edison demonstrated in Manhattan, New York a large-scale application of electric light invented by him a few years earlier. This demonstration foretold the demise of the oil lamp and the end of the use of kerosene as illuminating oil.

In respect of the year 1905. This year marked the start of the planning of the first mass-produced motor car, namely Ford's T-model, which in 1908 came on the market. It is of importance to note that this type of car was powered by a gasoline engine. The choice for the gasoline engine (rather than for the diesel engine which some years before (around 1893) had been invented by Rudolf Diesel) secured a demand and market for the lighter distillates at the expense of the middle distillates. The gasoline market would in the coming decades grow rapidly in line with the growing number of motor cars on the road in as much as these cars, in contrast with heavy duty vehicles, would continue to run on gasoline. To put the matter in its proper perspective, the real growth in the number of cars took place about fifty years later. Between 1945 and 1990 the number of registered cars would grow tenfold, from 40 million to about 400 million. The year 1905 is also the year in which the airplane established itself as a means of transport to be taken seriously, after the brothers Orville and Wilbur Wright on 17 December 1903 in Kitty Hawk (North Carolina) had made the first demonstration flight. The use and further development of the airplane opened a new market, initially for aviation gasoline, much later for kerosene (for the jet engine).

In respect of the year 1939. This year marks the year in which the Second World War started. The volume and geographic distribution of global oil production would prove to be a decisive strategic factor in the outcome of this war.

The year 1939 was also the last year of the period in which the US oil industry and the US oil market dominated the world of oil. Soon after the Second World War (in 1948) the US would become a net-importer of oil and the export/import of low-cost oil from the Middle East would since then forever dominate the international market.

In respect of the post-war years leading up to the year 1975. In those years, world oil production rapidly expanded. US oil production stabilized around 10 million barrels per day. Production from the fields in the Middle East Gulf area increased spectacularly, from 8.387 million b/d in 1965 to 19.733 million b/d in 1975.

In respect of the year 1975. The year 1975 was the year in which the global oil market started to function.

Table 2.2. Geographical Distribution and Volume of Oil Production (million b/d) 1975-2006

	1975	*1979*	*1989*	*1998*	*2002*	*2006*
US	10.0	10.1	9.2	8.0	7.6	6.9
Canada	1.7	1.8	2.0	2.8	2.9	3.1
Mexico	0.8	1.6	2.9	3.5	3.6	3.7
Total North America	12.5	13.6	14.0	14.2	14.1	13.7
Venezuela	2.4	2.4	2.0	3.5	2.9	2.8
Total South & Central America	3.7	3.9	4.2	6.9	6.6	6.9
FSU/ Russian Federation	9.9	11.8	11.1	6.2	7.7	9.8
Azerbaijan	-	-	0.3	0.2	0.3	0.7
Kazakhstan	-	-	0.5	0.5	1.0	1.4
Iran	5.4	3.2	2.9	3.8	3.5	4.3
Saudi Arabia	7.2	9.8	5.6	9.5	8.9	10.9
Iraq	2.3	3.5	2.8	2.1	2.1	2.0
Kuwait	2.1	2.6	1.4	2.2	2.0	2.7
UAE	1.7	1.8	2.0	2.6	2.3	3.0
Total Middle East	19.7	22.0	16.4	23.0	21.6	25.6
Algeria	1.0	1.3	1.3	1.5	1.7	2.0
Libya	1.5	2.1	1.2	1.5	1.4	1.7
Nigeria	1.8	2.3	1.8	2.2	2.1	2.5
Angola	0.2	0.1	0.5	0.7	0.9	1.4
Total Africa	5.0	6.8	6.2	7.6	8.0	10.0
UK	0.03	1.6	1.9	2.8	2.5	1.6
Norway	0.2	0.4	1.5	3.1	3.3	2.8
China	1.5	2.1	2.8	3.2	3.3	3.7
Indonesia	1.3	1.6	1.5	1.5	1.3	1.1
Total Asia Pacific	3.8	5.1	6.5	7.7	7.9	7.9
OECD	13.7	16.5	18.8	21.5	21.4	19.4
EU 25	0.4	2.0	2.5	3.4	3.2	2.3
OPEC 11*	27.2	31.2	22.9	31.2	29.0	34.2
World	55.8	66.0	64.0	73.6	74.5	81.7

* OPEC without Angola.
Source: BP Statistical Review of World Energy 2007.

Background of the years selected for Table 2.2:

- *The year 1975* was the year in which the global oil market started to function.

- *The year 1979* was the year in which the Iranian revolution was completed and the year of the second price shock. The political upheaval caused a considerable drop in Iranian oil production (from the all time high of 6.0 million b/d in 1974 to 3.2 million b/d in 1979 to the all time low of 1.3 million b/d in 1981) and sparked off for the second time a hefty rise of the prices on the global oil market.
- *The year 1989* saw the fall of the Berlin Wall (10 November 1989); this ultimately led to the ending of the Cold War and the dissolution of the Soviet Union (December 1991). The ensuing improvement of the political relationship between the Western States on the one hand and the Russian Federation and the newly-formed independent States, in particular Azerbaijan and Kazakhstan, on the other made it politically and financially possible, to make a start with Western investments for the development of the oil and natural gas potential of those States mindful of future export to the EU. Of more immediate importance, however, was the prospect of further expansion of the export of natural gas from the Russian Federation to the EU.
- *The year 1998* saw a further weakening of the oil price, continuing a decline which had started a year earlier. In December 1998 the world market price reached its lowest level ever (viz. 10/b US dollars [USD]). This price weakness was caused by an expansion of non-OPEC (high cost) production that was made possible by the favourable price levels that OPEC had set and tried to maintain.
- *The year 2002* witnessed a renewed surge of the international oil price, which would continue until today.
- *The year 2006*, the latest year about which statistical information is available, seemed to signal an end to the price surge at the very elevated level that was reached in that year but in the second half of 2007 prices started to climb again, in November 2007 the Brent price was nearing the USD 100/b mark. At the same time however the US dollar considerably weakened against the euro (EUR): USD 1.0 equalled EUR 0.69. These trends continued in the first quarter of 2008. In said quarter, the Brent price regularly exceeded USD 110/b. Concurrently the US dollar further dropped to EUR 0.64. The year 2006 confirmed the trend of increasing world production levels, but there are important differences between the oil producing countries: production declined in the US, the UK and Norway but increased in the Russian Federation and Angola. Interestingly, the latter joined OPEC as its 12th Member State as of 1 January 2007.

2.2.2 Natural Gas Production

Table 2.3. Geographical Distribution and Volume of Natural Gas Production (billion cubic metres) 1975-2006

	1975	*1985*	*1990*	*1995*	*2000*	*2006*
US	545	466	504	527	543	524
Canada	75	84	109	159	183	187
Mexico	15	27	27	27	36	43
Total North America	634	577	640	712	762	754
Total South & Central America	24	46	58	73	98	144
Netherlands	76	72	61	67	57	62
Norway	-	26	25	28	50	88
FSU/Russian Federation	270	431	598	555	545	612
Turkmenistan	-	78	82	30	44	62
United Kingdom	34	40	45	71	108	80
Uzbekistan	-	32	38	45	53	55
Iran	20	15	23	35	60	105
Qatar	2	5	6	13	24	49
Saudi Arabia	3	19	33	43	50	74
UAE	2	13	20	31	38	47
Total Middle East	33	64	101	149	207	336
Algeria	6	34	49	59	84	84
Nigeria	-	3	4	5	12	28
Total Africa	12	46	67	83	127	180
Australia	6	13	21	30	31	39
China	8	12	15	18	27	59
Indonesia	2	32	45	63	68	74
Malaysia	-	10	18	29	45	60
Total Asia Pacific	35	106	150	213	272	377
EU 27	194	199	186	213	232	203
OECD	806	786	851	972	1,070	1,078
World	1,203	1,667	1,992	2,135	2,425	2,865

Source: BP Statistical Review of World Energy 2007.

In the period preceding the discovery of important natural gas deposits in Western Europe, i.e., before 1959, the year in which the Slochteren gas field was discovered in the Netherlands, natural gas production and consumption were concentrated in the US and Canada. After the Second World War, large-scale natural gas production was developed in the Soviet Union. In all these countries, gas producing fields happened to be located in the vicinity of the places of consumption ensuring that the costs of transportation were manageable, which in turn allowed to keep the cost of supply within the range of those of competing coal and oil products. In these areas non-associated gas fields could commercially be developed and produced at an early stage. Nonetheless, the consumption and production of natural gas even in North America really became significant only after the Second World War. This development owed much to improved pipeline technology that then became available and which made it technically possible to build long-distance gas pipelines capable of withstanding the relatively high pressures under which natural gas has to be moved through the pipeline. The construction of the so improved pipelines opened up new local and regional markets and consumption could grow as is illustrated by the following figures: in 1935 US natural gas production sold amounted to 54 bcm; in 1950 and in 1960 this quantity had risen respectively to 178 bcm and 360 bcm. Forty-six years later, in 2006, US gas production stood at 524 bcm (almost ten times the level reached in 1935), but the country's gas consumption in 2006 (620 bcm) still exceeded that year's production. The deficit, which must be considered to be permanent, even growing, is covered by imports from Canada and Mexico.

Starting with the very substantial gas discovery made in 1959 near Slochteren, Groningen province in the Netherlands, many more gas deposits were found in the 1960s on the North Sea continental shelf. These deposits were sufficiently large to allow the development of a gas market in that region. On a global scale, however, the commercial prospects of natural gas definitively improved only after and as the (unexpected) result of the first oil price shock (1973/74), which raised the prices of the competing oil products (e.g., low-sulphur fuel oil and heating oil) dramatically. Export/import of natural gas on a commercial basis became possible due to the necessary improvement of the margins, allowing the recovery of the cost of long distance gas transportation whether by pipeline or by ocean-going tanker. It became also commercially possible to produce natural gas in high cost areas (e.g., the deeper waters). Since 2000, in response to and as a direct result of the spectacular increase of the international oil price this development has gained further strength. The overall effect becomes apparent when comparing the world natural gas production in 1975, which amounted to 1.083 million toe with the level reached in 2006, which was 2.578 million toe. World oil production increased also over this period, namely from 2.851 million tonnes in 1975 to 4.173 million tonnes in 2006, but not at same rate (Table 2.3). It is expected that in the coming decades mainly due to government policies in respect of climate change and security-of-supply, all favouring the use of natural gas as fuel above that of coal and oil, the gap between oil and natural gas production in terms of million toe will continue to narrow assuming that oil and gas prices are stabilizing at current levels.

2.3 PROVED RESERVES AND ULTIMATE RECOVERY

2.3.1 Conventional Oil

Table 2.4. Size and Geographical Distribution of Proved Conventional Oil Reserves (billion barrels) at End 2005 and 2006 and Reserves/Production (R/P) Ratios (years)*

	Proved Reserves at end 2005	*Proved Reserves at end 2006*	*R/P Ratio*
US	29.9	29.9	12
Canada*	17.1	17.1	15
Mexico	13.7	12.9	10
Total North America	60.7	59.9	12
Venezuela	80.0	80.0	78
Total South & Central America	103.2	103.2	41
Azerbaijan	7.0	7.0	29
Kazakhstan	39.8	39.8	76
Russian Federation	79.1	79.5	22
Iran	137.5	137.5	87
Iraq	115.0	115.0	>100
Kuwait	101.5	101.5	>100
Saudi Arabia	264.2	264.3	67
United Arab Emirates	97.8	97.8	90
Total Middle East	742.7	742.7	79
Algeria	12.3	12.3	17
Libya	41.5	41.5	62
Angola**	9.0	9.0	18
Nigeria	36.2	36.2	40
Total Africa	117.2	117.2	32
United Kingdom	3.9	3.9	6
Norway	9.6	8.5	8
China	16.2	16.3	12
Malaysia	4.2	4.2	15
Indonesia	4.3	4.3	11
Total Asia Pacific	40.5	40.5	14
World	1,209.5	1,208.2	40
European Union 27	7.3	7.1	8
OPEC 12***	914.5	914.5	70
Canadian oil sands****	163.5	163.5	n.a.

* Canadian proved oil reserves include 10.2 billion barrels contained in oil sands under active development.
** Angola joined OPEC as the 12th Member State as of 1 January 2007.
*** OPEC 12 includes Angola.
**** Official estimate of remaining established bitumen reserves in oil sands not under active development.
Source: BP Statistical Review of World Energy 2007.

The above listed proved oil reserves concern conventional oil and include natural gas liquids (NGLs) and condensates, except that the Canadian proved oil reserves include an element of non-conventional oil, viz. 10.2 billion barrels of bitumen contained in oil sands under active development. The official estimate of economically recoverable bitumen from the three Alberta oil sands areas at current prices using current technology, including 10.2 billion barrels contained in oil sands under active development amount to 174 billion barrels which is equivalent to about ten per cent of the estimated 1.700 billion barrels of bitumen in situ.

The R/P ratios of the listed countries/regions indicate to what extent these countries or regions are over- or under-producing their reserves. Examples of countries/regions with a low R/P ratio and thus over-producing their (collective) reserves are: North America, which include the US, Canada and Mexico, showing a R/P ratio of twelve years; China and Indonesia with R/P ratios of respectively twelve and eleven years; the United Kingdom with a R/P ratio of six years and Norway with a R/P ratio of eight years. Conversely, the Middle East shows a R/P ratio of seventy-nine years and in this region, which includes the prolific Gulf area, the reserves are under-produced.

Table 2.4 also shows a geographical mismatch between the places where the proved conventional oil reserves are located, i.e., the places of future conventional oil production, and the areas of major oil consumption as shown on Table 3.2, e.g., the US, EU, China and Japan. Significantly, close to sixty-two per cent of the world proved oil reserves are located in the Middle East (including non-OPEC-countries such as Syria, Oman and Yemen). Furthermore, the Member States of OPEC including the new Member State Angola, collectively referred to as OPEC 12, possess close to seventy-six per cent of the world reserves, while their current collective share in the world oil production is currently no more than 43.5 per cent or 35.6 million b/d. Future large scale exploitation of non-conventional oil, such as in particular the Canadian oil sands may change the geographical mismatch a little in favour of the major consuming countries.

Remains to be investigated how many recoverable reserves may be added to the world proved conventional oil reserves at end 2006 (168 billion tonnes, excluding any bitumen from Canadian oil sands) on account of new discoveries and reserves growth.

In a study from energy analysts Cambridge Energy Research Associates (CERA) in Massachusetts with base year 2000, ultimately recoverable reserves of conventional oil, including historical production up to the base year of 140 billion tonnes, are estimated to amount to 410 billion tonnes (source: *Nature*, vol. 445, 4 January 2007.) After correcting for the volumes produced between 2000 (the base year) and end 2006, i.e., 24 billion tonnes, and subtracting the historical production the remaining reserves at end 2006 of conventional oil amount to 246 billion tonnes. Subtracting the proved reserves at end 2006 (i.e., 168 billion tonnes excluding any bitumen from Canadian oil sands) the resulting figure of 78 billion tonnes represent the expectations regarding new discoveries and reserves growth in conventional oil in the coming years, at least according to the CERA analysis.

Note 1. By contrast, the US Geological Survey, as reported in IEA's publication *World Energy Outlook 2006*, also using the year 2000 as the base year, estimated the remaining ultimately recoverable reserves of conventional oil (referred to as 'light' oil) to be 322 billion tonnes. After correction for the volumes of oil produced between 2000 (the base year) and end 2006, i.e., 24 billion tonnes, the remaining ultimately recoverable reserves at end 2006 are 298 billion tonnes.

Subtracting the proved reserves conventional oil at end 2006 (i.e., 168 billion tonnes excluding any bitumen from Canadian oil sands) the resulting figure of 130 billion tonnes represent the US Geological Survey's (USGS) expectations regarding new discoveries and reserves growth in conventional oil in the coming years. As mentioned above, the corresponding, more conservative CERA figure is 78 billion tonnes, a substantial difference of 52 billion tonnes. To remain on the safe side, the CERA figure will be used for further estimates.

Note 2. New discoveries of conventional oil may be expected from the use of improved geophysical surveys. Reserves growth may be the result of the application of advanced extraction, production, drilling techniques to known reservoirs that had before been produced in a conventional manner or not yet at all. In both instances the aim is to improve the oil recovery factor that characterizes the reservoir. Apart from exceptional circumstances, such as a low viscosity oil (light oil and condensate) produced from a reservoir with a satisfactory permeability, an oil recovery factor of fifty per cent (which means that fifty per cent of the oil reserves in situ is left behind in the reservoir) is currently considered the maximum attainable, when the reservoir is produced in a conventional manner. If advanced recovery methods are applied to important oil fields that never passed the primary stage of exploitation (see Chapter 1) the reserves growth can be significant. In view of the costs involved, all this can only be realized if the international oil price stays at its current high level. If this requirement is met, and there is for the moment at least no reason to assume that this will not be the case, it would allow an optimistic estimate regarding the growth of the reserves.

2.3.2 Non-conventional Oil

Non-conventional oil is oil that at reservoir temperature will not flow through the reservoir and that therefore cannot be produced in the conventional manner (see Chapter 1). There are two categories of non-conventional oil: heavy oil with an API gravity of less than 22 degrees, including extra-heavy oil with an API gravity of less than 10 degrees, and natural bitumen, very dense and viscous oil that is mixed with sand and clay minerals.

Table 2.5. Geographical Distribution of Estimated Technically Recoverable Heavy Oil and Natural Bitumen Reserves (billion barrels)

	Heavy Oil	*Natural Bitumen*
North America	35	531
South America	266	-
Western Hemisphere	301	531
Middle East	78	-
Eastern Hemisphere	133	120
World	434	651

Source: US Geological Survey Fact Sheet FS-070-03, August 2003.

Table 2.5 shows that the Western Hemisphere has sixty-nine per cent of the world's technically recoverable heavy oil and eighty-two per cent of the technically recoverable natural bitumen. By contrast, the Eastern Hemisphere has about eighty-five per cent of the world's conventional oil reserves (see Table 2.4).

Total technically recoverable reserves of heavy oil and natural bitumen, viz. 1.085 billion barrels, are about equal to the proved reserves of conventional oil, viz. 1.198 billion barrels (see Table 2.4).

The largest extra-heavy oil accumulation is the Venezuelan Orinoco heavy oil belt, which contains 1.8 trillion barrels extra-heavy oil in situ of which thirteen per cent or 234 billion barrels are considered to be recoverable. The largest natural bitumen accumulation is found in Alberta, Canada. These oil sands are estimated to contain 1.7 trillion barrels in situ, of which thirty-two per cent or 544 billion barrels (76 billion tonnes) are estimated to be recoverable. Together the two deposits hold 778 billion barrels (109 billion tonnes) recoverable oil.

Currently, only the Canadian oil sands are under active exploitation. The Canadian proved oil reserves include an official estimate of 10.2 billion barrels for this part. Remaining established reserves not under active development are estimated at 163.5 billion barrels. But, as mentioned above, these oil sands are actually expected to hold 544 billion barrels (76 billion tonnes) of bitumen (by coincidence about twice as much as the proved reserves of conventional oil of Saudi Arabia, see Table 2.4). Since the Canadian oil sands are under active development and contribute to the current and future supply to the global oil market, it is justified to add these 76 billion tonnes to CERA's reserves figure of 246 billion tonnes of conventional oil. The result is an estimated 322 billion tonnes of ultimately recoverable oil generally, the sum of conventional and non-conventional oil.

For the moment there are no plans for a large-scale development of the Venezuelan extra-heavy oil deposits. This should not come as a surprise: it would make no economic sense to produce these reserves parallel with the much less expensive production of domestic conventional oil.

2.3.3 Oil Shale

Oil shale is a special type of sedimentary rock which contains kerogen (to be considered a proto-hydrocarbon) in a finely dispersed form. As a matter of fact oil shale is a source rock that through geological processes has been uplifted and become exposed at the surface. Because of this geological uplifting the natural hydrocarbon generation processes were interrupted and did not go beyond the intermediate stage of forming kerogen (refer Chapter 1).

Table 2.6. Geographical Distribution of Oil Shale Reserves in situ (in billion barrels)

	Shale Oil Reserves in situ
US	2,587
Democratic Republic of the Congo	100
Morocco	53
Russian Federation	247
Italy	73
Australia	32
Brazil	82
World	3,328

Source: Factual information obtained through Wikipedia, the free encyclopaedia, with further references.

The Green River Formation in Colorado, Wyoming and Utah, US, is the largest single oil shale deposit in the world, holding 1,466 billion barrels in situ. Since in no country serious exploitation of oil shales is contemplated and recovery factors are largely unknown the potential contribution to the supply to the global oil market must for the moment be left aside.

2.3.4 Natural Gas

Table 2.7. Size and Geographical Distribution of Proved Natural Gas Reserves (trillion cubic metres) at end 2005 and 2006 and Reserves/Production (R/P) Ratios (years)

	Proved Reserves at end 2005	*Proved Reserves at end 2006*	*R/P Ratios*
US	5.79	5.93	11
Canada	1.63	1.67	9
Mexico	0.41	0.39	9
Total North America	7.83	7.98	11
Total South & Central America	6.85	6.88	48
Kazakhstan	3.00	3.00	>100
Netherlands	1.39	1.35	22
Russian Federation	47.66	47.65	78
Turkmenistan	2.86	2.86	46
United Kingdom	0.48	0.48	6
Iran	27.58	28.13	>100
Iraq	3.17	3.17	>100
Qatar	25.36	25.36	>100
Saudi Arabia	6.82	7.07	96
United Arab Emirates	6.07	6.06	>100
Total Middle East	72.49	73.47	>100
Algeria	4.50	4.50	53
Nigeria	5.15	5.21	>100
Total Africa	14.08	14.18	79
Australia	2.61	2.61	67
China	2.45	2.45	42
Indonesia	2.48	2.63	36
Malaysia	2.48	2.48	41
Total Asia Pacific	14.66	14.82	39
European Union 25	2.49	2.43	13
Total World	180.20	181.46	63

Source: BP Statistical Review of World Energy 2007.

Table 2.7 shows that North America and the EU 25 have the lowest R/P ratios, namely eleven and thirteen years respectively. Within these regions the UK shows the lowest figure, namely six years. These regions and country are over-producing their gas reserves. If no new reserves are found, the time will come that natural gas must be imported (North America) or that the already existing import of natural gas must be increased (EU). By contrast, the Russian Federation has a R/P ratio of seventy-eight years and Total Middle East a R/P ratio of over one hundred years. These figures suggest that the natural gas reserves of these two areas are at the moment under-utilized and possess a great production and export potential. In the Russian Federation, a great effort is being made to realize this potential but the Middle East region is lagging behind in this respect. Currently, LNG export takes place from UAE and Qatar which is directed at Japan. In Qatar are also LNG export schemes under development directed at the West European market. In this market Qatar's liquefied natural gas will have to compete with natural gas that is exported from the Russian Federation by long-distance pipelines. The potential for increased gas exports from Qatar to Japan is not very great also due to future competition from Russian LNG in that market.

Note. Any estimate of recoverable natural gas reserves is dependent on the level of future gas prices. As long as the gas price remains linked with prices of comparable oil products and good quality coal and as long as these prices or some of them stay at the currently prevailing high level, the prospects for utilizing associated natural gas and for development of non-associated gas fields far removed from the markets will improve and more gas can be added to the recoverable reserves.

According to the USGS (2000), remaining ultimately recoverable gas reserves, including proved reserves, reserves growth and as-yet undiscovered resources, could total 283 billion toe. This was estimated using the year 2000 as the base year. Subtracting from this figure the volumes that have in the meantime been produced, i.e., about 15 billion toe, and subtracting further the proved reserves at end 2006, i.e., 163 billion toe, results in a figure of 105 billion toe. The latter figure represents the contributions that are expected from new discoveries after 2006 and, to a much smaller extent, from reserves growth. The cumulative production to date amounts to only fifteen per cent of total resources.

Note. The recovery factor for natural gas, permeability permitting, ranges already between eighty per cent and ninety per cent, when produced in a conventional manner. However there is still some scope for the application of recovery enhancing techniques, such as CO_2 injection into the reservoir. But compared with oil, there is overall less scope for reserves growth. It should be added that under circumstances of poor permeability, horizontal drilling techniques may help to overcome this gas flow restricting factor and restore the recovery efficiency to normal values, but the natural gas, so produced may turn out to be non-commercial (see Chapter 1).

There is also non-conventional natural gas. This non-conventional gas is represented by gas hydrates. The latter are crystals of a methane/water compound which are formed when natural gas surfaces at the bottom of the deep ocean and reacts with the ocean water (see Chapter 1).

The gas hydrates deposits are at the present moment not seriously investigated by the industry, except in Japan and in the US (Alaska North Slope). This will not change in the future. There exists considerable concern about the possibility that, as a result of any large scale exploitation, the stability of the hydrates deposits will be disturbed and cause the release of huge amounts of methane in the ocean and from there into the atmosphere. Such release of methane is very unwelcome because methane is a potent greenhouse gas and its presence in the atmosphere will have a very serious negative effect on global warming (see Chapter 1). Because of this risk it is not very likely that gas hydrates will ever be produced; these non-conventional reserves can therefore not be added to the ultimately recoverable conventional reserves.

2.3.5 COAL

For comparison purposes (see Chapters 3 and 4), the status of the coal reserves is indicated in the following Table.

Table 2.8. Geographical Distribution and Size of Proved Coal Reserves at end 2006 (billion tonnes) and Reserves/Production (R/P) Ratios (years)

	Proved Reserves Hard Coal	*Proved Reserves Lignite and Brown Coal*	*Proved Reserves Total Coal*	*R/P ratio*
US	111.3	135.3	246.6	234
Columbia	6.2	0.4	6.6	101
Kazakhstan	28.2	3.1	31.3	325
Germany	0.183	6.6	6.7	34
Poland	14.0	-	14.0	90
Russian Federation	49.1	108.0	157.0	>500
Ukraine	16.3	17.9	34.2	424
South Africa	48.7	-	48.7	190
Australia	38.6	39.9	78.5	210
China	62.2	52.3	114.5	48
India	90.0	2.4	92.4	207
OECD	172.4	200.9	373.2	177
World	478.8	430.3	909.0	147

* One tonne hard coal equals 0.66 toe; one tonne sub-bituminous coal equals 0.33 toe.
Source: BP Statistical Review of World Energy 2007.

Total proved coal reserves (hard coal and sub-bituminous coal) amount to 458 billion toe. The countries with the largest reserves of the economically more important hard coal are in order of magnitude: US, India, China, the Russian Federation, South Africa and Australia. For sub-bituminous coal is the ranking: US, the Russian Federation, China and Australia. The total reserves at the end of 2006 (458 billion toe) compares with the world proved conventional oil reserves (i.e., excluding any non-conventional oil such as the Canadian tar sands) of 168 billion tonnes and the world proved natural gas reserves of 163 billion toe. To make the fossil fuel reserves picture complete, the reserves of the Canadian tar sands should be added. For the moment the established reserves under active development and those not under active development are reported to amount to 1.4 billion tonnes and 23 billion tonnes respectively.

2.4 POLITICAL RISK

2.4.1 Historical Background

The global expansion of the petroleum industry brought IOGs into contact with more and more non-Western countries and their respective governments. These IOGs were confronted with different cultures, different political organizations and aspirations. Before the Second World War most non-Western countries were dependent States, mostly under Western domination, if not under European colonial rule. When trying to acquire and operate under petroleum authorizations in these countries, and in dealing with the host governments, IOGs could count on the political support and diplomatic protection of their home countries. IOGs could also count on the internal political stability of the host country or, if failing, could count on intervention of a Western power restoring order and bringing back stability. By contrast, governments of dependent States had mixed feelings about the arrival of IOGs in their country. On the one hand there was the expectation that the oil company would discover oil and make the country (or at least its ruling class) wealthy and on the other there was the fear that the new arrival would try to interfere and meddle in the domestic political situation. These feelings are reflected in a few provisions incorporated in the petroleum agreements, which in the 1930s were concluded between IOGs and the Middle East Gulf governments (see Chapters 5 and 6).

After the Second World War the international political situation changed rapidly. Two tremendous political developments were set in motion. In the first place there started a so-called 'Cold War' between the Soviet Union and its communist allics on thc onc sidc and the Western Powers on the other. The Cold War began in 1947, on the occasion of the adoption of the Marshall Plan for Western Europe and one year before the Universal Declaration of Human Rights was proclaimed by the General Assembly of the UN (GA Resolution 217 (III) of 10 December 1948). The Cold War definitively ended in 1991 but, as long as it lasted, the political world had been divided in two camps and any non-Western country being in the Western camp or prepared to join this camp was only judged on its political trustworthiness and internal political stability, without attention being paid to the way the country was governed or to how its citizens were being treated by the ruling class. Indeed, military regimes were welcome as allies because they could be relied upon to provide the desirable political stability and keeping any communist political activity in check. For these reasons, the said Universal Declaration of Human Rights, which was annexed to the Charter of the United Nations, could not be politically activated and remained dormant.

In the second place, decolonization processes took place that saw the creation of many new independent States. These new States, while politically independent from the Western countries, were faced with an economy that was still dominated by Western companies. They started a struggle to gain dominance over their economies and, as far as petroleum operations were concerned, sought to achieve this by state participation or outright nationalization or expropriation of IOG-held petroleum authorizations and other assets. At the same time and in the same context, the new States supported in this by other developing countries started a political campaign within the UN General Assembly in order to get international recognition and blessing of their struggle for economic independence, which focused on the concept of the permanent, inalienable sovereign powers of a State over its natural resources and on the introduction and establishment of a New International Economic Order (NIEO) (see Chapter 10).

When the Cold War gradually ended (1989-1991), Western governments got and used the political freedom to focus their attention on other States' attitude with respect to fulfilling their international obligations under the Charter of the United Nations. In particular attention was paid to any State's obligation to protect, respect and maintain the human rights of its own citizens in accordance with the Universal Declaration of Human Rights (see Chapter 10). In recent years, attention has shifted somewhat to those obligations under the Charter of the United Nations which require from any State that it will support and promote international peace and security and within this framework will abstain from supporting international terrorism, refrain from the threat or use of force against the territorial integrity or political independence of any State and not violate any Treaty aimed at prohibiting the manufacturing or proliferation of weapons of mass destruction of any kind. In a situation that a State does not fulfil the latter obligations, the UN Security Council, if all its permanent Members are in agreement, may impose trade sanctions, and if a petroleum producing country is involved, prohibit the export of oil and any foreign investment in petroleum operations. If a violation of human rights is judged to take place, trade sanctions may also be imposed, but if there is no international security aspect to such violation, trade sanctions, if deemed required, cannot be imposed by the UN Security Council, but must be put in place and maintained by a group of interested States, in which case Western States always take the lead.

Note. The imposition of trade sanctions or the taking of measures to hinder or prevent investments and the flow of capital on or with respect to any State that is considered not to fulfil any of its international obligations under the Charter of the United Nations, has generally devastating effects on the well-being of such State's population. If the offending State belongs to the category of capital importing developing countries, the above-described effects are reason enough to consider the imposition of trade/investment sanctions to amount to a violation of the human rights of their peoples and therefore not legitimate. But there is more. States that are parties to the New York Convention on Civil and Political Rights of 19 December 1966 must be considered to be prevented from imposing or from cooperating with other States in imposing trade/investment sanctions with respect to any State if the effect of these sanctions would be that the people of such State would be deprived from their means of existence. In the said Convention (Part I, Article 1.2), this is explicitly so stipulated.

2.4.2 Description and Scope

Political risk, as understood and experienced by an IOG-company operating in a host country, means the risk that for purely political reasons, not sanctioned by or envisaged under applicable law or the petroleum authorization concerned, said IOG-company will lose its petroleum authorization and its other assets in the country or is forced to stop its operations and to abandon its venture.

The most common political reasons underlying political risk are:

(i) a partly politically, partly economically motivated change in attitude of the host government towards IOGs, leading to nationalization or expropriation of such IOG's rights and assets;
(ii) a breakdown of the internal political stability of the host country, causing civil unrest, civil war or even a de facto break-up of the country and the disappearance of a central government (in such a situation, the country is referred to as a failed State). Such breakdown of internal political stability may make it for an IOG physically or for safety reasons impossible to continue its (authorized) petroleum operations in the country; or
(iii) a breakdown of the host country's foreign relations, either with respect to the international community, or with respect to the home country or to one or more other Western States.

With reference to (i). In case of politically or economically motivated nationalization, the IOG's main concern is to receive adequate and prompt compensation for the losses sustained by it. Under customary international law, any oil company whose petroleum authorization is cancelled and whose physical assets are nationalized or expropriated by the host government for other reasons than those based on applicable law or on the petroleum authorization that had been granted has the right to receive aforesaid compensation. This principle has been confirmed in many bilateral treaties for the promotion and protection of foreign investments.

The issue of nationalization and expropriation of the assets of foreign multinational corporations and of paying compensation if such measures were taken by a developing country had been raised by those countries in the context of formulating and declaring a NIEO and adopting a Charter of Economic Rights and Duties of States (GA Resolution 3281 (XXIX) of 12 December 1974 (see Chapter 10).

More recently, the issue of nationalization and the compensation that has to be paid in such event has been addressed in the 1994 European Energy Charter Treaty (Article 13). In the said article, it is stated that nationalization, where allowed under the terms of the Treaty, should be accompanied by the payment of prompt, adequate and effective compensation. Moreover according to the Treaty (Article 12), a foreign investor has a right to be treated on the same footing as a national investor or any other investor in respect of receiving compensation for losses due to political instability affecting the host country. The compensation has to be paid by the affected host country (see Chapter 10).

With reference to (ii). A breakdown of the internal political stability of a host country may make it, for any IOG-company operating in that country physically or for safety reasons, impossible to continue its operations. In extreme cases the oil company involved may be forced to abandon its operations altogether and to leave the country.

With reference to (iii). In the post Cold War era, any IOG could find itself operating in a host country that has a dispute with the UN or with a group of concerned Western States on account of not fulfilling its international obligations and, that as a consequence thereof, is subjected to the imposition of trade sanctions. Any IOG operating in such a country will suffer heavy losses as long as the sanctions are maintained and may be forced to abandon its petroleum venture. It is also possible that the government of a home country on its own decides to take action against the offending host country and orders the parent company of the IOG to stop investments, to shelve further development or to abandon the venture altogether. A new twist in the matter of human rights can be noted. Apart from taking such an extreme measure of urging the parent company of an IOG to suspend its operations and/or postpone any new investments, a home government may make use of the presence of the IOG-affiliate in the country and ask that this affiliate is directed to take action against such violation, e.g., by registering a protest, by refusing to cooperate, etc. Obviously, if the affiliate would become involved in political action, such action could easily have disastrous consequences for those individuals among its personnel, who are citizens of the host country. Moreover the existence of the affiliate itself (and thereby the future of its personnel) would be put at risk since a host government could decide to retaliate by nationalizing the politically active affiliate on account of it having meddled in the internal affairs of the country.

Note 1. As a transnational enterprise, any IOG is manoeuvred in an untenable position. On the one hand, the IOG (as represented by its locally established affiliate) is expected to respect host government's economic and social policies and to abstain from intervention or from taking sides in local political disputes, all in accordance with generally accepted and internationally recognized standards of conduct applicable or deemed applicable to multinational companies as explicitly laid down in the 1974 Charter of Economic Rights and Duties of States. On the other hand, and in accordance with the provisions of the aforesaid Charter (Chapter IV, Article 33 read together with Chapter I paragraph (j)), any government social policy should be in conformity with the country's international obligations, in particular with those formulated in the Charter of the United Nations. This means that any government's social policy is subjugated inter alia to that government's international obligation to protect, respect and maintain human rights and, to the extent that this would not be the case, an IOG-company's general obligation to respect a host government social policy is no longer binding. Nonetheless, an IOG should not be compelled by its home government to direct its local company to undertake political action if by such action investments made before the violation took place are put at risk and/or the safety of its national personnel is endangered. Besides, if a sovereign State fails to observe its international obligations, it should be held responsible and be censured, not by private sector enterprises but directly by the other States that share the same treaty or convention and the same obligations with the host country. For instance, any State which does not protect or respect human rights should be censured by its co-Member States of the United Nations or, if applicable, by its co-Member States of the New York Convention on Civil and Political Rights of 19 December 1966, but anyway not by an IOG-company whose parent falls under any such co-Member State's jurisdiction. Similarly, if a State does not respect international peace and security and is for instance actively supporting international terrorism it should be censured by UN Security Council and not by the Western private sector, provided always that the local IOG-company should have applied for and be granted its current petroleum authorization and started work and acquired assets in the country before the regime became repressive and before it began to ignore its international obligations.

Note 2. In stark contrast with the international rules that apply to paying/receiving compensation in case of nationalization/expropriation measures taken by a host government, an IOG-company suffering losses as a result of the imposition of trade sanctions by the UN Security Council or by a group of Western States, or as a result of having to respond to a request from its home government to abandon or suspend its venture in the host country is not offered and may not expect any such compensation.

2.4.3 Political Insurance

Private sector enterprises however have the possibility to take out insurance covering against certain elements of political risk. Such an opportunity is offered by the Overseas Private Investment Corporation (OPIC), the US government agency responsible for insuring US private investors against political risks in developing countries. OPIC was established in 1969 through an amendment of the Foreign Assistance Act of 1961. It started operations in 1971 as a distinct corporate entity. Risks insured are risks directly related with the situation in or action taken by the host country, such as currency inconvertibility, outright confiscation, creeping expropriation, nationalization, war, revolution, insurrection or civil strife up to a fixed amount in US dollars (USD 100 million or more) per project. The political risk of being ordered by the home government to leave the host country and abandon the venture is not covered. Such a situation does not fit in with an insurance scheme offered by a US government agency (not surprisingly, considering that in the past on several occasions the US government has called its citizens [and US oil companies] back from abroad). US investors seeking OPIC's insurance need to obtain from the host government a special approval for OPIC to issue insurance on their investment in accordance with the agreement which OPIC has with such a government. US and non-US oil companies may also use the facilities offered by the Convention establishing the Multilateral Investment Guarantee Agency (MIGA) which went into effect on 12 April 1988. The MIGA Convention was an initiative of the World Bank within the wider framework of the Bank's efforts in the field of encouraging the flow of investments to developing countries. This Convention is based on the same principles as, and has much in common with, the insurance schemes offered by OPIC. The Convention is intended to guarantee eligible investments against losses resulting from non-commercial risks, among which the transfer risk resulting from host governments' restrictions on currency conversion and transfer, the risk of loss from expropriation and confiscation measures emanating from the host government, the repudiation of government contracts in the cases where the investor has no access to a competent judicial or arbitral forum, and the risk of outbreak of armed conflict and civil unrest. Before concluding a contract of guarantee, the host government must have approved the investment and its insurance by the Agency against the risks mentioned in the guarantee contract. Similarly as the insurance offered by OPIC, the risk to be ordered back by the home government and be compelled to withdraw from the insured venture is not covered.

Chapter 3
Petroleum and the Economy

3.1 GENERAL REVIEW

3.1.1 Oil and Natural Gas

When emerging at the wellhead, petroleum is not suitable for use in the economy; as such it is not a marketable product. Before it can be moved to the market, it has to be treated in the field as described in Chapter 1. After this treatment, oil, natural gas and natural gas liquids (NGLs), if any, are each transported by pipeline and/or by ocean-going tanker to their respective destinations. Associated natural gas for which there is no market or which cannot be utilized in the oil field concerned for one purpose or another will be flared, but not so before NGLs are recovered from the gas if it proves to be a rich or wet gas.

Note. Before natural gas can be transported by an ocean-going tanker it must first be liquefied (LNG) at the port of export. At its destination the LNG is gasified and the gas is delivered to a gas company for further conditioning and delivery to the end-user.

Oil's destination is the refinery. In the refinery, a range of products is manufactured from it amongst which lubricants, all kind of fuels for generating heat, electricity or mechanical power, and feedstock for the petrochemical industry. The sectors using the oil products as fuel comprise the general industry, power stations, households and the transport sector.

In contrast to oil, most natural gas needs only conditioning, including the recovery of NGLs, and can thereafter, directly or through a distribution entity, be sold to the end-user mainly for heating purposes (industrial plant, power stations, households), less for further processing in refineries. Again in contrast to oil, there does not exist a global market for natural gas, there are only local or regional markets which are supplied by the gas producer through fixed long-distance pipelines or LNG-tanker on the basis of long-term sales contracts.

Oil and natural gas as such have partly differing, partly overlapping applications and markets; albeit in theory, oil in all applications and uses can be replaced by natural gas or by products made from natural gas through catalytic conversion. These catalytic conversions are costly in themselves. Moreover, due to the relatively low efficiency of the available conversion techniques, a loss of more than fifty per cent of the heat content of the natural gas being converted should be reckoned with.

3.1.2 Oil

The economic use of oil has a long history. As of the days that liquid or semi-solid natural hydrocarbons such as tar and asphalt were found at the surface, they were recognized to be useful in all sorts of applications. Oil found dripping or oozing from rocks was used as a lubricant, or as medicine for curing all sorts of illnesses, while burning natural gas played a role in religion. In the pre-drilling days, oil that leaked or seeped into pools or creeks was skimmed from the surface of the water with cloths and rags and subsequently bottled and sold. In some places, e.g., in Pechelbronn, France, shafts and galleries were dug in order to obtain better access to the oil or tar and increase their recovery. Tar found in pits or mixed with sand was utilized in the ancient world as a stone binding material in buildings and in walls. Tar or products distilled from tar (pitch) were widely used as a sealing material, more specifically for caulking seams of ships and boats. The native people of Alberta, Canada, used the tar sands found in outcroppings along the Athabasca River to waterproof their canoes. Much later, tar (coal-tar) was also obtained as a by-product of the manufacturing of town gas, and this coal-tar was applied in a similar fashion (in addition to its use in the chemical industry for the production of aniline, an organic base material).

Note. As could be expected, oil found its use also in warfare. Naphtha, presumably distilled from tar, was, in combination with sulphur and saltpetre, the main ingredient of Greek fire. The latter mixture had been invented by the Greek Kallinikos and was brought by him to Constantinople at the time this city was besieged by the Saracen fleet (673-678). By pouring this mixture out of tubes into the Saracen ships it set them alight and drove them off, saving the city. Since the slightest shock caused the mixture to explode, it could not safely be transported over land and its use was eventually restricted to naval battles, where it proved, as it did on the first occasion in 678, to be very successful and decisive. In fact, this flame-throwing weapon gave the Byzantine fleet for a long time control over the Mediterranean.

About one hundred years after the start of the Industrial Revolution (1750), oil became used as feedstock for the distillation of kerosene (a middle distillate). Kerosene was used for burning in oil lamps. Around 1850, the lamp was technically improved, an innovation made in Europe which led to an increased demand for kerosene and thus in turn for oil. The above described methods to recover oil from the surface proved inadequate to meet this increased demand. As more fully described in Chapter 1, entrepreneurs took notice and sought to increase production capacity by boring holes for reaching and extracting oil from the subsoil, trying in fact to copy the way in which in those days salt water was bored for and produced. The first oil well was bored in 1859 (Titusville, Pennsylvania) marking that year as the beginning of the petroleum industry.

Kerosene was obtained from the oil by means of a primitive distillation process. In this process kerosene formed the middle distillate. Unfortunately, there was no use for the lighter distillates (gasoline) and heavier distillates (fuel oil and residue) that were also obtained in the process and which therefore had to be discarded.

Note. Before the technically improved kerosene-burning lamp was introduced, light on the streets and in houses and offices was provided by gas lamps, in which inflammable gas (town gas) was burned. Town gas was manufactured from coals containing a high percentage of volatile matter. Apart from gas lamps, there were lamps burning sperm oil or suitable fats, and candles. In this competition, the kerosene burning lamp had several advantages: kerosene was easy to transport (compared to town gas which had to be piped), it was not expensive (it was cheaper than town gas as well as sperm oil) and the brightness of the light produced was satisfactory, anyway much better than that of candles, although not as good as that of gas light. Under these circumstances, the kerosene burning lamps succeeded in becoming the most important source of light till this lamp was gradually displaced due to the arrival of the electric lamp.

Due to the invention of the electric light, a large scale application of which was in 1885 demonstrated by its inventor, Thomas Edison, in Manhattan, New York, the dominant position of the kerosene-burning lamp and with that the demand for kerosene came to an end. The invention of the internal combustion engine (in 1893 Rudolf Diesel applied for a patent) and the use of this type of engine in cars in preference to the steam engine, came to the rescue and opened and created a new market, in particular for the lighter distillates, such as gasoline, for which previously there had been not much use. In this connection, it was of great importance that the first mass-produced car, viz. Ford's T-model introduced in 1908, used a gasoline engine rather than a diesel engine. Meanwhile crude oils were discovered that yielded less kerosene but more of the heavier fractions than the crudes produced so far (e.g., the light heavy crude oil that was discovered in 1901 at Spindletop, near Beaumont in Texas). The US industry, railways and shipping became interested in these heavier fractions (fuel oil) and recognized their potential for the generation of electricity and steam, a market that hitherto was the exclusive domain of coal.

Note. For a long time however coal succeeded in maintaining its dominant position in the market of thermal power generation. As a matter of fact, only in Western Europe and as from the 1970s did oil and natural gas succeed in making an inroad into coal's special domain.

Finally, some of the heavier refined products proved suitable as a basis for making lubricating oil, whereas the ultra-light refined products, in particular naphtha, became the most important feedstock for the petrochemical industry.

There are many types of oil, each type having its own specific value. The specific value depends on its quality and specific gravity. The quality depends on the relative concentration of the different types of hydrocarbon molecules as well as on the presence and concentrations of associated substances. The specific gravity is expressed in degrees API (API stands for American Petroleum Institute). The range 25 to 42 degrees API corresponds with specific gravities from 0.904 to 0.816. Generally, the lower the specific gravity and the higher the API-number, the higher the economic value of the oil. The aforementioned associated substances consist of sulphur and metallic compounds and are mainly found in the heavier oils, whereas the lighter oils, viz. condensates and NGLs, are relatively clean oils. Any sulphur present in the oil must be removed therefrom in order to prevent its ultimate release into the atmosphere after combustion. Furthermore, sulphur has a damaging effect on catalysts used in car exhaust systems for the purpose of reducing nitrogen oxide emissions. The metallic compounds may cause problems in the refinery processes as they may hinder or interfere with the catalysts that are being used in such processes. Oils containing sulphur or interfering metallic impurities are less valuable because of the costs that have to be made to remove these substances. These costs are mitigated in case the products that have to be removed possess a commercial value of their own.

Note. Sulphur offers a good example. The removal of sulphur takes place in a direct chemical process whereby gypsum ($CaSO_4$) is formed; or by treating the oil or oil fraction with hydrogen (H_2), whereby H_2S is formed; or by concentrating the sulphur in a small residue fraction from which sulphur in its elementary form can be extracted. Elementary sulphur has commercial value; hence the costs of the removal process can partly be recovered. Gypsum may sometimes also have a commercial value depending on whether it can be obtained without too many impurities.

3.1.3 Natural Gas

The economic use of and drilling for natural gas lagged far behind that of oil, at least in the world outside the US and Canada. Natural gas' relatively low energy density makes its transportation expensive relative to that of oil, which implies that natural gas must find a market close to the field where it is produced. In the first half of the twentieth century, those geographical conditions were only satisfied in the US and Canada. Outside the US and Canada, the search for hydrocarbons was focused on oil and not on natural gas. Not before natural gas in commercial quantities was discovered in Western Europe itself (starting in 1959 with the discovery of the Slochteren gas field in the Netherlands) was it commercially warranted to produce and consume natural gas in that region.

As fuel, natural gas is used in power stations, industrial plants and households. In these sectors it is used respectively for generating electricity, heat, space heating and cooking. Its use in North America started in the last decade of the nineteenth century as a substitute for town gas as discussed above. Substituting town gas, a product made from coal, natural gas started right from the beginning the competition with coal. Now that town gas is a thing of the past, the real gas-to-coal competition takes place in the market of thermal power generation. In this market natural gas has also competition from fuel oil and from alternative energy products (i.e., nuclear fission energy and renewables, as will hereinafter be further discussed).

Currently, a small part of the natural gas production is used as feedstock for conversion into more valuable chemicals or, alternatively, for conversion into hydrogen (H_2) or transport fuels.

Note. The abovementioned conversion process starts with the conversion of methane into synthesis gas or syngas. The latter is a mixture of carbon monoxide (CO) and hydrogen. It is obtained by a process in which methane is mixed with steam and oxygen, with oxygen alone or with air. Further processing of said syngas in accordance with the Fischer/Tropsch principle leads to the synthesis of larger hydrocarbon molecules. In a third stage these hydrocarbons are refined into transport fuels in the diesel/gasoline range. It is also possible to convert syngas into methanol (CH_3OH). Methanol as such is suitable for use as a transport fuel, but it is also used as feedstock for manufacturing a range of chemicals as well as the standard transport fuels, in particular gasoline. It should be added that research is going on into methods to convert methane directly into methanol.

Furthermore, as a low cost source for making hydrogen, methane may in the future become still more important if there will be a demand for hydrogen as transport fuel (see hereinafter).

There are many types of natural gas, each having its own specific value. When used directly as fuel, the specific value of the natural gas depends on its calorific value or heat content. When used as feedstock in catalytic conversion processes, its value derives from the value of the products into which the natural gas is converted, e.g., transport fuels, methanol, hydrogen and chemicals. If natural gas contains non-hydrocarbon gases, whose presence in the gas is a human health risk and/or will cause corrosion damage to pipelines and associated equipment, these gases need to be removed from the natural gas before the latter can be transported to its market, but its value is correspondingly lowered by the amount of the cost of removing these harmful gases.

Best example in this respect is hydrogen sulphide (H_2S). H_2S is not only toxic (a human health concern) but also extremely corrosive with the potential of causing great damage to pipelines and valves through which the natural gas has to be conducted. Therefore, H_2S has to be removed from the natural gas in the field of production at the earliest stage of treatment possible. The presence of non-inflammable gases, mainly nitrogen (N_2) and carbon dioxide (CO_2), reduces the calorific value (heat content) per cubic metre of the natural gas and hence its commercial value (as a fuel). If natural gas contains a significantly high percentage CO_2, say more than ten per cent, it may be considered worthwhile to separate the CO_2 from the natural gas and return it into the subsoil before the natural gas is transported and sold.

Note. As an option to dispose of associated CO_2, the underground storage of CO_2 has received considerable attention. Underground storage of such associated CO_2 is accomplished either by:

- injecting the CO_2 into the natural gas reservoir from which it was produced, by piping the CO_2 to a oil reservoir, into which it will be injected for the purpose of stimulating and improving the oil recovery; or
- piping the CO_2 to the location of a depleted gas field, for subsequent injection into such field.

On the other hand, the presence of inert gases, such as argon and helium, may be a commercial advantage provided the concentrations are large enough to warrant their sequestration and recovery.

Finally, the location of an oil or gas field has a decisive effect on the value of the natural gas that is or can be produced from that field, since (as explained in the beginning) the cost of transportation of gas to its markets is an important component of the cost of its supply. If the area of operation is located at a long distance from any gas market, the gas transportation possibilities will weigh heavily in taking a decision whether or not to develop a non-associated natural gas discovery, or how to dispose of any associated natural gas that is unavoidably produced in an oil field.

3.2 FACTORS DETERMINING ECONOMIC USE AND VALUE

3.2.1 The State and Requirements of Prevailing Technology

There are two kinds of technology at work here: demand-side technology and supply-side technology.

Demand-side technology is technology that requires for its application or operation the utilization of oil products and/or natural gas, and thus stimulates the demand for these products. The first example of technology that required an oil product for its application was the kerosene-burning lamp. When around 1850 in Europe after many experiments, an adequate, suitable oil lamp was constructed, i.e., a lamp, in which the kerosene would burn without smoke or smell, the demand for kerosene and hence for oil rose quickly, but there was still no use for the lighter and heavier distillates that were co-produced in the kerosene distillation process. Next example was the invention in 1893 Rudolf Diesel of the internal combustion engine and the use of this type of engine in cars. This created a new market for oil products, for the lighter distillates to be precise, which fraction had hitherto been considered as a nuisance. Present day examples are gas turbines, all kind of combustion engines, oil or gas-fired heating installations, etc.

Supply-side technology is technology that aims to improve the efficiency of petroleum operations, in particular that of drilling holes and extracting petroleum from the subsoil reservoir; that is involved in the long-distance transportation of petroleum to its markets; and that made and makes offshore operations in difficult and risky surroundings possible.

Note. Drilling operations became much more efficient as a result of the invention of rotary drill pipe technique, which allowed drilling much deeper holes as had previously been possible by means of cable-tool drilling (see Chapter 1). Long-distance pipeline transportation of natural gas became possible when pipelines could be made that were capable to withstand the high pressures under which the natural gas had to be transported over such distances. The latter achievement opened a very important gas market in North America. A more recent technological achievement in the field of long-distance transportation of natural gas concerned the technology involved in the liquefaction of natural gas. Such liquefaction makes the gas suitable for transport by ocean-going tanker.

3.2.2 The Cost of Supply

3.2.2.1 The Components

The cost of supply of oil and natural gas comprises all and any cost incurred by the producer from finding petroleum up to and including the cost of production and the cost of bringing the production to the market place. This cost must be recovered from the price (the producer price), which the producer receives in the (global) market place. Hence the cost of supply consists of:

(i) the cost of exploration work preceding discovery including an allocation of the cost of any failed search for new reserves;
(ii) the cost of extraction of petroleum from the discovered reservoirs and the cost of treating and storing the extracted petroleum in the field;
(iii) any production taxes (royalties), export and other duties, fees, area rentals and cash bonuses payable to the host State; and
(iv) the cost of transporting the oil and natural gas production from the field storage facilities and export terminals to the market.

With reference to (i). As far as the individual producer is concerned, the costs that have been spent on discovering the reservoirs from which the oil or natural gas are produced are part of the cost of supply and have to be recovered from the producer price. Usually the total expenditure spent on exploration work is written off over part of the lifetime of the field which has been discovered as a result of the exploration work. Furthermore, any producer is constantly searching worldwide for petroleum reserves, which are needed for replacement of the reserves that are being produced, but not all exploration work wherever undertaken leads to the discovery of new reserves. The costs of any such failed search have to be distributed over the producer's producing ventures.

With reference to (ii). The technical cost of extraction, treatment and storage in the field consists of capital expenditure and operating costs. Capital expenditure represents the cost of acquisition of a capital asset (e.g., a producing well, a production ship, any field treatment and storage facility) and is written off in accordance with a depreciation schedule based on the lifetime of the capital asset in question. Operating costs include the cost of leasing or hiring capital assets. The latter may be substituted for capital expenditure.

The level of the cost of extraction depends on (a combination of):

(i) the characteristics of the geographical location of the oil/gas field (offshore: Arctic waters, deep water or by contrast shallow water; onshore: desert area or other empty areas or by contrast environmentally sensitive areas such as densely populated areas in general or tropical rain forests in particular);
(ii) the depth and characteristics of the reservoir; and
(iii) the efficiency of the extraction process and the well productivity (production rate per well).

With reference to (iii). If production taxes (royalty), export and other duties, fees, area rentals and cash bonuses have to be paid to the host State the cost of production are correspondingly increased (see Chapter 8).

With reference to (iv). The cost of transportation consists of capital expenditure, operating costs and tariffs. Tariffs are payments made for the use of storage and transportation facilities, including such facilities used at export terminals. Capital expenditure represents the cost of acquisition of a capital asset (e.g., a pipeline) and is written off over the lifetime of the capital asset bought. The level of the cost of transportation from the field of production to the market place depends on distance and on whether oil or natural gas is being transported. Only the costs of long-distance transportation matter. Due to its low energy density, long-distance transportation of natural gas is expensive. It takes place by means of ocean-going tanker (after the natural gas is liquefied in a special plant) or high-pressure pipeline. Long-distance transportation of oil is not expensive, provided the oil can be moved by ship (oil tankers). Long-distance transportation of oil by pipeline is much more expensive, albeit always less so, than such transport of natural gas. It follows that when an oil field is situated near port facilities allowing the oil to be carried by tanker, the cost of supply is not much higher than the cost of its production. Natural gas however is always at a great disadvantage if its markets are far removed from the place of production and long-distance transportation by ship or pipeline is required.

3.2.2.2 High and Low Cost Unit-Production

When the costs of production (capital and operating costs of extraction and treatment in the field facilities plus the payments that have to be made to the State (production taxes, export duties, etc.) are compared with the price that the producer can obtain in the market (the producer price), distinction can be made between high cost unit-production (leaving a small margin) and low cost unit-production (leaving a large margin per unit of production).

Note. Obviously, any margin should be large enough to pay for the cost of final transport. In theory, a producer should stop producing (and close-in its producing wells) as soon as freight paid operating costs (i.e., operating costs plus transportation costs) equal the producer price. But, when it is expected that after a short while producer prices will recover, and considering the adverse consequences and costs of closing-in producing wells, any producer will always try to continue its operations for as long as possible. In the end, governments will always try, by fiscal or other measures, to protect the domestic high cost producer against imports of low cost production wherever coming from.

High-cost-production means oil or natural gas that:

(i) must be extracted from reservoirs with low to extreme low productivity characteristics, i.e., reservoirs with a low porosity and/or permeability, requiring the application of advanced drilling and extraction techniques;
(ii) must be extracted from reservoirs containing oil that cannot be satisfactorily recovered by using the conventional techniques but need the application of enhanced oil recovery (EOR) techniques to significantly improve the recovery factor;
(iii) must be extracted from reservoirs containing heavy oil, including very heavy oil, or natural bitumen; or
(iv) must be produced in technically challenging areas, such as the deep offshore generally, and some extreme offshore areas, such as the Arctic waters (occurrence of drifting icebergs) in particular.

Note. In respect of production cost per unit of volume, no region or country in the world has ever been able to compete with the oil producing regions in the Middle East. The deciding factor was and still is the impressively high well productivity (production rate per well), a result of large, extensive reservoirs with excellent reservoir characteristics making oil technically very easy and cheap to produce. Moreover the major producing fields are all situated in the vicinity of port facilities (Ras Tanura, etc.) and can be moved over sea at reasonably low cost (causing [but only in a situation of over-supply] problems for the high cost domestic producer).

By stark contrast, the extraction of natural bitumen from the oil sands located in Alberta, Canada, and the upgrading of the same to usable oil products is easily the most expensive production operation in the world. In mid-2006 the National Energy Board of Canada (NEB) estimated the operating cost of a new mining operation in the Athabasca oil sands to be USD 9/b to USD 12/b, while the cost of an in situ steam-assisted gravity drainage (SAGD) operation was estimated to be USD 10/b to USD 14/b. This compares negatively to operating costs for conventional oil wells; the latter range from less than USD 1/b in Iraq and Saudi Arabia to USD 6/b and higher in the US and Canada. The capital costs involved such as the costs of the equipment used in the open pit mining operations (shovels, dump trucks, crushing machines), and the costs of the conveyor belts or pipelines to transport the sand to the separation and processing plant may according to the NEB raise the total cost of production to USD 18/b to USD 20/b. For a SAGD operation, total costs up to processing are estimated to be USD 18/b to USD 22/b costs. Final costs include the cost of upgrading the bitumen and amount to USD 36 to USD 40/b. As a result of the recent oil price increases whereby most recently prices were nearing USD 100/b, the latest estimate of the capital expenditures that is earmarked to be invested in the exploitation of the oil sands up to 2015 is USD 200 billion. However due to an acute shortage of skilled workers which has developed in Alberta, it is not possible that such a huge amount of money can be spent and all planned projects can be realized.

3.3 MARKETS, PRICES AND PRICING SYSTEMS: OIL

3.3.1 The Early Kerosene Market (1860-1900)

In the nineteenth century, oil yielded after distillation only a single product of commercial interest, viz. kerosene, because of its use as lamp oil. Oil was refined near the place of production in order to avoid incurring cost for having to transport partly useless material. In those early days, it was a refined product, in this case the kerosene, and not the oil itself, which was shipped around and traded on the international market. In this manner, US kerosene was exported to Western Europe and the Far East (in 1880, fifty per cent of US kerosene production was exported); kerosene refined in Sumatra went to China and Japan; and Russian kerosene, refined in Baku, Caucasus, was similarly exported to Western Europe and the Far East. As a matter of fact, in 1892 a tanker shipped for the first time Russian kerosene from Batum on the Black Sea coast through the Suez Canal to Singapore and Bangkok.

3.3.2 The Pre-Global Market (1900-1975)

As a result of technical innovation, oil became in demand not only for its kerosene but also and more so for its lighter and heavier distillation fractions. The market for kerosene turned into a market for oil and, over the period 1900-1975, this market gradually developed into a global market. To meet the demand of this new market, the petroleum industry moved out and away from the continental US in search for new oil deposits. In the process the petroleum industry became dominated by a few major oil companies operating on a global scale (the international oil groups also referred to as 'the majors'). The majors' global activities began with supplying the markets in Europe and Asia, but, already from the start, the companies had to cope with an over-supply of oil. In the 1920s the situation got worse; too much oil was being exported to the said regions, in particular by the producers in the US and the Soviet Union. The resulting price cutting competition prompted the leading international oil companies of that time to negotiate and conclude among themselves (on 28 October 1928) an agreement (subsequently known as the Achnacarry or 'As-Is' Agreement) that provided for mutual cooperation in markets outside the

US in order to put an end to the then prevailing unbridled competition. The agreement however was never signed in view of US antitrust regulations. The principal participating companies were:

(i) Anglo-Persian Oil Company, a company formed in 1909 in connection with the famous discovery at Masjid-i-Suleiman, Persia, and which at the time of the 'As-Is' Agreement (as a matter of fact since August 1914) was owned for 51.7 per cent by the UK. Anglo-Persian changed its name in 1935 to Anglo-Iranian Oil Company after its original concession in Persia now Iran in 1933 had been renegotiated, and again changed its name, to British Petroleum (BP), on the occasion of the restructuring of this renegotiated concession in 1954;
(ii) Royal Dutch/Shell, a 60/40 British/Dutch partnership formed in 1907, currently a unified UK company under the name of Royal Dutch Shell; and
(iii) Standard Oil of New Jersey (later named Exxon, currently merged with Mobil, both companies originating from the dissolution of the Standard Oil Trust in 1911).

Note. Also participating were a number of smaller US oil companies, such as Standard Oil of Indiana (one of the companies emerging from the dissolution of the Standard Oil Trust, later Amoco and finally merged with BP), Standard Oil of New York (another company emerging from the dissolution of the Standard Oil Trust, later merging with Vacuum Oil Company to form Socony-Vacuum, known as Mobil. The latter finally merged with Exxon); The Texas Company (Texaco), a US company founded in 1902, and Gulf Oil, a US company established in 1907 (this company was in 1985 bought by Chevron, the former Standard of California, the latter also one of the companies that emerged from the dissolution of the Standard Oil Trust.

Under the terms of the (informal) 'As-Is' Agreement, the participating companies shared among and between themselves the markets in Europe and Asia based on the situation prevailing in 1928. In addition, the companies agreed to a pricing system for crude oils traded in the global market, where US crudes competed with crudes produced from outside the US. Under the agreed system, the price to be charged to the consumer was based on the oil prices posted by the US producers in the ports in the Gulf of Mexico area. Transportation costs were charged to the consumer as if the crude oil had been shipped from these ports to the actual place of delivery, irrespective of the real place of shipment. This pricing system reflected the fact that, before the Second World War, the US was not only the biggest oil producer in the world but also the biggest exporter of oil. Production was controlled to the extent that each company's production was only allowed to exceed the demands of its allocated market share if any such surplus would be sold to the other companies. The 'As-Is' Agreement was imperfect in the sense that it did not include independent US producers capable of exporting any surplus production to the markets covered by the Agreement. After production in the US was brought under control by the authorities, the effectiveness and relevance of the Agreement and its amendments improved. Of even greater importance for the control of markets and prices was the struggle for obtaining exclusive petroleum rights in the Middle East Gulf area. In this struggle the same major oil companies were involved as those participating in the 'As-Is' Agreement. By the end of the 1930s, a few major international oil companies, since then sometimes referred to as the seven sisters (five US and two European companies), operating within the framework of four different incorporated joint ventures, had succeeded in securing exclusive petroleum agreements in the Middle East Gulf countries. As later became clear, these petroleum agreements dealt with the world's largest reserves of oil and very important reserves of natural gas.

Meanwhile in 1922, oil was discovered in Venezuela, a country that after the Second World War would develop into a major exporter of oil and in this context would play a leading role in the establishment of the Organization of Petroleum Exporting Countries (OPEC) (see Chapter 4). By the middle of the 1930s, three of the seven sisters, viz. Royal Dutch/Shell, Gulf Oil and Standard of New Jersey (Exxon), had become the main concession holders in that new oil exporting country, which put these companies in control of exports that in the 1950s and 1960s were competing on the West-European market with their own exports from the Middle East Gulf area.

After the Second World War two important trends became visible. First, the US very soon became (as from 1948) a net-importer of oil; and secondly, the pre-war oil discoveries in the Middle East Gulf area were rapidly developed, yielding fast growing quantities of low cost oil for which a market had to be found. In fact throughout the 1950s and the 1960s, the Middle East Gulf governments were always urging their concessionaires to increase production and exports, a policy which would bring these governments into conflict with Venezuela, competing with its exports on the same West-European market. For the companies, it became necessary to adapt the pricing system to the new realities. A second price basing point was established at Ras Tanura, the Saudi Arabian port of export. The FOB (free on board) price posted by the companies Ex Ras Tanura was so related to the US FOB posted price in the Gulf of Mexico area that at a certain selected port of destination, known as the equalization point, the delivered price (FOB posted price plus freight charges to the equalization point) would be equal. Initially, the port of Genoa was selected as the equalization point, and later, in 1949, this point was shifted to New York. As from the introduction of this pricing system the international price of oil was set and controlled by the aforementioned majors. It was clear that the pricing system adopted tended to protect the US producers against competition from the low cost oil that was exported in increasing volumes from the Middle East Gulf area. Due to US government intervention (this government imposed on 11 March 1958 the Mandatory Oil Import Control Program), the US market became a market partly closed off to imports of Middle Eastern oils. The struggle for outlets between the exporting oil producers then concentrated on Western Europe. In that region the majors, holding the major oil producing concessions in the Middle East, were confronted with each other's exports, with oil exported from Venezuela and, as from 1955, with oil exported from the Soviet Union. To counter the competition, the concessionaires were forced to sell their crude oils below the then applicable posted prices.

Note. In 1955 the Middle East petroleum agreements of the 1930s were amended to the effect that the concessionaires were obliged to publish the price at which they were ready to sell their production FOB port of export to all comers, and this price was referred to as the posted price. As a result of this amendment, government revenues (derived from the then prevailing 50/50 profit-sharing arrangement) were henceforward calculated on the basis of the official posted price. In turn, the now adopted fiscal method caused concessionaires, if and when selling their exports below the applicable posted price, to lose the corresponding fiscal relief. Unable to continue with this practice, the concessionaires decided in February 1959 to adjust their posted prices to market realities and to reduce these prices by USD 0.18/b. As a result thereof, Arabian Light ex Ras Tanura was posted at USD 1.90/b. About one year later, in August 1960, the concessionaires again decided to lower their posted prices, initially by USD 0.14/b, later brought back to USD 0.10/b. Arabian Light was correspondingly posted first at USD 1.76/b, and later at USD 1.80/b. The following Table 3.1 illustrates the progression of the posted price during the 1950s up to and including 1960.

Table 3.1. Posted Prices (USD/b)

	1953	*1954*	*1955*	*1956*	*1957*	*1958*	*1959*	*1960*
Arabian Light ex Ras Tanura (34°)	1.82	1.93	1.93	1.93	2.00	2.08	1.90	1.80
Irani crude ex Bandar Mashur (34°)		1.91	1.91	1.91	1.98	2.04	1.86	1.77
Iraqi crude ex Fao (34°)	1.79	1.90	1.90	1.85	1.92	1.97	1.80	1.72
Kuwaiti crude ex Mina-al-Ahmadi (34°)	1.96	1.72	1.72	1.72	1.78	1.85	1.67	1.59

3.3.3 OPEC and Its Early Pricing Policies

Since any reduction of the posted prices meant to the governments concerned a corresponding loss of fiscal revenues, the posted price reduction that was effectuated in August 1960, was a matter of great concern to the governments of five major oil exporting countries, viz. Saudi Arabia, Iran, Iraq, Kuwait and Venezuela. Within one month after the latest posting, they met in Baghdad and succeeded at that meeting in reaching agreement on the establishment of an organization of mutual cooperation and assistance named OPEC. At the time, the concessionaires, being the exporting oil producers, were themselves not directly involved with the organization. The governments of the OPEC Member States could therefore only act (in marketing and pricing matters) through and by means of giving or withholding approval to the prices posted by their concessionaires. In fact the concessionaires kept control over prices throughout the 1960s. The posted prices were maintained at the level fixed in 1960, although the concessionaires had to continue their practice of selling below the posted price, in the light of the competition experienced in the West European market from Libyan and Middle Eastern crudes produced by other companies (referred to as the independents), which had replaced the exports from the Soviet Union.

On 16 October 1973, ten days after the outbreak of the Yom Kippur War (also known as the Ramadan War), OPEC finally succeeded to take away control over the international price from their concessionaires. On the aforesaid date the posted price of Arabian Light ex Ras Tanura was unilaterally raised by no less than seventy per cent, from USD 2.898, the price which in June 1973 had been agreed with the concessionaires, to USD 5.119. Soon thereafter, with effect from 1 January 1974, OPEC raised the posted price again, this time by almost 130 per cent to USD 11.651. These price movements have since been referred to as the first oil price shock or as the first oil crisis, although there was absolutely no physical lack of oil.

Until the end of 1974 posted prices were maintained at their level as fixed on 1 January 1974. As of 1 January 1975 the posted price as such disappeared but continued as the non-published fiscal or export price. Meanwhile a new published price was introduced under the name of 'official government selling price'. This price was fixed at a level of ninety-three per cent of the fiscal price and indicated the price at which concessionaires, which meanwhile had lost a sixty per cent share in their respective concessions, could buy back that so lost share of production. As of 1 January 1975, the official government selling price became the only price that was published, this time by the governments themselves.

Note. To remain in control of prices and pricing, it was necessary for the OPEC governments to bring their state participation policies to a conclusion. The wish to acquire a stake in the petroleum agreements of the 1930s was, for the first time, brought to the foreground at an extra-ordinary meeting of the OPEC Conference in September 1971. Soon afterwards, negotiations with the concessionaires began. After initially having agreed to some intermediate stages of state participation, in the course of 1975 the governments involved finally acquired one hundred per cent – and in some instances fifty-five to sixty per cent – of the long-term concessions (Gulf area) and other concessions/long-term leases (Nigeria, Libya, Venezuela) held by the majors in their respective countries. The majors lost thereby control over a large part of their production capacity and oil reserves outside the US, forfeiting forever their capability to dominate the international market and to control the prices. This dramatic turnabout was illustrated by the fact that in 1970, immediately before the governments started the negotiations on participation, the majors possessed about seventy per cent of the then proven world oil reserves outside the US and about the same percentage of the world oil production outside the US.

3.3.4 The Global Oil Market (as of 1975)

In the aftermath of the political disturbances that took place in the Middle East in the first half of the 1970s and as a result of OPEC acquiring during that period the major world oil reserves and oil production capacity, an open, freely accessible global market for internationally traded oil came into operation.

On this global market through the intermediary of oil traders, oil can be freely bought from exporting producers by the refineries outside and independent from the networks and channels of major integrated oil companies. However, this market, although freely accessible to producers and buyers alike, is not a perfect market. Producer prices are not arrived at as a result of freely conducted negotiations between independent selling and buying parties. In reality, the price is determined and established by the actions and behaviour of the Member States of OPEC, regulating and controlling the supply of crude oil to the market within the limits of their prevailing production capacity (see Chapter 4).

Note. In the process of the forming of a global oil market, a world spot and short-term market price for oil has been established which is represented by the spot or short-term prices of regional benchmark crudes, such as Brent Dated, Mexican Isthmus, OPEC export crudes or West Texas Intermediate (WTI) and Alaska North Slope (ANS). These prices are quoted in the trade journals and serve as reference prices for incorporation in sales price formulas used in term sales contracts. Oil producers operating in oil importing countries usually sell their production to inland refineries at import parity price, i.e., at the level of the prices, that have to be paid by those refineries for their imports, provided the production offered is of a type and quality that can be handled by the refinery concerned. Non-OPEC exporting oil producers, such as Norway and the Russian Federation, will generally charge for their exports a price which matches the price of comparable and relevant OPEC export crudes. On this basis, Brent Dated is linked to the price at which Nigeria's Bonny Light is traded. In its turn Brent assumes the role of marker crude for the crude oils produced or traded in the North Sea area.

Table 3.2. Geographical Distribution and Volume of Oil Consumption (million b/d) 1975-2006

	1975	*1980*	*1990*	*1995*	*2000*	*2005*	*2006*
US	16.3	17.1	17.0	17.7	19.7	20.8	20.6
Japan	4.8	4.9	5.3	5.8	5.6	5.4	5.2
China	1.4	1.7	2.3	3.4	4.8	7.0	7.4
India	0.5	0.6	1.2	1.6	2.2	2.6	2.6
FSU/Russian Fed.	7.1	8.5	5.1	3.0	2.6	2.6	2.7
EU 25/27	13.5	14.2	13.4	13.8	14.4	15.2	15.2
OECD	38.7	41.0	41.4	44.5	47.7	49.4	49.0
World	55.0	61.7	66.8	69.8	76.3	83.1	83.7

Source: BP Statistical Review of World Energy 2007.

Table 3.2 shows that the consumption of oil is concentrated in the OECD-countries, which include the US, the European Union and Japan. In 2006 the OECD Member States accounted for almost fifty-nine per cent of the world oil consumption. Some refineries however wholly or partly export their products, so statistically the country's registered consumption of oil may exceed its domestic or inland consumption. Since 2005 China has joined the ranks of the big consumers, even now surpassing Japan.

Table 3.3. Comparison of Geographical Distribution of Oil Production and Consumption in 2006 (million b/d)

Country/Region	*Oil Production*	*Oil Consumption*
US	6.9	20.6
North America	13.7	24.8
Venezuela	2.8	0.6
Total South & Central America	6.9	5.1
Norway	2.8	0.2
The Russian Federation	9.8	2.7
Saudi Arabia	10.9	2.0
Middle East	25.6	5.9
Nigeria	2.5	n.a.
Africa	10.0	5.9
Japan	-	5.2
China	3.7	7.4
India	0.8	2.6
Asia Pacific	7.9	24.6
OECD	19.4	49.0
OPEC 11*	34.2	n.a.
EU 25**	2.3	14.9
World	81.7	83.7

* OPEC 11 excludes Angola which joined the organization as its 12th Member State as of 1 January 2007.

** EU 25 became EU 27 only after 1 January 2007 when Bulgaria and Rumania joined the Organization.

While accounting for almost fifty-nine per cent of the world oil consumption, the OECD countries' share of world oil production is only twenty-four per cent. To put it differently, seventy-six per cent of the world oil is at the moment produced outside the OECD area. Of this share, fifty-six per cent is being produced by the Member States of OPEC.

Note. The geographical discrepancy between areas of production and consumption requires that oil must be transported over considerable distances. Fortunately, distance is not a problem in a commercial or technical sense since the oil can be transported over sea and the costs of such transport is relatively low compared to the cost of production.

Where OPEC-production must be qualified as low cost to very low cost production, it could (if OPEC would be willing to do so) be supplied to the major consuming areas at very competitive prices. In the past such marketing strategy has more than once caused problems for the high cost domestic producer operating within a country that is importing oil from any of the OPEC Member States, e.g., the US, or that is exporting its high cost oil in competition with OPEC oil, e.g., Norway.

3.3.5 LIMITS TO PRICES AND PRICING

On the global oil market, oil is sold by the producer through the intermediary of an oil trader to the refinery in accordance with the terms of a sales contract comprising type of oil, price, quantity, quality and delivery date. If the sale does not concern a spot sale but a short, medium or long-term sale, the producer price is determined in accordance with a formula including references to the spot price or short-term price of one or more comparable marker crudes. In accordance with these contracts, oil of the chosen type, quality and quantity is shipped to and delivered at the refinery concerned. In the refinery, the oil in question is converted into specific products which will then be delivered to the end-user in the market for that particular product.

But as mentioned before, the world market price is subject to, and depends on possible negative or positive OPEC intervention on the supply side by means of reducing or increasing production volumes and/or expanding production capacity. OPEC's powers of intervention are bound and limited on the downward side by the cost of supply of oil produced by the marginal producer amongst the Member States (lower limit), and on the upward side, by the price that the refineries are willing or can afford to pay (upper limit).

3.3.5.1 The Upper Price Limit

The price that the refineries are willing or can afford to pay for a particular type of oil on the global market is determined by the prices which the end-users are prepared to pay for the various end-products that have been made from that oil.

Note. The inland market prices, i.e., the prices to be paid by the end-users, are prices comprising the price paid by the refinery for the feedstock, including import duties, as the case may be, the cost of processing, distribution costs and all kind of taxes and fiscal duties and charges. By manipulating the fiscal charges, a government can regulate the consumption of the end-products (see Chapter 4).

The end-user of oil products is found in the following sectors of the economy: transport (road transport, shipping and aviation), thermal power generation, petrochemical industry, general industry, including the refineries themselves and households (domestic sector). In each of these sectors, the products from the refinery have to compete with one or more alternative products, all of which have their own price and availability. If there is no competition or very restricted competition due to non-availability of alternative products, the upper limit of the price in each sector (the sector price) is determined by the readiness of the end-user in the sector to pay the price that is demanded. Putting it differently, the upper limit of the sector price will be reached when the end-user starts to limit its consumption of the refined products and/or is switching to readily available alternative products, if any. In the latter case, the cost of supply of the readily available alternative products represents also an upper price limit.

Note. A review of the alternative products available or potentially available in the above-mentioned sectors of the economy is presented hereinafter.

3.3.5.2 The Lower Price Limit

In theory, the lower price limit is formed by the cost of supply of the oil produced and delivered by the marginal producer among the OPEC Member States. The lower limit has been tested in 1986 in the context of the application of the 'flooding-of-the-market' strategy, but has since then never been repeated. There is no reason for OPEC to use this strategy again and anyway OPEC has simply not enough production capacity to do this.

3.3.5.3 The Situation from 2000

The year 2000 can be seen as a real turning point in the global oil market. Compared to the foregoing years, OPEC no longer had to contemplate production cut backs to support a chosen price level. In fact, prices moved without its intervention at a comfortable level of around USD 25/b, as a matter of fact within a price band (22-28 USD/b) that OPEC in March 2000 had set for the price of the OPEC Reference Basket for the purpose of pursuing pricing policies by means of production adjustments. As from 2005, the international price reached spectacular levels: in November 2007 the spot price of Brent was nearing the USD 100/b mark whereas the yearly average of the same spot price in 1998 had been as low as USD 12.72. Even when correcting for the current weakness of the US dollar, this is an astonishing development. Nonetheless, these trends continued in the first quarter of 2008. In the last part of said quarter, the Brent price regularly exceeded USD 100/b, whereas the USD further weakened to around EUR 0.66. At the moment, the Member States of OPEC are producing at full capacity which means that their supply to the market can only be increased if new investments in expanding production capacity are made. Prices may come down in the process, but OPEC will always have the power to stop any dramatic fall. In this context, it is interesting to note that in the beginning of 2007 when the oil price fell back to around USD 50/b for a short while, OPEC indicated not to allow a fall back to pre-2006 price levels, a target that can easily be achieved, if need be, by some restrictive production adjustment. In the light of the current price level, it may safely be assumed that OPEC has moved this lower limit upwards, to say USD 90/b or even USD 100/b.

Note. For a long time, OPEC has resisted making new investments in expanding production capacity, being of the opinion that the high oil prices are the result of speculation reflecting the weakness of the US dollar and fear for political unrest involving some Member States in the Middle East and Africa. Then on 22 April 2008, OPEC suddenly announced at an Energy Conference held in Rome to be planning new investments in production capacity over a period of four years that should result in an increase of the daily production by 5 million barrels in 2012. What the effect on the price level will be remains to be seen, but fact is that OPEC will never allow the international price to drop below their chosen lower limit.

Table 3.4. A Review of Oil Prices (USD/b) 1975-2006*

Year	*USD money of the day*	*USD 2006*
1975	11.53	43.38
1980	36.83	90.46
1985	27.56	51.71
1990	23.73	36.76
1995	17.02	22.74
1998	12.72	16.22
2000	28.50	33.93
2004	38.27	40.83
2005	54.52	56.27
2006	65.14	65.14

* 1975-1983 Arabian Light posted at Ras Tanura; 1984-2006 Brent Dated.
Source: BP Statistical Review of World Energy 2007.

Note. In November 2007 the Brent Dated price surpassed the till then highest yearly average price of USD (2006) 90.46 attained in 1980. The lowest yearly average price: USD (2006) 16.22 in 1998 (USD 12.72).

3.4 MARKETS, PRICES AND PRICING SYSTEMS: NATURAL GAS

3.4.1 GENERAL

Due to its inherent physical constraints, natural gas cannot at low cost be shipped around the world and brought to the places where there is a need for the gas. This lack of mobility has prevented the forming of a global gas market. Instead natural gas is sold to dedicated markets connected with the producing fields by a long-distance pipeline or by ocean-going LNG tanker. Against this background, a gas producer sells its gas (that can be either associated natural gas collected from an oil field or non-associated natural gas produced from a gas field) either:

(i) directly to the end-users located in the market concerned; or
(ii) to a gas transport and trading enterprise (a gas company), whose business it is to condition the gas bought from the producer and to transport and deliver the conditioned gas to the final consumer.

In situation (i), it is the gas producer who is responsible for the transportation of its production all the way from the field of production, wherever located, to the intake point of the end-user. In situation (ii), it is the gas company who operates a pipeline system, treatment plant and even one or more underground storage facilities. Such a company may have sales contracts with several customers and several producers and therefore be better able to balance supply and demand and to cope with seasonal fluctuations. In particular, such a balancing act between seasons will be possible if the gas company is in possession of and operates underground storage facilities.

Except in markets where there truly exists gas to gas competition, the price to be paid by the end-user, which comprise general industry, power stations and the gas companies supplying households, is a formula price derived from and tied to the price actually paid or assumed to have to be paid by said end-users in the local or regional market on a calorific basis for competing or potentially competing fossil fuels, such as low sulphur coal, low sulphur fuel oil and heating oil. The prices of these fossil fuels are incorporated in the price formula used in the usually long-term sales contract. In addition the end-users may have to pay eco-taxes. In long-term gas sales contracts, it is stipulated that the sales price is reviewed every six months.

Note 1. Fuel prices incorporated in any gas price formula should exclude any eco-tax levied on those fuels. Even if no eco-tax is applicable but only a simple energy tax, imposing a levy on the heat content of the fuel (or electricity) consumed, the formula prices should exclude such energy tax, provided the same energy tax will be levied on the gas.

Note 2. The maximum price, that a power plant is prepared to pay to a gas producer, is a price that allows this power plant to produce electricity at the same cost as when the plant had to use low-sulphur coal or low-sulphur fuel oil. In comparing gas prices with the prices of competing fuels, due account should be taken of differences in efficiency of converting heat into electricity and the capital and operating costs of the conversion facilities. If for instance by using natural gas instead of coal, a greater heat conversion efficiency can be achieved, say forty-five per cent instead of thirty-five per cent (which means that compared to coal a greater part of the heat content of the gas is converted into electricity), correspondingly more can be paid for the natural gas on a heat content basis. Any imposition of discriminatory eco-taxes to be paid by the power plant will give natural gas a commercial advantage.

Note 3. Households are using fossil fuels for cooking and space heating purposes and are supposed to buy the cheapest fuel on offer. In practice however, coal is no longer a competing fuel in this sector. Its use in households has too many disadvantages where handling and storage are concerned. For the greater convenience, households are prepared to pay a premium in favour of natural gas or heating oil. Hence, heating oil and not coal is the competing or potentially competing fossil fuel. Therefore, the price charged by the gas company to households for their natural gas deliveries must stay at the same level as the price which is charged or potentially could be charged to households for heating oil. In contrast to the situation where natural gas is sold to power plants, no further adjustment to the said price level is required in as much as there are no significant differences between heating oil and natural gas in relation to the efficiency of heat conversion. Only on account of any imposition of discriminatory eco-taxes would preference be given to the use of natural gas.

Note 4. If the natural gas producer sells its gas to the chemical or fuel processing industry where it will serve as feedstock for conversion or transformation into chemical products or fuels the price paid by the plant is derived from the price that is paid in the market for the manufactured products (e.g., chemicals, transport fuels, methanol, hydrogen, etc.).

Note 5. If a gas producer does not sell directly to the end-user but to a gas company, the gas producer's price is derived from the price received by the said gas company from its customers. Between the two prices, there will exist a difference representing the capital and operating costs incurred by the gas company in conditioning and distributing the gas among its customers plus a suitable fee for off take services rendered, such as utilizing underground storage capacity.

Table 3.5. Geographical Distribution and Volume of Natural Gas Consumption (billion cubic metres) 1975-2006*

	1975	*1985*	*1990*	*1995*	*2000*	*2006*
US	553	489	543	629	661	620
Canada	48	58	67	80	83	97
Mexico	15	27	27	28	38	54
Australia	5	14	17	20	24	29
Japan	9	39	50	61	75	85
China	10	13	15	17	24	56
Asia Pacific	35	111	159	219	298	438
FSU/Russian Federation	270	361	420	378	377	432
EU 25	187	256	290	341	420	467
OECD	819	882	999	1,182	1,343	1,420
World	1,197	1,662	1,982	2,145	2,428	2,851

* 1 billion cubic metres NG is equivalent to 6.25 million barrels or 0.9 million tonnes oil, 1 billion cubic metres per year is equivalent to 17,123 barrels/d oil.
Source: BP Statistical Review of World Energy 2007.

Table 3.6. Comparison of the Geographical Distribution of Natural Gas Production and Consumption in 2006 (billion cubic metres)

Country/Region	*Production*	*Consumption*
US	524	620
Canada	187	97
North America	754	770
The Russian Federation	612	432
Middle East	336	289
China	59	56
Japan	-	85
India	32	40
Asia Pacific	377	438
EU 25	190	467
OECD	1,078	1,420
World	2,865	2,851

The major areas of consumption of natural gas are the North American region including the US and Canada, the European Union (EU 25), the Russian Federation and Asia Pacific (pre-dominantly Japan, lately joined by China). As shown on Table 3.5 above, the North American region is practically self-sufficient in natural gas, i.e., indigenous production is neatly in balance with the indigenous consumption. Within this region, the US is the largest consumer and producer. The production shortfall is made up by imports mainly from Mexico and Canada. In the EU 25, indigenous natural gas production falls far short of satisfying demand. Roughly forty-two per cent of consumption is covered. The shortfall is made up by importing natural gas by pipeline from Norway and the Russian Federation and by LNG tanker from Algeria and Libya. In the future, gas may be imported by LNG tanker from Qatar. The Russian Federation is an important consumer country itself, but its current gas production being supported by favourable gas prices is expanding rapidly, leaving enough room for export to the EU by various long-distance pipelines. Furthermore, in the near future, LNG will be delivered to Japan. Asia Pacific taken as a region is beginning to consume more gas as it produces and appears no longer to be self-sufficient. Within the region, Japan is the largest natural gas consumer, closely followed by China. Having no indigenous production at all, Japan needs to import gas from gas-exporting countries elsewhere in the region such as Indonesia, Brunei, Malaysia, Australia and, in the near future, the Russian Federation. China intends to import gas from Qatar and Australia. Asia Pacific is also the region in which the natural gas trade involves transportation exclusively by LNG-tanker. The LNG trade in this region is dictated by the respective geographical locations of the importing and exporting countries involved and by the interregional distances that have to be coped with.

Assuming that the international oil price in the long term will stabilize at levels of around USD (2008) 90/b to USD 100/b (OPEC permitting) and assuming too that gas prices will remain linked to oil product prices and in this way can benefit from the high level oil price, natural gas development and production in combination with long-distance transportation either by pipeline or by LNG tanker have a bright future. In particular the LNG trade will get the opportunity to expand. Important LNG exporting countries include: Trinidad, Algeria, Nigeria, Australia, Brunei, Indonesia and Malaysia. In the coming years, they will be joined by Qatar, Oman, Libya and the Russian Federation.

Table 3.7. Recent and Current Natural Gas Prices 2004-2006 (USD per million Btu)*

	LNG Japan cif	*Natural Gas EU cif*
2004	5.18	4.56
2005	6.05	6.28
2006	7.14	8.77

* 1 million Btu is the calorific equivalent of 28 cubic metres NG or 0.025 tonnes oil.

3.4.2 Marketing and Use of Associated Natural Gas

A geographical discrepancy exists between the places of final gas consumption, the location of the producing fields and the gas reserves (the places of future gas production). In contrast to oil, the geographical mismatch in the case of natural gas is not easily corrected or overcome by shipping the natural gas around the world: the inherent physical constraints of natural gas make this impossible. While the development of non-associated gas fields may be put on hold, in anticipation of marketing opportunities and the solving of the transportation problems, huge quantities of associated gas, which are unavoidably produced in the oil fields, may have to be flared, i.e., burned off in the field, if the gas cannot be utilized in a plant in the vicinity of the oil field concerned or re-injected either into the original reservoir or into another reservoir for one purpose or another.

Note. Despite the recent increase of gas prices, considerable flaring of associated natural gas still takes place in the oil fields of Nigeria, Siberia and the Middle East. Even after extraction of NGLs, which is a possibility if the associated gas happens to be a rich or wet gas, flaring of the gas must be considered to be a regrettable waste of a valuable natural resource. It is generally agreed by governments and industry that this should be avoided if at all economically possible. Meanwhile solutions have been found and put in operation. These involve building of industrial facilities in the immediate vicinity of the oil field, in fact at the collection point of the gas, in which the associated natural gas is utilized either as fuel or as feedstock for an industrial process. Suitable facilities are for instance gas-fired power plant in combination with the industrial plants that use the electricity produced by such power plants; petrochemical plant in which the gas is converted into a whole range of chemicals, or a small, compact refinery (referred to as a gas-to-liquids [GTL] plant) in which the natural gas is converted into one or more commercially attractive liquids, e.g., petroleum liquids, methanol; or industrial plant, which uses the gas in a direct manner as fuel. Frequently applied solutions envisage the building of a gas-fired power station in combination with an aluminum smelter or a water-desalination plant. In all such cases, the associated natural gas can be supplied at the cost of no more than that of collecting the gas in the oil field. Associated gas (and non-associated gas for that matter) may also be converted into liquids or electricity in order to facilitate its transportation (in the form of liquids or electricity) to a suitable market. Such transportation is easier and less expensive than the transportation of the natural gas itself. In this situation, the producer's gas price is derived from the price that is paid in the end-market for the liquids or electricity. A drawback of any conversion is that the conversion process has a relatively low efficiency in as much as more than fifty per cent of the heat content of the natural gas will be lost thereby.

3.5 TAXES

3.5.1 Export Duties

An export duty is a duty that is levied on the value of the exported production at the port of export. Unless such a tax is levied on all major export crudes, in which case the global producer price will be raised by the amount of the tax, the cost of the tax must be absorbed by the producer.

3.5.2 Import Duties

Import duty is a duty that is levied on the value of the imported oil at the port of import. Such duty has to be paid by the refinery that is importing the oil. The main purpose of an import duty is to protect domestic oil producers, which sell their production to an inland refinery and get paid a price derived from the prices paid by the refinery for imported crude oils after adjustments on account of differences in quality and gravity. If such price turns out to be too low, i.e., if and when world market prices show a tendency to fall below the level at which these producers could still operate commercially, the government of the importing country may contemplate to impose an import duty. Such a measure is a powerful fiscal instrument and had been contemplated by the US government in 1986, at the time OPEC was pursuing a 'flooding-the-market' strategy which caused the world market price to tumble. Domestic oil producers operating within an oil-importing country and selling their production to an inland refinery are paid a price derived from the prices paid for imported crude oil after adjustment on account of differences in quality and gravity. If an import duty is levied, the domestic oil producer will benefit therefrom.

3.6 ALTERNATIVES FOR OIL PRODUCTS AND NATURAL GAS

3.6.1 Introduction

Oil (products) and natural gas have over the decades become essential ingredients for a satisfactory performance of the world economy. In fact the latter has become completely dependent on their availability. Already for this reason alone, it is opportune to search for or develop other products that can take over the functions of oil products and natural gas at the appropriate time in all or any sector of the economy; these other products are hereinafter referred to as alternative products or alternatives and relate exclusively to oil and natural gas. From this perspective, the search for and development of alternatives should start when the depletion of the recoverable oil and gas reserves is approaching. If the end is in sight, alternatives should be introduced in the relevant sectors of the economy at the same rate as the depletion of the reserves is progressing.

Note. For a proper assessment of the status of the reserves, those of oil (including non-conventional oil) and natural gas should be added together since oil and gas are mutually interchangeable, in the sense that all products manufactured in the refinery from oil can also be manufactured from natural gas by means of the appropriate catalytic conversion processes, albeit at considerable cost.

However the introduction of suitable alternatives could be speeded up in order to prolong the lifetime of the reserves, and/or for security-of-supply reasons, or in case alternatives or some of them might offer a real technological advantage over the use of oil or natural gas.

Apart from all these considerations, the use of oil products and natural gas as fuel has the disadvantage that such use lead to the generation of CO_2 that, without taking precautionary measures, will be emitted into the atmosphere and thereby increase the atmospheric concentration of CO_2 which in turn will contribute to global warming and climate change (see Chapter 1). These environmental concerns form another reason for looking for alternatives.

Note. Besides looking for alternatives, the government may also try to discourage the consumption of, for instance, transport fuels by imposing fiscal charges and making the fuels more expensive for the end-user (see Chapter 4).

3.6.2 A Review of Alternatives

3.6.2.1 Coal

Coal (comprising hard coal and sub-bituminous coal) is and remains, for the time being and technically speaking, the only readily available alternative for oil and natural gas. It has been demonstrated that oil and natural gas can be replaced by coal or by products made from coal in all sectors of the economy where currently the former are used. Even as feedstock for the refineries as well as feedstock for the petrochemical industry, coal can replace both oil and natural gas. When used as such (in the form as delivered from the mine), coal can replace oil products and natural gas as fuel in power plants, in industrial plants for steam generation and even in households for space heating. In the past, coals which contained a high percentage of volatile matter were processed into town gas. Town gas was used in gas lamps, lighting up the streets of the town and buildings, and was used for space heating and cooking. For lighting purposes, town gas (and therefore coal) was finally replaced by electric light; in households, town gas was replaced for cooking and space heating purposes by natural gas as soon as the latter became available (in the vicinity of the consumer) and suitable pipelines could be constructed for its transport over longer distances.

Coal can also replace oil and natural gas as feedstock for manufacturing diesel oil and hydrogen.

Note. Starting point for both processes is the gasification of the coal. In this process a mixture of carbon monoxide (CO) and hydrogen referred to as syngas is produced. If syngas is further treated with steam, the CO part thereof is converted into hydrogen and CO_2. After removal of the latter, the syngas will then have been completely transformed into hydrogen. In the next step, hydrogen can be delivered as such to the end-user (e.g., a refinery where it is used for upgrading hydrocarbons or for removing sulphur from the heavy oil products) or used as fuel to drive a gas turbine generator. But syngas can also be further processed for manufacturing hydrocarbons, which in turn can be refined into diesel oil, a process that is referred to as the coal-to-liquids process. The coal-to-liquids process is a catalytic conversion process, which is a further development of the Fischer-Tropsch process invented in 1925. In the past, the Fischer-Tropsch hydrocarbon synthesis using coal as base material yielded poor quality transport fuel due to imperfect catalysts and had only been economically justified and warranted in countries that, having no oil or gas reserves themselves, were unable due to adverse political circumstances to import the necessary oil or natural gas supplies. In this context, large scale application of the hydrocarbon synthesis based on coal had only taken place in Germany during the Second World War and further in South Africa during the period that this country was subjected to an oil embargo in connection with the existence in that country of an apartheid regime. Currently, the elevated level of oil prices in combination with improvements of the catalysts have revived interest in this process, and factories are being built for testing the new catalysts.

Coal has the advantage of being abundantly available. Its proven reserves are large and geographically speaking widely spread which makes coal less risky from a geopolitical point of view (see Chapter 2). Moreover the costs of production are reasonable in particular when open pit mining is possible. Coal can also be shipped at a moderate cost and can thus be exported at a very competitive price when compared to today's prices of exported oil. In short, coal would be the perfect alternative to oil and natural gas if not for its many practical and environmental drawbacks. The latter put it at a disadvantage; so much so, that replacing oil and natural gas should only be considered if the recoverable oil and natural gas reserves are getting depleted or if a government wishes to stimulate the use of coal in the context of pursuing a policy of security-of-supply.

Note 1. Handling, storage and transportation of coal are cumbersome and compare in these respects unfavourably with oil and natural gas. But its most prominent drawback (when viewed in the context of climate change) is the generation of CO_2, which arises when coal as such is used as fuel or when coal is gasified. When compared with oil products and natural gas, coal generates (per unit of heat produced) respectively twenty-five per cent and seventy-five per cent more CO_2. This drawback can be eliminated by installing at the plant facilities for the capture and underground storage of CO_2 (CCS-facilities), so as to prevent the CO_2 generated from being emitted into the atmosphere. The future of coal and its usefulness to replace oil and natural gas may depend on to what extent these measures prove to be practical and economically justified.

Note 2. It should be recalled that environmental concerns raised in connection with the burning or combustion of fossil fuels are not restricted to CO_2 generation alone. One environmental drawback shared by all these fuels is the oxidization of the nitrogen in the air that is mixed with the fuel in the combustion process, with the result that nitrogen oxides (NOx) are formed and released into the atmosphere together with CO_2 and with potentially damaging effects (smog). Furthermore, some types of coal contain a high percentage of sulphur. This substance has to be removed but it must be pointed out that the heavier types of crude oil as well as natural gas may also contain sulphur compounds up to a significant percentage, which also have to be removed.

3.6.2.2 Biomass

Biomass (plant material) can be described as a kind of proto-coal, since on a geological timescale coal is generated out of biomass after this material as a result of geological processes becomes deeply buried and exposed to high temperatures and pressure. From this point of view, biomass can be seen and treated as coal and solely or mixed with coal serve the same purpose, i.e., the generation of heat and electricity. Like coal, biomass can be gasified and converted into syngas, which can further be processed as described above in relation to coal. Some types of biomass can be converted into ethanol and diesel oil (see hereinafter).

3.6.2.3 Non-fossil Electricity

Non-fossil electricity is electricity that is generated by non-fossil energy sources such as hydropower in various forms, nuclear fission energy and, on a much smaller, local or even regional scale, biomass, wind turbines, photovoltaics, solar power and high temperature geothermal energy (the latter collectively referred to as renewables). With exception of biomass, all renewables can also be described as non-carbon energy sources.

In 2005, total power generation amounted to 18,197 TWh divided between non-fossil electricity and fossil electricity in a 33/67 proportion. Non-fossil electricity is generated by nuclear energy (fifteen per cent), hydroelectric (sixteen per cent) and by renewables (two per cent). Fossil electricity is generated by coal (forty per cent), oil (seven per cent) and natural gas (twenty per cent). From these figures, the conclusion can be drawn that if non-fossil electricity were ever to replace fossil electricity (and that seems at the moment the only possibility to push back the generation of CO_2), the capacity of its energy sources will have to be expanded in a drastic manner. Since a lack of space and suitable sites prevents any meaningful expansion of the electricity-generating capacity of the renewables as well as any further expansion of the currently established capacity of hydroelectric, the generation of non-fossil electricity appears to become a matter of expanding the capacity of nuclear energy (with a current market share of fifteen per cent). Such expansion should first take place at the expense of coal as being the biggest contributor to CO_2 generation and then at that of oil. It follows that a maximum reduction of CO_2 emissions could be achieved if electricity would exclusively be generated in nuclear power stations but a mix of 80/20 with natural gas would also be very effective for getting CO_2-emissions under control (see Chapter 4 for further discussion).

Note with reference to nuclear fission energy. This source of energy is based on the fission of uranium-235 or plutonium-238 atomic nuclei in a reactor by a bombardment with neutrons. Uranium-235 is naturally occurring in uranium ore, but only for a very small part (about 0.7 per cent). The rest is made up of uranium-238. Before being usable in a reactor, the uranium-235 content of the fuel intake (uranium oxide) must be increased to about three to four per cent. This treatment of the fuel intake through the intermediary of the gaseous uranium hexafloride (UF6) is referred to as 'enrichment'. After enrichment, the reactor fuel contains about four per cent uranium-235; the remainder consists of uranium-238. Because the fission creates new neutrons, the process is self-sustaining. Apart from splitting uranium-235 nuclei, high-speed neutrons may also transform uranium-238 into plutonium-238. The fission of uranium-235 as well as the transformation of uranium-238 can be controlled by 'moderating' the speed of the neutrons. If the neutrons are not slowed down, all uranium-238 may be transformed into plutonium-238 that in turn can be split and produce energy. In this way, the reactor is 'breeding' its own fuel. This type of reactor is referred to as a 'fast breeder' reactor. But depending on the moderation of the neutrons, plutonium-238 will also be formed in the conventional reactor. In a closed cycle process, this plutonium-238 will be extracted from the spent fuel and subsequently mixed with the primary fuel that is going into the reactor. However by using the aforesaid fast breeder reactors as described above, the world usable uranium reserves would increase a hundred fold, providing the feedstock for non-fossil electricity for thousands of years. From this point of view, nuclear fission energy could rightly be called a renewable form of energy on a par with solar power, etc.

Note with reference to nuclear fusion energy. Still, hopes are pinned on the possibility that, in the not too distant future, non-fossil electricity will be generated by nuclear fusion energy. In contrast to nuclear fission energy, this type of nuclear energy derives its energy from the fusion of two hydrogen isotopes deuterium and tritium into helium. Deuterium is produced from seawater and tritium from lithium. In the reaction, one atom of deuterium fuses with one atom of tritium to form one atom of helium, releasing thereby one high-speed neutron. The most promising design for a fusion reactor is the tokamak. The latter is a toroidal chamber in which intense magnetic fields keep the atomic nuclei (plasma) confined as it heats up electrically the latter to more than 100 million degrees Celsius. In the next step, microwaves are beamed into the hot plasma to increase the efficiency of the fusion reaction. All by all, the reactor consumes huge quantities of energy and the challenge is to extract from the fusion reaction significantly more energy as is consumed by it. As far as the neutrons are concerned, they escape the magnetic confinement and hit the shielding material around the reactor. These collisions create heat that could be utilized to produce additional steam. More importantly, the high-speed neutrons could be used to breed tritium by shooting them into shielding material consisting of lithium. Other uses for the fast neutrons are also under investigation. To bring the development of the nuclear fusion option forward, construction has started of the International Thermonuclear Experimental Reactor (ITER) in Cadarache in southeastern France. ITER is supposed to start operations in 2016. Costs are estimated to amount to USD 10 billion. If successful, ITER is expected to generate ten times the energy it consumes. The ITER-project is a joint undertaking of the EU, the Russian Federation, the US, Japan, China, India and South Korea. It was first proposed in 1985 at the summit meeting between Reagan and Gorbachev in Geneva. If all goes well and allowing fifteen years for carrying out experiments, fusion energy could be confirmed as a commercial proposition as from 2030, perhaps right in time to replace nuclear fission energy and fossil fuels in power plants.

Note with reference to hydropower. Hydropower consists in using the energy of flowing water to drive turbines which in turn generate electricity. Hydropower comes in various forms. The most simple and widespread technique of creating the right conditions for converting hydro-energy into electricity consists in building dams in rivers and creating behind the dam an artificial lake that serves as water reservoir. Water from the reservoir is then taken in by the turbines which are placed inside the dam. Less common is the technique that seeks to make use of offshore tides and waves. How attractive this all may sound, significant expansion in hydropower is blocked by a lack of sites suitable for building the necessary dams and creating artificial lakes or for building tidal or wave-power stations.

Note with reference to high temperature geothermal energy. Large-scale application of high temperature geothermal energy is restricted to sites located in a volcanic region, e.g., Iceland.

Note with reference to wind turbines. Even when wind turbines are used with a capacity of 2 MW, large-scale application of this non-carbon energy source entails the setting up of wind farms comprising of at least 250 turbines. Wind farms of this scale require much space and can therefore not be built in densely populated areas. More suitable for this purpose are semi-desert areas or coastal waters. This raises immediately the problem of how the electricity that is generated by the farm must be transported to the market. Apart from that, wind power suffers from the fundamental flaw that power generation is not stable or reliable, dependent as it is on wind speed which, as is well known, can vary between zero and more than 250 km per hour and all that within a relatively short period. Unfortunately, if wind speed exceeds a certain maximum, wind farms must be shut down. But as could be expected, serious research is undertaken to come up with technical (not necessarily commercial) solutions to this problem of fluctuating wind speeds. Most proposals amount in essence to pumping around water. Whatever is proposed however, it will certainly require large investments and again much space. From this point of view, it can be counted upon that plans for large-scale wind farms will be discarded as soon as public opposition to nuclear fission energy fades away, and nuclear power stations are allowed to be built.

Note with reference to solar radiation. Solar radiation can be converted into heat or electricity. Only the conversion into electricity (photovoltaics) offers scope for a meaningful expansion of this source of energy. But large-scale application of photovoltaics as envisaged today requires just like wind farms much space. The required space would of necessity have to be claimed in the major desert or semi-desert areas of the world, because only these areas, due to their geographical location and cloudless skies, receive sufficient solar radiation for a meaningful generation of electricity. For instance, the average annual solar radiation in the desert areas of Northern Africa, the Arabian Peninsula, US and Australia is more than 2,200 kWh per square metre, which figure compares with a modest 1,000 kWh per square metre for Northern Europe. Except for the US and Australia, places of consumption are far removed from the suitable areas, which means that any electricity generated and collected in these areas must be transported over large distances. In fact 'electricity from the desert' (EFD) faces the same marketing problems as has to be faced by natural gas and wind power. However this may (technically) be, before EFD can ever become reality, some serious 'security-of-supply' issues must be addressed and resolved.

3.6.2.4 Alternative Fuels for Road Transport

Currently alternative fuels for road transport comprise biofuels and hydrogen. At the moment, two types of biofuel are being manufactured: ethanol and biodiesel. Ethanol is made from corn, wheat and barley (US and other OECD countries) or from sugar cane (Brazil). Biodiesel is produced mainly from oil-seed crops, including rapeseed, palm oil and sunflowers.

Note with reference to ethanol and biodiesel. Ethanol (C_2H_5OH) can be blended up to high percentages (eighty-five per cent) with gasoline, but at a higher percentage than thirty per cent, the car engine has to be adjusted. Biodiesel is comparable with conventional diesel oil and makes it an attractive blending component. For both, biofuels hold that their production has a negative environmental impact and disastrous consequences for food prices. From this perspective research is focused on the possibility to produce ethanol from ligno-cellulosic crops such as eucalyptus, poplar or willow trees and grasses or from waste (e.g., straw). This so-called second generation ethanol should ideally replace corn-, wheat- and barley-based ethanol considering the many environmental and societal drawbacks associated with its use as mentioned above.

Note with reference to hydrogen. Hydrogen is presently used as rocket fuel and in the refineries for the upgrading of hydrocarbons and for the removal of sulphur from the heavy fractions of the distillation process. However in the context of government policy to reduce CO_2-emission, the option of using hydrogen as fuel for car engines is being researched. In this option, hydrogen can be used in two ways: either directly fuelling the car engine in substitution of gasoline or diesel oil (whereby the internal combustion principle can be maintained), or through the intermediary of a fuel cell supplying electricity to an electric motor. But replacing carbon transport fuels such as gasoline and diesel oil by hydrogen faces big problems of a technical nature when the safety aspects involved in the fuel replacement are taken into account. Due to its inherent physical constraints, i.e., its low energy density, hydrogen in its gaseous state must during transportation to the filling stations and when stored in the car be kept under high pressure. On the other hand, when stored in a motor car in its liquid state, the hydrogen must be cooled to temperatures below minus 253 degrees Celsius which in turn requires a very special tank. It is very doubtful whether a motor car that is equipped for the use of hydrogen in its gaseous or liquid state can ever satisfy the standard safety regulations and pass any crash tests. It should be added that research is going on into the potential that metal hydrides offer for storing hydrogen in cars. Much lower pressures are involved which improves the safety, but the problems involved with distributing hydrogen to the filling stations still remains (not to mention the difficulties that motorists will have to cope with when trying to fill up the hydrogen storage devices installed in their cars). Whether hydrogen is directly or indirectly fuelling cars in both instances no CO_2 is generated and thus no danger of an emission of CO_2. Whether in the total picture no CO_2 is generated depends on the way in which hydrogen itself is manufactured. Hydrogen can be produced in many ways. The most common way is to produce hydrogen from methane (extracted from natural gas). Hydrogen can also be produced from syngas that in turn is obtained from the gasification of coal or biomass. In both instances CO_2 is produced as a by-product. Finally, hydrogen can be manufactured in a direct manner by the electrolysis of water. Whether or not CO_2 is generated in this process depends on the feedstock intake of the power plant that is supplying the electricity needed for the electrolysis and on how this plant handles the CO_2 that may be generated in the plant. From this perspective, the electricity needed for the electrolysis should either be supplied by a nuclear power plant or by a coal-fired power plant that is equipped with CCS-facilities. Finally hydrogen can be generated in a thermochemical reaction. The required heat is supplied by high-temperature steam, which in turn is generated in a Very-High-Temperature Reactor (VHTR) in a nuclear power station.

3.7 MARKET OUTLOOK

3.7.1 INTRODUCTION

In *World Energy Outlook 2007*, the latest publication of the International Energy Agency (IEA) two oil/gas consumption scenarios are presented which are respectively referred to as the Reference Scenario and the Alternative Policy Scenario. Both scenarios cover the period up to 2030. In projecting oil/gas demand over the Outlook period, the Reference Scenario takes into account those government policies and measures that have been enacted or adopted by mid-2007, regardless of whether they have yet been fully implemented. The Reference Scenario does not take into consideration possible, potential or even likely future policy actions. By contrast, the Alternative Policy Scenario was developed inter alia to highlight and assess the impact of future policy actions aimed at a stabilization of the atmospheric concentration of CO_2 at a level of 450 ppm. However no revolutionary developments on the energy scene, such as a large-scale switch from coal to nuclear energy in the power generating sector, coupled with the introduction of the hydrogen-powered or all-electric car (replacing diesel oil and gasoline in the transport sector), etc. have been taken into account. But according to both scenarios, the impact of new policies is less marked in the period up to 2015. In that period, the opportunities for change are limited because the cost of replacing capital stock prematurely is high, even when the new stock is more energy-efficient and/or contributes more to limiting CO_2-emissions. A case in point is the aforesaid possible replacement of coal by nuclear energy in the power generation market. However this may to assess the market outlook for oil and natural gas, the projected consumption has to set off against ultimate recovery. For this specific purpose the more pessimistic Reference Scenario is better suitable since it has a built-in margin for error. Moreover as far as oil is concerned, it is reasonably assumed that consumption in the major consuming regions, i.e., North America and Europe will stabilize at slightly above 2006 levels and only will significantly grow in the developing countries in general and in China and India in particular. As far as natural gas is concerned, the Reference Scenario envisages a modest increase in natural gas consumption in the major gas consuming areas and a strong growth in again China and India.

Underlying both scenarios is the projection of the world population growth. The latter determines in first instance the world primary energy consumption, including that of oil and natural gas. According to the United Nations' report, *World Population Prospects: The 2004 Revision* (UNPD, 2005), global population is projected to grow by one per cent per year on average, from an estimated 6.4 billion in mid-2004 to over 8.1 billion in 2030.

3.7.2 Market Outlook: Oil

Table 3.8. Projected Oil Consumption in the Reference Scenario (million b/d)

	2006	*2010*	*2015*	*2030*	*Annual % Growth 2006-2030*
North America	24.9	26.2	27.7	30.0	0.8%
Europe	14.3	14.5	14.7	14.7	0.1%
Russia	2.6	2.8	3.0	3.3	0.9%
China	7.1	9.0	11.1	16.5	3.6%
India	2.6	3.1	3.7	6.5	3.9%
Other Asia	5.5	6.2	6.9	8.9	2.0%
Middle East	6.0	7.0	7.9	9.5	1.9%
World	84.7	91.1	98.5	116.3	1.3%

Source: World Energy Outlook 2007.

If annual production proves indeed to be capable to satisfy the projected consumption, it implies that as of the end of 2006 up to and including 2030 124 billion tonnes will have been produced. Subtracting this figure from the 322 billion tonnes calculated for the remaining ultimately recoverable reserves (including Canadian oil sands) at end 2006, leaves 198 billion tonnes available for the years after 2030, i.e., for thirty-three years of production at an annual rate of 6 billion tonnes. Moreover, a considerable volume of oil is left that (if need be) could be produced from US oil shales or Venezuelan extra-heavy oil.

3.7.3 Market Outlook: Natural Gas

Table 3.9. Projected Natural Gas Consumption in the Reference Scenario (billion cubic metres)

	2005	*2015*	*2030*	*Annual % Growth 2005-2030*
North America	765	887	994	1.1%
Europe	550	639	771	1.4%
Russia	431	516	586	1.2%
China	51	131	238	6.4%
India	35	58	112	4.8%
Other Asia	177	262	360	2.9%
Middle East	261	394	639	3.6%
World	2,854	3,689	4,779	2.1%

Source: World Energy Outlook 2007.

If annual production will indeed be capable of satisfying the projected consumption, it implies that as of end 2006 and up to and including 2030 83 billion toe will have been produced. Subtracting this figure from the 268 billion toe calculated for the remaining reserves at end 2006 (see Chapter 2) leaves 185 billion toe available for the years after 2030, i.e., for forty-three years of production at an annual rate of 4.3 billion toe.

Chapter 4
Government Petroleum Policies

4.1 AN OVERVIEW

From the start of the industrial revolution (1750) till today the primary energy demands of the world economy have been met exclusively or to a large extent by coal, oil and natural gas. First of all came coal (*King Coal*), later it was joined and even partly replaced by oil and natural gas. More recently and at least for now, nuclear energy and hydroelectric make their smaller contributions. Statistics show that in 2006 coal, oil and natural gas jointly satisfied eighty-eight per cent of the world primary energy consumption (totalling 10.9 billion toe), whereas nuclear energy and hydroelectric each contributed six per cent. Expressed in volumes, oil's contribution amounted to 3.9 billion tonnes and that of natural gas 2.6 billion toe. Considering that oil is also needed and utilized to manufacture lubricants, bitumen and the basic feedstock for the petrochemical industry, it is clear that next to coal, oil and natural gas are for the present moment in one way or another of vital importance for the functioning of world economy and its further development in the foreseeable future. This is also true, albeit in different ways, forms and importance, for the economy of each individual country.

Note. From a global perspective (and apart from any geo-political considerations and environmental concerns), this dependency on the fossil fuels is by no means alarming. Worldwide there are ample recoverable reserves of coal, oil and natural gas still available: those of oil and natural gas seem sufficiently large to cover world consumption till the end of the twenty-first century, whereas the proved reserves of coal alone could if still required for the economy last much longer.

Against this background, any government (of a sovereign State) was and is required to formulate a policy with respect to the domestic consumption and production of petroleum and to make decisions in the matter of import or export of the same and most probably such decisions may turn out differently where oil or natural gas is involved. In principle, every government followed and follows a petroleum policy which is aimed at promoting its own interests in petroleum matters.

But in a few very important instances, sovereign States recognizing a common interest in petroleum matters have coordinated their policies in this regard within the framework of a formal Organization. The first such Organization and one which was solely dedicated to petroleum matters was the Organization of Petroleum Exporting Countries (OPEC). This Organization was founded on 1 September 1960 by five major oil exporting States, for the stated purpose of coordinating and unifying their petroleum policies and of determining the best means of safeguarding their interests individually, as well as collectively.

West-European countries formed in 1947 the Organisation for European Economic Co-operation (OEEC). This Organization was meant to administer American and Canadian aid under the Marshall Plan for the reconstruction of Europe after World War II. The OEEC was succeeded in 1961 by the Organisation of Economic Co-operation and Development (OECD). This new Organization comprises thirty free market economies among which the Member States of the former OEEC, the US, Canada, Australia, Japan and later on also South Korea. On 15 November 1974, in a direct reaction and response to the oil crisis of 1973/1974 which was triggered by the total or gradual oil embargo imposed by the Arab Member States of OPEC against the United States and some West-European countries, the OECD founded the International Energy Agency (IEA) and adopted three days later the International Energy Program (IEP). More broadly based economic cooperation but also including petroleum matters was intended by the European Economic Community (EEC) founded by the Treaty 'Establishing a European Economic Community' signed in Rome on 25 March 1957, effective date 1 January 1958. Much later the EEC was succeeded by the European Community (EC), which was created by the Treaty on the European Union (EU) as one of the three pillars envisaged for this Union (7 February 1992).

Apart from the countries that have joined one of the aforementioned Organizations (OPEC, OECD, EU), there are many other countries which on their own and independent from any interstate organization are promoting the search for and, if successful, the development of their domestic petroleum potential. To this non-organized group belong most of the developing countries but also countries with an economy in transition such as the Russian Federation, India and China. In all cases said countries aspire to establish a viable petroleum industry with enough reserves not only for their own needs (which as far as the developing countries are concerned may initially be small) but also for export to the major consuming Western countries. If a developing country is very successful, it may hope one day to join OPEC (as Angola did: this State joined OPEC as the 12th Member State on 1 January 2007).

In the 1990s the international community under guidance of the United Nations became aware of the fact that global warming was going on as a result of the emission of CO_2 into the atmosphere originating from the burning and combustion of the fossil fuels and that this global warming was projected to continue and reach unacceptable rates if the emission of CO_2 would continue at its present rate. This awareness and concern was shared by the governments of the Member States of the OECD and given concrete form in the protocol signed in December 1997 in Kyoto, Japan (Kyoto Protocol).

4.2 COMMON PETROLEUM POLICIES (NON-OPEC)

From the outset, but before the Kyoto Protocol was signed and Member States of the OECD had to focus their attention on ways and means to fulfil their CO_2 commitments, under this protocol the petroleum policies of said Member States had much in common with the policies pursued by other interested countries outside OPEC. These common objectives can be summarized as follows:

(i) to search for, develop and produce domestic petroleum resources. The volumes produced are in the first place meant to supply the domestic market but, if there is any surplus, the government concerned will certainly try and export any such surplus to countries which by lack of sufficient domestic resources are forced to import;

(ii) to establish state-owned companies for worldwide searching for and developing petroleum resources;

(iii) to stimulate energy saving and an efficient use of energy in order to reduce consumption of petroleum if the latter has to be imported;

(iv) to aid research into and the development of suitable fossil fuel alternatives in order to reduce dependency on foreign suppliers; and

(v) to make choices between different forms of energy, where such choice is possible. In this matter of choice, governments of OECD countries have less room to manoeuvre than governments of other countries because of the environmental policies pursued by the former. As a result of these policies, the OECD countries (with the notable exception of France, Finland and Japan) became cautious and hesitant, for reasons of public safety and health, about any expansion of the use of nuclear fission energy in the power generating sector, despite the fact that in this sector nuclear energy could adequately replace any of the fossil fuels; and

(vi) to stimulate fuel switching and to give preference to oil, oil products and natural gas over coal in the thermal power plants in order to increase energy transformation efficiency and to avoid the inherent disadvantages associated with its use.

Note concerning (i). In the recent and less recent past, before the international oil price reached such spectacular high levels, the government of any petroleum importing country could decide to stimulate and encourage the establishment of domestic production, even if the cost thereof would exceed the price payable for imports. When needed, expensive domestic production can be supported and protected by fiscal and/or administrative measures (e.g., abolishing royalty and/or lowering of the rate of the applicable income tax, and/or the imposition of quantitative restrictions on imports and/or levying an import duty). Best known example of a government measure to safeguard domestic oil production is the US Mandatory Oil Import Control Program imposed in 1958 and meant at the time to protect the US domestic oil producers against the low-priced imports from the Middle East. It should be added that in 1986, when as the result of the flooding-of-the-market-strategy pursued by Saudi Arabia, the world market oil price dropped to about USD 10/b, the US Bush-Administration seriously considered the option to introduce an import duty in order to protect the high cost US domestic production. During the brief period of declining world market prices (1997-2000), which decline started in the beginning of 1997 and reached for the time being its lowest level (viz. USD 10/b) in December 1998, high cost domestic producers everywhere immediately proposed to their respective governments that fiscal burdens be alleviated, e.g., by abolishing royalty, by relaxation of depreciation rules and by lowering of the rate of the applicable income tax(es).

Note concerning (ii). France and Italy established in 1953 their respective state-owned companies Entreprise de Recherches et d'Activités Pétrolières (ERAP) and its subsidiary Société Nationale ELF Aquitaine and Ente Nazionale Idrocarburi (ENI) and its subsidiary Agip. Both oil groups have now been privatized. Japan pursued the same policy and established state-owned Japan National Oil Corporation (JNOC). JNOC is expected to acquire petroleum rights in foreign countries in competition or preferably in cooperation with foreign oil groups. In the United Kingdom a state oil corporation was established under the Petroleum and Submarine Pipe-lines Act 1975. The corporation was named the British National Oil Corporation (BNOC). The new corporation was authorized in compliance with the directions of the Secretary of State to explore for and obtain petroleum in any part of the world. BNOC was also designated as the official government advisor in petroleum affairs and was given a preferential position in acquiring any licence interest that was intended to be transferred. Under the Oil and Pipelines Act 1985, the existence of BNOC was terminated (see Chapter 6).

Apart from these primary objectives described above, there were and still are secondary objectives, but these apply only in case a country possesses or is likely in the near future to possess its own petroleum resources. These secondary national objectives concern matters such as:

- (i) institution building, putting a petroleum law and regulations in place, becoming party to international conventions, multilateral or bilateral treaties regarding petroleum operations;
- (ii) petroleum taxation;
- (iii) state participation in petroleum operations;
- (iv) treatment of foreign enterprises who wish to invest in the search for and development of the domestic petroleum resources; and
- (v) fulfilling international obligations which derive from the international conventions or treaties to which the State has become a party.

Note concerning (iii). Often used arguments for state participation in petroleum rights and operations are profitability and control or on a combination of the two. If domestic production proves to be a profitable undertaking, state participation will also be profitable. But if such operations are not likely to be or become profitable, and therefore commercial petroleum enterprises show no interest in applying for the required authorization, the government concerned may wish that in the public interest those petroleum operations (in particular exploration) are nonetheless undertaken and will designate a state company to do so. Such approach may be made possible by including a sole risk option in a state participation agreement. As to control, a government may feel that it needs to stay very close to the licensee/contractor and its operations in order to:

(a) share in the decision making process, in particular concerning development planning and investments in development;
(b) gain first-hand experience with the technical, administrative and commercial aspects of petroleum operations; and
(c) supervise the implementation by the licensee/contractor of all non-technical and non-fiscal provisions of the authorization, among which national interest provisions, such as the employment and training of nationals, the use in the operations of locally manufactured products and of the goods and services of local suppliers.

But it will be appreciated that the argument of having to stay close to a holder of a petroleum right in order to achieve the objectives (a) through to (c) above is most convincing where exclusive licences (as opposed to contracts of work) are concerned. If a state enterprise enters with another, usually foreign oil enterprise, into a contract of work, the nature and form of this contractual relationship (e.g., a production sharing agreement) should guarantee sufficient scope for achieving the aforesaid objectives.

4.3 UN CLIMATE CONFERENCES AND IPCC'S ASSESSMENT REPORTS

4.3.1 World Climate Conference (February 1979)

At the level of the UN, serious interest in the possibility of climatic problems resulting from the burning and the combustion of fossil fuels can be traced back to the World Climate Conference convened by the UN World Meteorological Organization (WMO) in February 1979. At this Conference the following statement was issued:

> The causes of climatic variations become better understood, but uncertainty exists about many of them and their relative importance. Nevertheless, we can say with some confidence that the burning of fossil fuels, deforestation, and changes in land use have increased the amount of carbon dioxide in the atmosphere by about 15 per cent during the last century and is at present increasing at about 0.4 per cent per year . . . It is possible that some effects on a regional and global scale may become detectable before the end of this century and become significant before the middle of the next century. This time scale is similar to that required to redirect, if necessary, the operation of

many aspects of the world economy, including agriculture and the production of energy.

In May 1979 the Congress of the WMO approved the World Climate Programme (WCP) which consisted of four components, viz. Climate Data, Climate Applications, Climate Research and the Climate Impact Studies Programme for which the UN Environmental Programme (UNEP) took the responsibility. The next step was taken nine years later, in 1988, when WMO and the UNEP jointly established a new institution named the Intergovernmental Panel on Climate Change (IPCC). The panel formed three Working Groups, namely, I. Scientific Assessment of Climate Change; II. (Economic/Social) Impacts of Climate Change; and III. (Government) Policy Response to Climate Change. Working Group I (WG I) is charged with the task of reporting every six years on the latest scientific insights concerning global warming and its consequences for sea level rise and climate change. WG I issued its Fourth and latest Assessment Report in February 2007.

4.3.2 The IPCC WG I First Assessment Report (1990)

The IPCC WG I's First Assessment Report was issued in the summer of 1990 in preparation of the Second World Climate Conference held later in the year. In this report, the Working Group gave as its best estimate that by the year 2100 the atmospheric concentration of CO_2 would have risen to twice the pre-industrial level (which would entail a rise to about 550 ppm) and the average temperature to 3.5 degrees Celsius compared to the pre-industrial level. This meant that an additional rise of 3 degrees Celsius by the year 2100 was foreseen considering the 0.5 degrees C increase that already had taken place since the industrial use of the fossil fuels. As to the sea level, a rise of 45 centimetres was predicted.

4.3.3 The Second World Climate Conference (1990)

Against the background of WG I's First Assessment Report the Second World Climate Conference convened in Geneva (6-7 November 1990). One hundred thirty-seven countries took part therein. Just a few days before, on 29 October 1990, the joint Energy/Environment Council of the EC had decided to take actions aiming at reaching stabilization of the total CO_2 emissions by 2000 at 1990 level in the Community as a whole. This initiative of the EC was noted and welcomed by the participants and the Final Declaration of the Conference made reference to it. Since then the year 1990 has become the reference year for formulating CO_2 emission policies. The most important statements contained in the Final Declaration were the following:

1. We, the Ministers and other representatives from 137 countries and from the European Communities, meeting in Geneva from 6 to 7 November 1990 at the Second World Climate Conference, declare as follows:

2. We note that while climate has varied in the past and there is still a large degree of scientific uncertainty, the rate of climate change predicted by the Intergovernmental Panel on Change (IPCC) to occur over the next century is unprecedented. This is due mainly to the continuing accumulation of greenhouse gases, resulting from a host of human activities since the industrial revolution, hitherto particularly in developed countries. The potential impact of such climate change could pose an environmental threat of an up to now unknown magnitude; and could jeopardize the social and economic development of some areas. It could even threaten survival in some small island states and in low-lying coastal, arid and semi-arid areas.

5. Recognizing that climate change is a global problem of unique character, we consider that a global response . . . must be decided and implemented without further delay based on the best available knowledge such as that resulting from the IPCC assessment. Recognizing further that the principle of equity and the common but differentiated responsibilities of countries should be the basis of any global response to climate change, developed countries must take the lead. They must all commit themselves to actions to reduce their major contribution to the global net emissions and enter into and strengthen cooperation with developing countries to enable them to adequately address climate change without hindering their national developments goals and objectives.

7. Where there are threats of serious or irreversible damage, lack of full scientific certainty should not be used as a reason for postponing cost-effective measures to prevent such environmental degradation.

8. The potentially serious consequences of climate change, including the risk for survival in low-lying and other small island States and in some low-lying coastal, and arid and semi-arid areas of the world, give sufficient reasons to begin by adopting response strategies even in the face of significant uncertainties. Such response strategies include phasing out the production and use of GFCs, efficiency improvements and conservation in energy supply and use, appropriate measures in the transport sector, sustainable forest management, afforestation schemes . . . proper land use planning, the use of safe and cleaner energy sources with lower or no emissions of carbon dioxide, methane, nitrous oxide and other greenhouse gases and ozone precursors, paying special attention to new and renewable sources.

10. We agree that the ultimate global objective should be to stabilize greenhouse gas concentrations at a level that would prevent dangerous anthropogenic interference with climate.

12. Taking into account that the developed world is responsible for about 3/4 of all emissions of greenhouse gases, we welcome the decisions and commitments undertaken by the European Community with its Member States, Australia, Japan, Canada and other developed countries to take actions aimed at stabilizing their emissions of CO_2, or CO_2 and other greenhouse gases not controlled by the Montreal Protocol, by the year 2000 in general at 1990 level, yet recognizing the differences in approach and in starting point in the formulation of the above targets.

16. The specific difficulties of those countries, particularly developing countries, whose economies are highly dependent on fossil fuel production and exportation, as a consequence of action taken on limiting greenhouse gas emissions, should be taken into account.

28. We call for negotiations on a framework convention on climate change to begin without delay after a decision is taken by the 45th Session of the General Assembly of the UN. We urge all countries and regional economic integration organizations to join in these negotiations and recognize that it is highly desirable that an effective framework convention on climate change, containing appropriate commitments, and any related instruments as might be agreed upon on the basis of consensus, be signed in Rio de Janeiro during the UN Conference on Environment and Development.

29. We reaffirm our wish that this convention contain real commitments by the international community.

4.3.4 The Establishment of the UN Framework Convention on Climate Change (1992)

The EC CO_2 stabilization policy, referred to in the Final Declaration, was confirmed by the joint Energy/Environment Council on 13 December 1991 and was put forward as a proposal for global implementation at the UN Conference held in 1992 in Rio de Janeiro. At this conference, the aimed for UN Framework Convention on Climate Change was adopted. According to this convention, it was the intention to convene regularly climate conferences to monitor progress made and review measures taken in reducing the emission greenhouse gases. Annex I Parties to the Framework Convention, comprising the developed countries, committed themselves to adopting policies and measures to return greenhouse gas emissions to their 1990 levels by the year 2000.

Meanwhile the US government became also involved in the CO_2 debate. Its interest in the matter dates from 1989 in which year a meeting of experts was convened by the US President Bush in Washington to discuss the alleged environmental risk of CO_2 emissions. A few years later, in response to the Framework Convention on Climate Change, his successor, US President Clinton, declared on 21 April 1993 (Earth Day) as follows:

> We must take the lead in addressing the challenge of global warming that could make our planet and its climate less hospitable to human life. Today, I reaffirm my personal, and announce our nation's commitment to reducing our emissions of greenhouse gases to their 1990 levels by the year 2000. I am instructing my administration to produce a cost effective plan . . . that can continue the trend of reduced emissions. This must be a clarion call, not for more bureaucracy or regulation or unnecessary costs, but instead for American ingenuity and creativity, to produce the best and most energy-efficient technologies.

The first meeting of the Conference of the Parties to the UN Framework Convention took place in 1995 in Berlin. At this meeting the conclusion (referred to as the Berlin Mandate) was reached that the commitments made by the Annex I Parties would not be sufficient to meet the stated objectives and that further commitments beyond the year 2000 were needed.

4.3.5 The IPCC WG I's Second Assessment Report (1995)

In the same year (1995), the IPCC WG I issued its Second Assessment Report. In this report, the Working Group adjusted the lower limits of its forecasts slightly downwards. The average temperature was on this occasion estimated further to increase (i.e., in addition to the observed increase of 0.5 degrees Celsius over the past one hundred years) by between 1 and 3.5 degrees Celsius, if the atmospheric concentration of CO_2 were to rise to about twice its pre-industrial level. Sea level rise was put between 15 and 95 centimetres. The panel declared to be aware of the fact that there were many uncertainties and many factors that limited its ability to project and detect future climate change. As to the observed 0.5 degrees Celsius increase over the past one hundred years, the panel convincingly argued that this rise was unlikely to be entirely natural in origin since it appears to have been quicker than previous temperature changes. The panel concluded that the balance of evidence suggests a discernible human influence on global climate.

4.4 THE 1997 KYOTO PROTOCOL

On the strength of IPCC WG I's Second Assessment Report and mindful of the 1995 Berlin Mandate, the Parties to the UN Framework Convention on Climate Change (COP 3) convened again in December 1997 in Kyoto, Japan (1 to 11 December 1997). The intention was to come to binding commitments for the Annex I Parties with respect to reducing greenhouse gas emissions for the period after the year 2000. This time 159 countries participated in the conference. The final result was laid down in a Protocol to the Framework Convention. The basic principle is set out in Article 3. In accordance with this article, the Parties to the Conference shall ensure that their aggregate anthropogenic carbon dioxide equivalent emissions of the greenhouse gases listed in Annex A do not exceed their assigned amounts calculated pursuant to their quantified emission limitation and reduction commitment listed in Annex B, with a view to reducing their overall emissions of such gases by at least 5 per cent below 1990 levels in the commitment period 2008 to 2012. The greenhouse gases listed in Annex A comprise carbon dioxide (CO_2), methane (CH_4), nitrous oxide (N_2O), hydrofluorocarbons (HFCs), perfluorocarbons (PFCs) and sulphur hexafluoride (SF_6). However any Annex I Party may use 1995 as its base year for the three fluor compounds. The EU 15 accepted a reduction commitment of 92 per cent for itself and for each of its fifteen Member States, Norway: 101 per cent; the US: 93 per cent; Canada, Hungary,

Japan, Poland each: 94 per cent; Australia: 108 per cent; Iceland: 110 per cent; and the Russian Federation: 0 per cent. The EU overall reduction commitment of 92 per cent was subsequently in mutual agreement between the Member States re-distributed as follows: UK: 87.5 per cent; the Netherlands: 94 per cent; Austria: 87 per cent; Germany: 79 per cent; Denmark: 79 per cent; France, Belgium: 92.5 per cent; Italy: 93 per cent, whereas Portugal, Spain, Sweden and Greece are allowed to exceed their 1990/1995 emission levels. The Protocol was signed by eighty-four States, among which the US, on 11 December 1997. According to its rules, the Protocol would enter into force if it was ratified by fifty-five participating States accounting in the aggregate for 55 per cent of the 1990 CO_2 emissions. However the ratification procedures were delayed because the Parties to the Conference wished to await the outcome of further negotiations on detailed rules on how the Protocol should be implemented, in particular with respect to various schemes such as joint implementation, clean development mechanism and emissions trading. These negotiations took place at the UN Climate Conference in Buenos Aires (1998) and subsequently at the COPs held in Bonn (COP 6bis 2001) and in the same year in Marrakech (COP 7 2001). At these COPs, the rules regarding the way the Protocol should operate were worked out and agreed, thus opening the way for the ratification of the Protocol by the participating States. The EU 15 ratified the Protocol in 2002. On 23 October 2004 this was also done by the Russian Federation. With the Russian ratification, the conditions for entering into force of the Protocol were fulfilled. Accordingly, the Protocol entered into force on 16 February 2005. Up to now the US which did sign the Protocol has refused to proceed to ratification. As of December 2006, a total of 169 countries and other government entities have ratified the Protocol.

As agreed under the Framework Convention, no commitments were imposed on the developing countries, but on the other hand the Kyoto Protocol envisages three schemes set out in Articles 6, 12 and 17 respectively that allow the Annex I Parties to mitigate the financial consequences of fulfilling their commitments under Article 3. Under Article 6 an Annex I Party is allowed to transfer to or acquire from any other such Party emissions reduction units (ERUs) resulting from projects aimed at reducing anthropogenic emissions by sources or enhancing anthropogenic removals by sinks of greenhouse gases in any sector of the economy. Such acquisition of ERUs shall be supplemental to domestic actions for the purposes of meeting a Party's commitments under Article 3. This approach is referred to as 'Joint Implementation' (JI) but is not described as such in this article. In Article 12 a 'Clean Development Mechanism' (CDM) is defined. Under this scheme, Parties not included in Annex I (i.e., the developing countries) will benefit from project activities resulting in certified emission reductions (CERs) and Annex I Parties may use the CERs accruing from such project activities to contribute to compliance with part of their quantified emission limitation and reduction commitment under Article 3. The reason behind CDM is that the costs of projects falling within the scope of the Protocol are generally lower in developing countries than in industrial countries. Article 17 describes the principle of emissions trading. Annex I Parties may participate in emissions trading for the purposes of fulfilling their commitments under Article 3. Any such trading shall be supplemental to domestic actions for the purpose of meeting the commitments under that Article.

4.5 CLIMATE CHANGE PETROLEUM POLICIES POST-KYOTO (OECD, EU)

4.5.1 A Review of the Options

After the Kyoto Protocol and the associated agreements became effective (16 February 2005), the Member States of the OECD and the EU (the Annex I Parties or commitment-States) confirmed their serious intention to take all measures necessary to achieve the targets set by said Protocol. To this end they are investigating and stimulating the measures and actions that have to be taken jointly or individually to curb the emission of CO_2 in one way or another. The aforesaid commitment-States were thereby joined without accepting commitments however by many other States including the Russian Federation, India and China (the 'non-commitment countries').

For trying to achieve the Kyoto targets, the governments of the commitment States have the following five options at their disposal:

- The first option is to curb the consumption of the fossil fuels either by introducing fiscal measures to discourage the end-user from using these fuels; and/or by stimulating the efficient use of energy and/or the efficiency of energy transformation in which the fossil fuel concerned takes part. Implementing the first option includes as allowed under the Kyoto Protocol the setting up of emission cap-and-trading schemes whereby CO_2 emission limitations (caps) are allocated to power stations and industrial plants accompanied by the right to buy or sell emission rights if respectively risking to exceed or staying below the cap. To make the trading scheme more effective, initial caps should be rather restrictive and wholly or partly not given away free but be auctioned. This will of course create a competitive pressure on and between CO_2-emitting industrial plants and power stations, not only in the same region but also internationally. It remains to be seen whether this form of trading scheme will have an overall damaging economic impact. The first option could be expanded by measures aimed at the capture of the CO_2 that is unavoidably generated in the industrial plant or power station and the subsequent underground storage of such captured CO_2 in a suitable porous rock formation (e.g., a depleted natural gas reservoir).
- The second option concerns the electricity generating sector. In this sector, the use of non-fossil electricity could be expanded. The most practical way to do this is to expand the use of nuclear energy in view of the limitations attached to the use of the renewables.

- The third option concerns the transport sector. In this sector the use of alternative fuels for road transport could be introduced or further expanded. Available or potentially available alternatives under investigation or even actually manufactured are biofuels and hydrogen.
- The fourth option concerns the industrial sector. Fossil fuels generate heat and steam for use in industrial plants. No real alternatives are available except for the application of the first option and for the possibility of fuel switching, whereby for instance coal is replaced by natural gas or mixed with biomass. Such switching by its own will offer an interesting reduction of CO_2 generation.
- The fifth option concerns the domestic sector (space heating and cooking). In this sector, refined products such as heating oil compete with natural gas and coal. Apart from the first option, the only alternative that can be thought of is replacing the fossil fuels by electricity (the all-electric house). Whether this contributes to a reduction of CO_2 generation depends (as in the case of hydrogen) again on the way in which the electricity that is delivered is being generated. If generated by nuclear energy, there is a big reduction to be expected.

4.5.2 The Final Choice

Given the fact that pursuing the first option including schemes to capture CO_2 and store the same underground has its limitations which very soon will be reached, that in the transport sector there are no realistic or acceptable alternatives to oil products or natural gas, considering that in the industrial sector the only alternative consists in fuel switching which may even increase the demand for natural gas and considering further that in the domestic sector the only alternative appears to be electricity itself, it is clearly up to the use of non-fossil electricity (to be generated by preference by means of nuclear energy) to provide the means to tackle the problem of curbing anthropogenic CO_2 generation (and subsequent emission).

Note. See Chapter 3 for more details on the above discussed options.

However this may be, public opposition against nuclear fission energy and in particular against the fast breeder reactor (because of the production of plutonium) may hinder the expansion of non-fossil electricity that otherwise would be achievable in say the next ten to fifteen years. Public opposition follows from concerns about investment costs, disposal of radioactive waste, plant safety and nuclear weapons proliferation. Public opposition is in particular strong in the EU with exception of France and Finland. The utilization of nuclear fission energy meets less opposition in the Russian Federation and in the emerging economies such as China and India. In particular China seems to embrace nuclear energy in a significant manner, considering the announced intention to build a large number of nuclear power stations in the coming decades, but in this country as well as in the Russian Federation and India, nuclear power must compete with the abundantly available low cost coal. The same applies to the US which country happens to possess the largest reserves of coal. Australia is a case apart. This country possesses large coal, natural gas and uranium reserves. It exports uranium oxide but uses coal and natural gas itself for thermal power generation. Anyway, a replacement of coal by nuclear energy can only be a gradual process. In the meantime, coal-fired power stations, coal gasification plants and coal-fired industrial installations could either be equipped with CO_2 capture and underground storage facilities (referred to as CCS facilities) which of course will be a very expensive operation, and/or be made suitable for mixing with biomass and/or converted to gas-fired plants and installations.

Note. The first accident involving a nuclear reactor occurred on 8 October 1957 at Windscale, a few hundred metres from the Calder Hall nuclear power station. Windscale No. 1 plutonium production reactor caught fire releasing a plume of radioactivity that drifted far across northern Europe. Serious concern about the operational safety of nuclear plants came to the foreground on 28 March 1979 when one of the two reactors of the Metropolitan Edison Company's Three Mile Island nuclear power plant, near Middletown (Harrisburg), Pennsylvania, suffered a partial meltdown. But radiation levels outside the plant were normal and people living in the vicinity of the plant did not have to be evacuated. In May 1985 it was decided to restart the other, undamaged reactor, which all those years had been left idle. Whether or not there was a direct connection with the incident, fact is that since the year of the meltdown no nuclear power plant has been built in the US. The negative feelings about plant safety got a new impulse when on 26 April 1986 reactor number 4 of the nuclear power station at Chernobyl, Ukraine, exploded, contaminating tens of thousands of square kilometres in northern Ukraine, southern Belarus and Russia's Bryanks region. The fallout, whereby four hundred times more radioactivity was released than at Hiroshima, drove a third of a million people from their homes and triggered an epidemic of thyroid cancer in children. Total costs over the years have mounted into the hundreds of billion of dollars. The disaster is for a great deal responsible for public opposition against the use of nuclear energy in Western Europe (with exception of France and Finland), albeit recently this opposition has somewhat shifted its focus from plant safety to the matter of disposal of radioactive waste and the high construction costs.

Nonetheless, when in the near future the negative impact of increased atmospheric concentrations of CO_2 on surface temperatures and the consequences thereof for the climate become clearer, and a new generation of nuclear fission reactors becomes available, the nuclear fission option might get more and more support. A strong argument for such support will always be that the use of nuclear fission energy in the power generating market avoids the generation of CO_2, so if all worldwide required electricity would be supplied by nuclear power plants instead of by fossil fuel-fired power plants, the CO_2 emission problem would be approximately 40 per cent solved. But for coal-rich countries this substitution would mean that their existing coal-fired power plants must be converted to adjust to nuclear energy and that their coal mines would have to be closed. The substitution could only be averted if governments would require the owners of the coal-fired power plants to equip the latter with the aforementioned CCS facilities. About this possibility not much can be said for the moment: the techniques involved in capture, transport and underground storage are still being tested with respect to costs and energy conversion efficiency. How governments of the aforementioned coal rich countries try to solve this dilemma remains unclear, albeit

China has announced not to hesitate to start a vigorous programme of nuclear power station building. Governments of countries without domestic coal resources must decide with respect to fuelling their power sector between the use of natural gas and that of nuclear energy. If such country also happens to be a natural gas importing country, the government concerned has little choice; a switch to nuclear energy would have the additional benefit of improving the country's security of supply.

4.5.3 The Consequences for Oil and Natural Gas

Should governments decide to adopt a policy of replacing coal by nuclear energy in the power generation sector the overall consumption of oil and natural gas will not be affected thereby. In the said sector, the consumption of oil products is already modest and the latter's possible total elimination will make therefore not much difference thereto. The consumption of natural gas in the sector is more substantial but there is no reason to expect that all electricity will be provided by nuclear energy. To reserve a twenty per cent share for natural gas appears reasonable and desirable. In the other sectors not much will or can change as far as the demand for oil and natural gas is concerned, except for fuel switching in the industrial sector which will only increase the demand for natural gas, and except for strenuous efforts to reduce consumption of oil and natural gas in one way or another. Meanwhile research into the possibilities of an all-electric world economy should be started in time, hopefully in combination with the development of nuclear fusion energy.

4.6 THE IPCC WG I FOURTH ASSESSMENT REPORT (2007)

In February 2007, the IPCC WG I Fourth Assessment Report was issued under the title, Climate Change 2007: The Physical Science Basis.

In conformity with the previous six years' reports, this latest report contains on the basis of the latest insights facts, predictions and other information concerning global warming and its consequences for sea level rise and climate change.

Note 1. In the Fourth Report the following facts, observations and conclusions are presented. Total temperature increase from 1850-1899 to 2001-2005 is 0.76 [0.57 to 0.95] degrees Celsius and appears to have accelerated during the last fifty years 0.13 [0.10 to 0.16] degrees Celsius per decade. This is larger than the corresponding trend for 1901-2000 given in the Third Assessment Report, namely 0.6 [0.4 to 0.8] degrees Celsius. The global atmospheric concentration of CO_2 has increased from a pre-industrial value of about 280 ppm to 379 ppm in 2005. The latter value exceeds by far the natural range over the last 650,000 years (180 to 300 ppm) as determined from ice cores. The primary source of the abovementioned increased atmospheric concentration results from fossil fuel use, with land use change providing another significant but smaller contribution. Annual fossil CO_2 emissions increased from an average of 6.4 GtC (23.5 $GtCO_2$) per year in the 1990s to 7.2 GtC (26.4 $GTCO_2$) per year in 2000-2005. CO_2 emissions associated with land-use change are estimated to be 1.6 GtC (5.9 $GtCO_2$) per year over the 1990s, although these estimates have a large uncertainty. The global atmospheric concentration of methane has increased from a pre-industrial value of about 715 ppb to 1,732 ppb in the early 1990s, and is 1,774 ppb in 2005. This latter value exceeds by far the natural range of the last 650,000 years (320 to 790 ppb) as determined from ice cores. It is very likely that the observed increase concentration is due to anthropogenic activities, predominantly agriculture and fossil fuel use, but relative contributions from different source types are not well determined. There exists a very high confidence that the globally averaged net effect of human activities since 1750 has been one of warming, with a radiative forcing of + 1.6 W per square metre. The combined radiative forcing due to increase in CO_2, methane and nitrous oxide is + 2.3 W per square metre. Observations since 1961 show that the average temperature of the global ocean has increased to depth of at least 3,000 m and that the ocean has been absorbing more than eighty per cent of the heat added to the climate system. Such warming causes sea water to expand, contributing to sea level rise. Widespread decreases in glaciers and ice caps and losses from the ice sheets of Greenland and Antarctica have also contributed to sea level rise. All in all global average sea level rose at an average rate of 1.8 mm per year over 1961 to 2003. Even if all radiative forcing agents are held constant at year 2000 levels, a further warming trend would occur in the next two decades at a rate of about 0.1 degree Celsius, due mainly to the slow response of the oceans. About twice as much warming (0.2 degree Celsius per decade) would be expected if emissions are within the range of scenarios developed by the Working Group I for the purpose.

Projected Globally Averaged Surface Warming (degrees Celsius) and Sea Level Rise (metres) at the End of the Twenty-first Century

Case	*Temperature Change – Best Estimate*	*Temperature Change – Likely Range*	*Sea Level Rise*
Constant year 2000 concentration	0.6	0.3-0.9	NA
Low Scenario (B1)*	1.8	1.1-2.9	0.18-0.38
Very High Scenario (A1F1)*	4.0	2.4-6.4	0.26-0.59

* B1 and A1F1 Scenarios assume atmospheric CO_2 concentrations to reach in 2100 600 ppm and 1,550 ppm respectively.

Warming tends to reduce land and ocean uptake of atmospheric CO_2, increasing the fraction of anthropogenic emissions that remains in the atmosphere. Depending on the scenario used, the global warming at 2100 could be increased by more than 1 degree Celsius. Increasing atmospheric CO_2 concentration leads to increasing acidification of the ocean, hindering marine organism in forming of $CaCO_3$ for the building of their skeletons and shells and thus reducing the locking up of CO_2. Projected warming in the twenty-first century shows scenario-independent geographical patterns similar to those observed over the past several decades. Warming is expected to be greatest over land and at most high northern latitudes, and least over the Southern ocean and parts of the

North Atlantic Ocean. Sea ice is projected to shrink in both the Arctic and Antarctic under all scenarios. It is considered very likely that hot extremes, heat waves and heavy precipitation events will continue to become more frequent. Climate carbon cycle coupling is expected to add CO_2 to the atmosphere as the climate system warms, but the magnitude of this feedback is uncertain. Model studies suggest that to stabilize at 450 ppm CO_2 could require that cumulative emissions over the twenty-first century be reduced from an average of approximately 670 GtC (2,460 $GtCO_2$) to approximately 490 GtC (1,800 $GtCO_2$).

Note 2. In the geological past, Earth has experienced much higher atmospheric CO_2 concentrations (up to 2,000 ppm) with correspondingly higher surface temperatures without adversely affecting the biosphere, albeit the distribution of the land masses over the planet was then different and must have affected the resulting surface temperatures. From this perspective, it could be argued that the current surface temperatures are on the low side and that the climate generally speaking could benefit from some warming up. The problem is that apparently the atmospheric CO_2 concentration has moved over the last 650,000 years within rather narrow limits (180 to 300 ppm), which suggests that, from a geological time perspective, extremely rapid switch to a warmer, perhaps in several aspects more beneficial climate might pose a serious problem of adaptation for the biosphere, albeit the power of adaptation of the larger species must not be underestimated.

Note 3. CO_2 is not the only infrared absorbing trace element in the atmosphere. Water vapour as well as methane are also capable of absorbing and re-emitting infrared radiation, but the effect molecule for molecule differs. That of water vapour is much weaker and that of methane is much stronger.

Note 4. Higher atmospheric CO_2 concentration will have an important fertilizing effect on plants: it stimulates their growth and thereby the production of oxygen.

Note 5. Global warming in itself may have beneficial effects, in particular by warming the northern high latitudes in winter. Expansion of farm land in these regions becomes a possibility. The expected disappearance of summer ice at the North Pole may offer shipping in that region many new opportunities. From global warming, a lengthening of the plant growing season may also be expected. An added benefit may be mentioned, any strengthening of plant growth removes more CO_2 from the atmosphere on account of an intensified photosynthesis process.

4.7 THE ORGANIZATION OF PETROLEUM EXPORTING COUNTRIES

4.7.1 ORGANIZATION, STATUTES AND STATED OBJECTIVES

4.7.1.1 Organization

The Organization of Petroleum Exporting Countries (OPEC) embodies a formal cooperation with respect to petroleum policy between the major crude oil producing and exporting countries. Already in 1949 Venezuela had approached the four major oil exporting countries in the Middle East Gulf area, i.e., Saudi Arabia, Kuwait, Iraq and Iran, with suggestions for setting up an organization for discussion and exchange of views on the price competition that was going on in Western Europe and that was undermining the value of their export oils. The right moment for actually founding such an organization came when in August 1960 their concessionaires took the decision to reduce for a second time in a row the posted prices of their export oils. In a fast defensive reaction, the five major oil exporting countries founded OPEC on 1 September 1960 in Baghdad.

OPEC was established as a permanent intergovernmental Organization with an international status. In accordance with Article 102 of the Charter of the UN, the Agreement creating OPEC was registered with the UN Secretariat on 6 November 1962.

The Organization consists of the following organs:

(i) the Conference, meeting at least twice a year;
(ii) the Board of Governors;
(iii) the Secretariat; and
(iv) the Economic Commission.

In addition, the Conference will establish from time to time and as required ad hoc ministerial committees, composed of a variable number of petroleum ministers.

The Conference is the supreme authority of the Organization, responsible for the formulation of its general policy. It consists of the ministers of petroleum affairs in the governments of the Member States. The Conference decides upon reports and recommendations submitted by the Board of Governors or by the aforementioned ad hoc ministerial Committees. It meets at least twice a year. Binding decisions require unanimity. The Board of Governors directs the management of the Organization and is responsible for implementing the resolutions of the Conference. The Board consists of one Governor for each Member State. The Secretariat through the Secretary-General is responsible for maintaining contacts with governments and organizations, it possesses several divisions and departments for collecting information and carrying out studies in the field of energy, finance and economics. The Economic Commission is a specialized body operating within the framework of the Secretariat. It is responsible for assisting the Organization in promoting stability in international oil prices at equitable levels. Ad hoc ministerial committees were established as from the time the Organization was forced, in the light of weak markets and prices, to fix overall production ceilings and individual production quotas for the Member States. The committees were charged with the task to submit recommendations regarding questions of price and production levels and/or to monitor the implementation by the Member States of the resolutions adopted by or agreements made at the Conference, in so far these resolutions and agreements concerned prices, overall production ceilings and Member States' individual production quota. The most important of these committees is the still functioning Ministerial Monitoring Committee (MMC), which was established in 1982. At the time of its establishment, the MMC was composed of four petroleum ministers. Currently all Member States are represented in the Commission. A meeting of the MMC with all its members present functions therefore as an alternative meeting of the Conference and is in fact deliberately used as such to bring sensitive matters to a solution outside an extraordinary or ordinary meeting of the Conference.

4.7.1.2 The Statutes

According to its Statutes of November 1971, the principal aim of the Organization is the coordination and unification of petroleum policies of the Member States and the determination of the best means of safeguarding their interests individually as well as collectively. In particular the Organization should strive to ensure the stabilization of prices in the international oil markets with a view to eliminate harmful and unnecessary fluctuations. At all times, so is stated in the Statutes, the Organization must give due regard to the interests of the producing nations and to the necessity of securing a steady supply income to the producing countries. At the same time, the Organization must ensure to consuming nations an efficient, economical and regular supply of petroleum and to investors in the petroleum industry a fair return on their capital. The Statutes also formulate the requirements for membership of the Organization. A candidate-State, wishing to qualify for membership, must possess a substantial net export of crude oil, fundamentally similar interests to those of the Member States and must be accepted by a majority of three-fourths of the full members, including the concurrent vote of all five founder members. Soon after its founding, other oil exporting States joined the Organization: Qatar (1961), Indonesia (1962), Libya (1962), the United Arab Emirates (1967), Algeria (1969) and Nigeria (1971). Ecuador joined in 1963 but withdrew in 1992; Gabon joined in 1975 but withdrew in 1995. Much later Angola joined the Organization and as of 1 January 2007 became the 12th Member State. In October 2007, Ecuador rejoined the Organization.

4.7.1.3 The Solemn Declaration of 1975

A review of the stated objectives of the Organization took place at a meeting of the Sovereigns and Heads of State of the Member States (Algiers, 4 to 6 March 1975). This review was felt necessary in the light of the adverse reactions that OPEC's actions on prices and takeovers of concessions had caused in the OECD countries. At the end of their meeting, the Sovereigns and Heads of State made public a Solemn Declaration, in which they came out strongly in favour of the New International Economic Order, which was about one year earlier introduced by a Declaration and Programme of Action to that effect respectively contained in UN General Assembly's Resolutions 3201 (S-VI) and 3202 (S-VI) of 1 May 1974 (see Chapter 10).

But the Solemn Declaration had also much to say about oil prices. It was pointed out that an artificially low price for oil in the past has prompted overexploitation of this limited and depletable resource and that continuation of such policy would have been disastrous from the point of view of conservation and world economy. Very interestingly, an attempt was made to define what the 'right' price should be. The following determining factors were mentioned:

(i) the imperatives of the conservation of petroleum, including its depletion and increasing scarcity in the future;
(ii) the value of oil in terms of its non-energy uses; and
(iii) the conditions of availability, utilization and cost of alternative sources of energy.

With respect of the level of the then prevailing petroleum prices (1975 level), it was pointed out that in spite of the apparent magnitude of the readjustment, the high rate of inflation and currency depreciation had wiped out a major portion of the real value of price readjustment and that the current price was markedly lower than that which would result from the development of alternative sources of energy. Although the Sovereigns and Heads of State did not subscribe to the notion that their export prices were too high, they declared themselves prepared to negotiate the conditions for the stabilization of oil prices so as to enable the consuming countries to make necessary adjustments to their economies. As regards the supply of petroleum, a very sensitive subject in view of the embargoes and supply restrictions which between 20 October 1973 and 18 March 1974 had been imposed against the US and the Netherlands (full embargo) and against France and the UK (partial embargos), the Solemn Declaration, appearing one year after these supply interruptions had been terminated, stated that the Sovereigns and Heads of State reaffirmed their countries' readiness to ensure supplies that would meet the essential requirements of the economies of the developed countries, provided that the consuming countries should not use artificial barriers to distort the normal operation of the laws of demand and supply. In order to achieve this purpose, the Member States declared themselves willing to establish close cooperation and coordination among themselves in order to maintain balance between oil production and the needs of the world market. However this may be, the promised close cooperation and coordination between the Member States was effectuated only after a situation of over-supply had developed in the international oil market and prices started to decline (as from 1981).

4.7.1.4 State Participation

After having settled various fiscal and pricing issues within the framework of the Teheran Agreement of 14 February 1971 (see Chapter 8), OPEC's attention turned to the issue of acquiring interests in the petroleum agreements of the 1930s and other long-term leases or licences. At an extraordinary meeting of the Conference held on 22 September 1971, a resolution was adopted calling upon the Member States to open negotiations with their concessionaires and licensees in order to reach agreement on the desired participation. All Member States, with the exception of Iran and Indonesia, were involved.

4.7.1.5 OPEC's Long-Term Strategy as Seen in the 1980s

A few years after the Solemn Declaration, OPEC's ideas about what should be the 'right' price for oil in the international market, the price that should be maintained and supported by collective action, were further worked out by OPEC's six-member Long-Term Strategy Committee (LTSC). The LTSC presented its report to the 56th extraordinary meeting of the Conference (Taif, Saudi Arabia, 7 to 8 May 1980), where it was accepted by a majority of the Member States. The recommendations of the LTSC were based on the idea that oil prices should in the long-term approximate the level of the cost of alternatives and that a pricing floor should be established to guarantee continuous advance over time towards that target.

Note. In the view of the LTSC, the proposed pricing floor should combine three elements:

(1) an inflation adjustment based on the change in unit values of export of the OECD Member States;
(2) an exchange rate adjustment in the manner of the Geneva II Agreement (see Chapter 8); and
(3) a rise of the oil prices in real terms proportionately to the growth in real GNP of the aforesaid OECD Member States.

The link laid with the growth in real GNP of the OECD Member States was meant to lay a link with the OECD area's ability to absorb price increases and the tendency to greater energy consumption (i.e., a pricing policy based on what price the market can bear and how much the market needs the oil). The LTSC noted that as markets swing from surplus to shortage, the pricing strategy should be flexible enough to allow prices to go beyond the minimum floor price level, in the light of signals (spot prices or contract prices) received from the market.

The LTSC recommendations on pricing strategy have never been implemented.

4.7.2 OPEC's Interventions in the Global Oil Market

4.7.2.1 Pricing Policies from 1975 to 2000

After the oil crisis of 1973/1974 had run its course, OPEC remained focused on achieving even higher producer prices, apparently without regard to the global economic effects of such high prices and clearly in contradiction to the principles and objectives proclaimed in the Solemn Declaration. On 1 January 1979 in the aftermath of the Iranian Revolution (December 1978), the official selling price for Arabian Light ex Ras Tanura was fixed at USD 13.335/b, but already six months later, on 1 July 1979, this price was raised to USD 18.00/b. In the following months and years, the official selling price would be raised still more, to reach an all time high of USD 34/b on 1 November 1981. In fact the market price of Arabian Light ex Ras Tanura was still higher, reaching in 1980 an all time high of USD 36.83/b, which is USD 90.46/b in 2006 USD.

By trying to achieve even higher prices, OPEC became confronted with oversupply and a weak market. This situation was caused by a combination of three factors, namely a decline in world demand, a slow increase in non-OPEC production, and Saudi Arabia's insistence to maintain its production at the exceptionally high level of 10.1 million b/d. On the part of Saudi Arabia, this was a deliberate policy meant to force the other Member States to adopt a more restrained price policy. At a meeting in Vienna which in the course of the proceedings was converted into an extraordinary meeting of the Conference (19-20 March 1982), it was decided that as of 1 April 1982 total OPEC production would be bound to an upper limit. On this occasion, the ceiling was fixed at 18 million b/d, but after the meeting Saudi Arabia announced a further reduction of its production on a voluntary basis with 0.5 million b/d, thereby lowering its ceiling to 7.0 million b/d and the overall ceiling to 17.5 million b/d. But it was a temporary agreement. The ceiling was to be reviewed a few months later at the 60th meeting of the Conference which was scheduled for 20/21 May 1982 in Quito. At the meeting the Member States agreed with effect from 1 June 1982 to individual production ceilings. The price for the marker crude was confirmed at the level of USD 34/b. This was the first time in the history of OPEC that a majority of the Member States was prepared to accept a reduction of their output to avoid having to cut their prices. The Conference also decided to establish a Monitoring Ministerial Committee (MMC) which was expected to monitor the market situation and recommend to the Conference the necessary measures to be taken. The MMC would be responsible for checking each country's production level and selling prices.

In the years that followed, OPEC was continuously confronted with overproduction and oversupply and a declining demand for OPEC oil. Internally, OPEC was divided over the strategy that should be followed. The Arab Gulf States led by Saudi Arabia favoured a policy of regulating prices only, without any binding commitments as to how much oil each Member State would be allowed to produce. In contrast, other Member States, Venezuela, Algeria, Libya and Iran in particular, were in favour of a policy directed at obtaining the desired price level by means of regulating production and individual export quotas. Under this policy, the individual quota would be allocated in accordance with the economic needs of the Member State concerned. From this standpoint only, the Member States, which were running a surplus on their current account balance (e.g., Saudi Arabia, Kuwait and Abu Dhabi), should be subject to and accept any such restriction of their production as would be necessary to stay below an agreed overall production ceiling.

In its struggle to support the price structure, OPEC lowered in 1983 the price per barrel of the reference crude to USD 29/b. This measure had no effect. In this connection OPEC requested in the second half of 1984 the UK government to persuade BNOC, the UK national oil corporation, to maintain the price of Brent Blend, the oil which functioned as the North Sea reference oil and which was connected to the price of oil exported from Member State Nigeria, viz. Bonny-Light, at its prevailing level of USD 30/b in order to give Nigeria no cause to breach the OPEC price arrangements. However, BNOC experienced increasing problems in selling its oil against the price at which the company had bought the oil from its licensees (the latter receiving from BNOC the open market price). So when Statoil, the Norwegian state oil enterprise, started in October 1984 to sell its oil at a discount, BNOC reacted by lowering the price of its Brent Blend from USD 30/b to USD 28.65/b (meanwhile the open market price had dropped to USD 27/b). Nigeria felt obliged to follow suit and, in contravention of the OPEC price arrangements, lowered the price of its Bonny-Light by USD 2.0 to USD 28.0/b. In reaction to the price cutting of Statoil, BNOC and Nigeria OPEC met in Geneva, from 29 to 31 October 1984, which meeting was converted into an extraordinary meeting of the Conference. At this meeting, OPEC decided to support its price structure by lowering the overall production ceiling with effect from 1 November 1984 from 17.5 to 16 million b/d and by reconfirming the price of the reference oil at USD 29/b (as had been fixed in the previous year). The decisions taken at this meeting of the Conference were summarized as follows:

> The Conference reviewed with great concern the recent developments in the world oil market following the price cuts undertaken by Statoil of Norway and BNOC of the United Kingdom, as well as by member country Nigeria. The Conference believes that the present oil price structure should be maintained and market stability restored as a means to secure healthy world energy balances and to stimulate world trade. Beneficiaries of such stability are not OPEC countries alone, but the international community at large, producers and consumers alike. The Conference is, therefore, determined to defend the price structure of its marker crude (Arabian Light, 34 degrees API, ex Ras Tanura) at the level of USD 29/b and to consolidate market stability. For this purpose, the Conference decided to cut as from 1 November 1884 and on a temporary basis the global production ceiling of OPEC from 17.5 to 16 million b/d.

The largest reduction was accepted by Saudi Arabia (647,000 b/d). Zero reductions were granted to Iraq and Nigeria. The total production ceiling of 16 million b/d was distributed over the Member States in quotas ranging from 137,000 b/d for Gabon to 4,353,000 b/d for Saudi Arabia. The quotas allocated to the Member States producing more than 1 million b/d were as follows: Saudi Arabia: 4.353; Iran: 2.300; Venezuela: 1.555; Nigeria: 1.300; Iraq: 1.200; Indonesia: 1.189 (source: MEES, 5 November 1984).

In January 1985 OPEC decided to lower the price of the reference oil to USD 28/b with effect from 1 February 1985. Nevertheless the decline in demand for OPEC oil and in particular that for Saudi Arabian oil continued and became very pronounced in that year as is shown by the diminishing production figures presented on following Table 4.1.

Table 4.1. Decline in World and OPEC Production (million b/d)

	1979	*1980*	*1985*	*% change 1979-1985*
World	66.0	62.9	57.5	– 13.3
OPEC	31.2	27.2	16.7	– 46.5
OECD	16.5	17.1	20.0	+21.0
Former Soviet Union	11.8	12.1	12.0	+ 1.7
Saudi Arabia	9.8	10.3	3.6	– 63.3
Mexico	1.6	2.1	2.9	+81.1
Non-OPEC*	23.0	23.6	28.7	+24.8

* Excludes former Soviet Union.
Source: BP Statistical Review of World Energy 2007.

Note. The percentage increase (+24.8 per cent) of non-OPEC production (excluding production from the former Soviet Union) was due to production from Mexico, Alaskan North Slope and the North Sea continental shelf. The drastic reduction of Saudi Arabian production must be seen against the background of an overall declining world production.

At the end of 1985, OPEC decided to change its strategy and to concentrate on market share and to ignore price levels for the time being. In this context, it was decided at the 76th meeting of the Conference (Geneva, 7-9 December 1985) 'to secure and defend for OPEC a fair share in the world oil market consistent with the necessary income for [M]ember States' development' (MEES, 16 December 1985). Pricing was not discussed. In order to achieve the desired fair market share, Saudi Arabia started to produce at full capacity. As a result of this 'flooding of the market' strategy, the international producer's price started to collapse, beginning in February 1986 and reaching minimum values in June of that year. Such a sharp drop in the price had not been expected by OPEC. To stop further price erosion, OPEC decided later in the year to abandon the aforementioned 'flooding of the market' strategy. At the 79th meeting of the Conference (Geneva, 6-22 October 1986), Saudi Arabia declared that the price for Arabian Light should not be lower than USD 18/b and that each Member State should adhere to this price. The Conference confirmed, albeit on an interim basis, the return to a system of overall production ceilings and individual production quotas derived therefrom. Only Iraq was free to produce up to its capacity. The matter of a fixed price structure was referred to an already existing ministerial committee of three petroleum ministers which was freshly named the Ministerial Committee on Pricing (MCP). On 14 November 1986 the MCP submitted its recommendations. It was proposed to replace Saudi Arabia's Arabian Light as marker crude by a basket of seven crude oils, among which were also three non-OPEC crude oils, namely West Texas Intermediate, Mexican Isthmus and North Sea Brent Blend. According to the proposal, the prices of the OPEC export crude oils should be related to the price of the basket on the basis of quality and API gravity differences and geographic location. The calculated price differentials would result in individual prices above or below USD 18/b. At the 80th meeting of the Conference (Geneva, 20 December 1986), the Member States concluded for the first time in the history of the organization among themselves a formal agreement on pricing and production levels instead of adopting a Conference resolution about these matters, as had hitherto been the custom. In accordance with the recommendation of the MCP, it was agreed to return promptly to a fixed pricing system at a level of USD 18/b for OPEC's reference price. This price would be based on a basket of seven crude oils, consisting of six OPEC crude oils and only a single non-OPEC export crude oil, Mexican Isthmus. The export crude oils and their individual prices resulting in an average price of USD 18/b for the basket were agreed as set out in Table 4.2 below.

Table 4.2. 1987 Spot Prices of the OPEC Reference Basket (USD/b)

Basket	18.00
Arabian Light	17.52
Sahara Blend	18.87
Minas	17.56
Bonny Light	18.92
Dubai	17.42
Tia Juana Light	17.62
Mexican Isthmus	18.07

Source: MEES 1987.

Note. The allocation of the individual prices had been based on the decision that the difference in value between Arab Heavy (outside the basket) and Bonny Light (the lightest oil inside the basket) should not exceed USD 2.65/b.

The Agreement laid down the permitted overall and individual production levels for each quarter of 1987. The overall production ceiling averaged for the whole of 1987 amounted to 16.6 million b/d. The Agreement contained very typical provisions, which have been preserved in the later agreements. These typical provisions were the following:

(i) the Member States' production levels would be monitored by a group of high-level marketing representatives from all Member States under supervision of the President of the Conference;
(ii) all Member States should undertake to abide by and adhere strictly and seriously to the production levels as defined in the agreement;
(iii) in case any Member State would violate unilaterally the Agreement and exceed its temporary, allocated production level, in whole or in part, and such violation would be confirmed through the group of marketing experts, under the guidance of the aforesaid President, other Member States may consider themselves free of their obligation under the Agreement; and
(iv) the production levels agreed upon would be temporary and would not constitute a basis for any fair distribution of national quotas in any future negotiations. (MEES, December 1986.)

At the 81st meeting of the Conference (Vienna, 25-27 June 1987) OPEC established the production levels for the second half of 1987. The overall ceiling was fixed at 16.6 million b/d. The Conference stressed the necessity of cooperation of the oil exporting countries outside OPEC, for instance Norway and the United Kingdom, as an essential prerequisite for a lasting market stability.

As from 1988, world demand for oil started to increase and OPEC could more than proportionally benefit therefrom. This development led to regular upwards adjustments of the overall production ceiling and the national quotas as agreed at successive meetings of the Conference. During the first half of 1990, a new period of overproduction started. The UAE and seven other Member States, among which Saudi Arabia and Kuwait, were to blame for this overproduction by exceeding their quotas despite the fact that these quotas had regularly been increased. Similarly as in 1986 the overproduction led in April 1990 to sharp fall in prices. In June 1990 the calculated spot price of the basket reached a minimum of USD 14.05/b. In an attempt to reverse this trend, the Member States reached at the 87th ordinary meeting of the Conference (Geneva, 26-27 July 1990) a new agreement on production quotas and on an overall production ceiling of 22.491 million b/d, but also decided to raise the price of the reference basket from USD 18/b to USD 21/b.

Note. The agreement secured Kuwait's and the UA Emirates' acceptance of a quota which would be the same for both Member States, namely 1.5 million b/d. This long sought quota parity had earlier been achieved by Iraq in relation to Iran at a level of 3.140 million b/d.

A few days after the 87th meeting of the Conference, the Middle East was at war again. On 2 August 1990 Iraq invaded Kuwait and in response to this violation of the international peace, the UN Security Council imposed with effect from 6 August 1990, trade sanctions against Iraq as well as against occupied Kuwait. The July 1990 production and quotas agreement appeared to have lost all its relevance. The trade sanctions had as an immediate effect that the exports of oil and refined products from Iraq as well as from Kuwait came to a halt. The prices on the international oil market started to rise immediately. The calculated basket spot price rose accordingly, reaching its highest value (USD 34.58/b) in October 1990. In contrast to OPEC's pricing policies followed with respect to the two preceding price shocks (1973/74 and 1979/80 respectively), OPEC did in 1990 not attempt to maintain the price at the elevated level reached at the time. Instead, the majority of the Member States, amongst which Saudi Arabia, agreed at a meeting of the MMC (Vienna, 29 August 1990) to increase production in excess of the production quotas fixed in July 1990, in so far as needed to maintain market stability and a regular supply of oil to the customers (i.e., the western countries). In fact, OPEC solved the issue by going back to its basic objectives as stated in the Solemn Declaration of March 1975. OPEC showed thereby to have appreciated the negative economic effects on the world economy of high oil prices. A price explosion was averted that otherwise surely would have taken place. In the months that followed, OPEC proved very successfully in coping with the loss of contributions from Iraq and Kuwait. Already by the first quarter of 1991, OPEC production (without Iraq and Kuwait) had exceeded the July 1990 overall production ceiling (viz. 22.5 million b/d) by one million b/d. Ironically, the production increase achieved by OPEC was so effective that the target price for the basket (USD 21/b) remained unachievable.

In the next years OPEC managed to cope without the contribution of Iraq. So successfully in fact that, in spite of a steadily growing demand for oil, which became apparent since 1995, problems of overproduction on the part of OPEC Member States kept the target price most of the time out of sight. At the 11th meeting of the MMC (Geneva, 25-29 September 1993), the Committee was confronted with a basket price that had fallen below USD 15/b. In order to halt this slide and to reverse the downward trend, the overall production ceiling for the next six months was fixed at 24.5 million b/d. This overall ceiling was distributed amongst the Member States in the form of temporary allocations. At the 95th meeting of the Conference (Vienna, 23-24 November 1993), it was decided to uphold the September 1993 agreement. The November 1993 calculated spot quotation for the basket was USD 14.47.

Price weakness continued unabated because non-OPEC (high cost) production (not bound by production restrictions) was expanding under the umbrella of the favourable price levels which OPEC had set and tried to maintain. Price weakness manifested itself during the last part of 1997 and the whole of 1998. On 31 March 1998, the OPEC Member States finally agreed between them to reduce their overall production by 1.26 million b/d with reference to the production actually achieved in February 1998 (namely 31.7 million b/d), without altering the individual production quotas that had been established in November 1997. Remarkably, five non-OPEC States, namely Norway, Mexico, Egypt, Oman and Yemen, joined OPEC in this action and pledged to reduce their production in the aggregate by 0.27 million b/d (Norway and Mexico each pledged 0.1 million b/d). Nevertheless, the producer price continued its decline. The spot quotation for the OPEC reference basket reached in December 1998 its lowest level (USD 9.96/b). After the Member States succeeded at their meeting in March 1999 to agree amongst themselves to bring down the overall level by 1.72 million b/d based on the production in February 1999 (i.e., a reduction of 6.4 per cent), the world market price has ever since moved upwards.

Table 4.3 below illustrates the impact of OPEC's intervention in the global oil market over the period 1975-2000.

Table 4.3. World/OPEC Production (million b/d) and Spot Prices (USD/b) 1975-2000

	1975	*1980*	*1985*	*1990*	*1995*	*1998*	*2000*
World	55.8	62.9	57.5	65.5	68.1	73.6	75.0
OPEC	27.2	27.2	16.7	24.6	27.7	31.2	31.5
% World	48.7	43.2	29.0	37.6	40.7	42.4	42.0
Y.Avg. Price*	11.53	36.83	27.56	23.73	17.02	12.72	28.50
Y.Avg. Price**	-	-	-	-	16.86	12.28	27.6

* 1975-1983: Arabian Light posted at Ras Tanura, 1984-2000: Brent Dated.
** OPEC Reference Basket.
Sources: OPEC Briefings; BP Statistical Review of World Energy 2007.

4.7.2.2 Pricing Policies from 2000 Onwards

The year 2000 can be considered as a real turning point in the global oil market. The demand for oil started to increase quite rapidly as a result of an expanding world economy. In particular the emerging economies (China, India, etc.)

increased their consumption. In first instance only the international price reacted to the increased demand, production volumes would follow in 2003. OPEC did no longer have to contemplate production cut backs to support a chosen price level, although up to July 2005 at successive meetings of the Conference production quotas for its Member States were established which however were largely ignored production capacity permitting (see Tables 4.5 and 4.6 hereinafter). In fact, prices moved without its intervention at an elevated level of around USD 25/b, as a matter of fact within a price band (22-28 USD/b) that OPEC in March 2000 had set for the price of the OPEC Reference Basket in the context of pursuing pricing policies by means of production adjustments. Meanwhile, the political situation in the Middle East started to become unsettled again. It began with the destruction of the WTC Twin Towers in New York on 11 September 2001 which event led to the War on Terror in Afghanistan. The turmoil increased when on 20 March 2003 Iraq was invaded by a coalition of willing Western States for the purpose of disarming the country and to remove its leader Saddam Hussein. Later tensions developed around Iran and its uranium enrichment programme, which prevented that the political storms could calm down. Remarkably, all this time the export of crude oils from the Gulf area was never interrupted, not even temporarily. There never was a physical shortage of crude oil from the Gulf area. Nevertheless the financial markets reacted strongly, driven by fears about the political future of the Middle East and the possibility that access to its oil production facilities and oil reserves would at some time in the immediate future be denied to consumers in the OECD countries. As from 2002, prices surged further upwards to levels that had not been seen since the second price shock (1979-1985). As from 2005 the price exceeded USD 50/b. In July 2006 the Brent Dated price peaked at USD 73.66/b. In the second half of 2006 and the first half of 2007, prices came down a little and even touched briefly the USD 50/b, the level that had been passed in 2005. At this point OPEC reacted immediately, making it known that they would consider re-introducing production cutbacks if the international price would decline still further, which however did not happen.

Table 4.4. World/OPEC Production (million b/d) and Spot Prices (USD/b) 2000-2006

	2000	*2002*	*2003*	*2004*	*2005*	*2006*
World	75.0	74.5	77.1	80.2	81.2	81.7
OPEC	31.5	29.0	30.9	33.2	34.1	34.2
%World	42.0	38.9	40.1	41.4	42.0	41.9
Y. Avg. Price*	28.50	25.02	28.83	38.27	54.52	65.14
Y. Avg. Price**	27.6	24.36	28.1	36.05	50.64	61.08

* Brent Dated.
** OPEC Reference Basket.
Sources: OPEC Briefings; BP Statistical Review of World Energy 2007.

Note on the OPEC Reference Basket. As mentioned before, Saudi Arabia's Arabian Light was replaced with effect from 1 January 1987 as marker crude for price policy and monitoring purposes by a basket of six OPEC crude oils viz. Arabian Light, Sahara Blend, Minas, Bonny Light, Dubai and Tia Juana Light, and one non-OPEC crude stream, Mexican Isthmus. At the 135th meeting of the Conference held in Isfahan, 16 March 2005, it was decided to change the composition of the Basket (ORB) from the seven crudes as hitherto used to all OPEC 11 crude streams. The Basket was now made up of the following: Arabian Light, Sahara Blend, Minas, Iran Heavy, Basra Light, Kuwait Export, Es Sider, Bonny Light, Qatar Marine, Murban and BCF 17 (Venezuela). The Basket has a gravity of 32.7 degrees API, slightly heavier than the previous seven crudes Basket (34.6 degrees API). The change was implemented as of 16 June 2005. As of 10 September 2007 the composition of the Basket was changed again, this time to include 'Girassol', the export crude of its new Member State, Angola. As of October 2007, the Basket price includes the Ecuadorean crude 'Oriente'.

Up to July 2005, despite the price surge and tight market conditions that in that period were prevailing (and that are still continuing) OPEC maintained its policy to establish, at successive meetings of the Conference, production quotas for its Member States.

Table 4.5. Production Quotas (thousand b/d) April 2000-September 2001

	2000 (eff. April)	*2001 (eff. April)*	*2001 (eff. Sept.)*
Algeria	788	773	741
Indonesia	1,280	1,255	1,203
Iran	3,623	3,552	3,406
Iraq	-	-	-
Kuwait	1,980	1,941	1,861
Libya	1,323	1,296	1,242
Nigeria	2,033	1,993	1,911
Qatar	640	627	601
Saudi Arabia	8,023	7,865	7,541
UAE	2,157	2,113	2,025
Venezuela	2,845	2,786	2,670
OPEC 10*	24,692	24,201	23,201
OPEC 10**	28,898	28,334	28,334
OPEC 11**	31,512	30,857	30,857

* Excludes Iraq.
** Actual annual production, ex/includes Iraq.
Sources: BP Statistical Review of World Energy 2007; MEES 2007.

Table 4.6. Production Quotas (thousand b/d) June 2003-July 2005

	2003 (eff. June)	*2003 (eff. Nov.)*	*2004 (eff. Aug.)*	*2004 (eff. Nov.)*	*2005 (eff. July)*
Algeria	811	782	830	862	894
Indonesia	1,317	1,270	1,347	1,399	1,451
Iran	3,729	3,597	3,817	3,964	4,110
Iraq	-	-	-	-	-
Kuwait	2,038	1,966	2,087	2,167	2,247
Libya	1,360	1,312	1,392	1,446	1,500
Nigeria	2,092	2,018	2,142	2,224	2,306
Qatar	658	635	674	700	726
Saudi Arabia	8,256	7,963	8,450	8,775	9,099
UAE	2,217	2,138	2,269	2,356	2,444
Venezuela	2,923	2,819	2,992	3,107	3,223
OPEC 10*	25,400	24,500	26,000	27,000	28,000
OPEC 10**	29,540	29,540	31,145	3,1145	32,235
OPEC 11**	29,540	30,884	33,175	33,175	34,068

* Excludes Iraq.
** Actual annual production, ex/includes Iraq.
Sources: BP Statistical Review of World Energy 2007; MEES 2007.

The Tables above show that during the period for which production quotas were established, i.e., up to July 2005, actual OPEC 10 production constantly exceeded the OPEC 10 collective quota. In 2006 the annual OPEC 10 oil production exceeded the collective OPEC 10 July 2005 quota by no less than 4.8 million b/d. This difference confirms that for some years now Member States are unofficially allowed to produce at full capacity.

But the international community worrying about the price levels that meanwhile have been attained (see Chapter 3) is constantly pressing OPEC to start investing in new production capacity. Up to very recently, OPEC has resisted any such demands with the argument that the oil supply was sufficient and that prices near or above the USD 100/b level were due to the weakness of the US dollar and speculation in the oil market inspired by fear for political unrest and conflict involving major oil exporting countries such as Nigeria, Iraq and Iran. But then, apparently yielding to the international pressure, OPEC's Secretary General announced on the 11e International Energy Forum held in Rome (20-22 April 2008) that the Organization is planning new investments in production capacity over a period of four years with the aim to increase collective production by 5 million b/d in 2012. Whether OPEC's intentions will convince the market and lead to somewhat lower prices remains to be seen. Fact is that OPEC will never allow the international price to drop below a level that it feels to be justified under the given circumstances and that presumably ranges between USD (2008) 90/b and USD (2008) 100/b.

4.8 THE EUROPEAN COMMUNITY

4.8.1 Background

Economic cooperation between European States started when six West-European States, i.e., France, West-Germany, Italy, Belgium, Netherlands and Luxembourg, founded in 1951 the European Coal and Steel Community (ECSC) by the Treaty of Paris. Said Treaty entered into force on 23 July 1952 for a limited duration of fifty years. Shortly thereafter, the same six Member States founded the European Economic Community (EEC) by the Treaty 'Establishing a European Economic Community' signed in Rome on 25 March 1957, effective date 1 January 1958. Much later and after Denmark, Ireland, Portugal, Spain, Greece and the United Kingdom had joined the EEC, the latter was formally replaced by the European Community (EC). This substitution was effectuated by the Treaty on the European Union (EU) which was signed on 7 February 1992 in Maastricht and entered into force on 1 November 1993. In 1995, 2004 and 2007 respectively three, ten and two more European States joined the EU expanding the membership to twenty-seven States. The EC is one of the two European Communities still remaining after the ECSC on 23 July 2002 had ceased to exist, the other is the European Atomic Energy Community (Euratom). The former's responsibilities and assets have been assumed by the EC.

The original EEC Treaty created a common market for all goods, persons, services and capital, with exception always of coal and nuclear products which were and are governed by their own specific Treaty i.e., ECSC and Euratom. In December 1985 the EEC Treaty was amended by the adoption of the Single European Act. This Act established an internal market intended to be functioning by 1 January 1993. The internal market is described as a place without frontiers, in which the free flow of goods, persons, services and capital is safeguarded in accordance with the provisions of the EEC Treaty. The Act also introduced a new Title (Title VII) in the EEC Treaty containing a statement of the objectives of the Community in environmental matters. Generally, it was laid down that Community environmental policies shall be integrated with Community energy policy.

4.8.2 Restrictions on Community Level Petroleum Policies

Member States of the EU are sovereign States and while prepared to cooperate with one another with respect to the production and consumption of energy products, including petroleum within the territory under their jurisdiction, they do not wish to surrender their sovereign rights over petroleum resources within said territory, including those located on the continental shelf and wish to decide for themselves which type of energy product to use and consume, whether it should be coal, oil, natural gas, nuclear energy and/or renewables. Finally, Member States insist on the right to independently select the countries from which they wish to import their energy products, be it electricity, coal, natural gas, oil or uranium ore, from inside or outside the Community. In particular the last point may cause friction at Community level considering that for covering the consumption of fossil fuels and nuclear energy, the EU as a whole is very much dependent on foreign suppliers, a situation that would suggest a uniform approach under the umbrella of the EU, if any supply source is subject to geopolitical risk as seen from a EU point of view. Where coal is supplied by politically reliable countries such as the US, Canada and South Africa, it can safely be assumed that no such geopolitical risk is involved. But the supply of oil and natural gas is a different matter. Oil is for eighty-five per cent supplied by foreign countries, which are almost all Member States of OPEC. Natural gas is for fifty-eight per cent (a share that will rapidly increase in the future), supplied by foreign countries among which the Russian Federation and Norway (by pipeline) and Algeria and Libya (by LNG tanker). In the future the LNG exporters will be joined by Qatar and Oman. It follows that compared with coal, the EC runs with the supply of oil and natural gas a greater geopolitical risk. Still on this issue, the Member States wish to draw their own plans, to pursue their own policies and to take their own responsibilities, e.g., the gas sales and transport agreements between Germany and the Russian Federation involving the building of a pipeline (called Nordstream) through the Baltic Sea (see Chapter 9 for the environmental aspects of this project).

4.8.3 Policy Objectives Realized

4.8.3.1 EC Directive 94/22/EC of 30 May 1994

On the production side, the EC has introduced legislation aimed to provide the legal instruments for mobilizing and developing the Community's own oil and natural gas resources. It concerns the EC Directive 94/22/EC of 30 May 1994 which related to the conditions for granting and using authorizations for the prospecting, exploration and production of petroleum in the Member States. The Directive was aimed at ensuring the non-discriminatory access to and pursuit of petroleum operations under conditions which encourage greater competition in this sector (see Chapter 6)

4.8.3.2 European Energy Charter and Charter Treaty

A few years before the introduction of the EC Directive 94/22/EC, the European Commission on behalf of the European Communities had been the driving force behind the creation of the European Energy Charter. This Charter was on 17 December 1991 signed by forty-nine countries among which the Member States of the EC, the other Member States of the OECD including the US, the countries in central and eastern Europe, the Baltic States and almost all Member States of the CIS (former Soviet Union), including the Russian Federation itself. The Charter was a non-binding political declaration, setting out the objectives, principles and actions which the signatory States declare willing to undertake in the fields of trade and investment in energy products and energy production. A few years later the European Commission this time acting on behalf of the newly created EC succeeded in bringing about a Charter Treaty (signed on 17 December 1994) in which the aforesaid principles and objectives of the Energy Charter were worked out in detail and turned into commitments for the contracting States (see Chapter 10).

4.8.3.3 Internal Market in Natural Gas

On the consumption side, the EC succeeded in creating an internal market in natural gas. The completion of such market took longer than was expected at the time the Single Act was adopted. The completion process took place in two phases. The first phase was formed by two Council Directives, one on the transit of natural gas through grids (Council Directive 91/296/EEC of 31 May 1991) and the other concerning a Community procedure to improve the transparency of gas (and electricity) prices charged to industrial end-users (Council Directive 90/ 377/EEC of 29 June 1990) as amended by Commission Directive 93/87/EEC. The final phase was formed by the European Parliament and Council Directive 98/ 30/EC of 22 June 1998, the final text of which was established at a meeting of the Energy Council on 8 December 1997. The Directive 98/30/EC of 22 June 1998 establishes common rules for the transmission, distribution, supply and storage of natural gas. Upstream pipeline networks, i.e., any pipeline or pipeline network operated as part of an oil or natural gas production project are not involved and do not form part of the internal market, but LNG facilities owned and/or operated by a natural gas undertaking do.

The internal market in natural gas is not complete in as much as the Commission is working on arrangements that would take away ownership and/or control regarding the gas transportation system from the gas suppliers/ producers. The effect of creating the internal market on the consumption of gas is one of lowering the prices to be paid by the end-user. This effect may be expected to lead to an increase of the consumption which in itself is in conflict to the overall post-Kyoto policy of restricting in one way or another the consumption of fossil fuels.

4.8.4 The Shaping of an Energy Policy

4.8.4.1 Up to the Oil Crisis of 1973

Up to the outbreak of the first oil crisis in October 1973, Community energy policy was focused on economically protecting and safeguarding the existence of the coal mines and coal production of the Member States, whereby, due to the worsening economic circumstances, of necessity the objective of a common market for hard coal had gradually to be discarded. Due to their relatively low prices, both crude oil and imported hard coal succeeded to penetrate the West-European energy market. The depressed level of the oil prices was caused by the severe competition that went on between and amongst the international oil companies on the West-European market. This competition was also viewed with concern by the governments of the oil exporting countries, in particular by Venezuela. The selling of their crudes in the US had become difficult as a result of the implementation of the Mandatory Oil Import Control programme which became effective in March 1958. To meet this competition, the concessionaires in the Middle East Gulf area were forced in August 1960 for the second time in a row to lower their posted prices which prompted five major oil exporting countries to set up with effect from 1 September 1960 their Organization of Petroleum Exporting Countries (OPEC) (see Chapter 3).

4.8.4.2 From the 1973 Oil Crisis up to 1986

During the oil crisis of 1973/1974, the Arab Member States of OPEC either stopped or gradually reduced their exports of oil to the western countries. This embargo resulted in a drastic increase of the price of oil on the world market. An immediate reaction came not from the EEC, but from the side of the OECD, since not only West-European States were involved. As described before, the OECD established on 15 November 1974 a subsidiary organization by the name of the International Energy Agency (IEA) and adopted three days later the International Energy Programme (IEP).

Note. Every Member State of the OECD may become a member of the IEA. The IEA is charged with the task to execute the IEP. This Programme envisages cooperation between the members of the IEA which is aimed at taking measures to reduce its Members' dependency on the import of oil and to promote a gradual replacement of oil by other energy sources or products. The IEP provides also for a crisis management system that starts functioning as soon as the supply of oil in any significant degree is interrupted. Members are obliged to keep and maintain oil emergency stocks sufficient to substitute for ninety days' imports. During a crisis the members are obliged to share still existing supplies with one another in accordance with the allocation procedures established as part of the IEP. The emergency allocation system is activated either with respect to all participating countries or in respect of any one country (selective trigger) when available supplies are seven per cent or more below consumption during a base period.

The Community was prompted by the events to formulate a coherent energy policy. In 1974 the Council formulated the following policy objectives:

- the need to reduce the dependency on imported energy products, in particular oil;
- the development of nuclear fission electricity;
- the development of the Community's own coal and petroleum resources; and
- stimulating and promoting technical research in the field of energy.

Note. In the aftermath of the oil crisis of 1973, the Community policy makers were of the opinion that hard coal should play an important role in the implementation of the policy to reduce the dependence of the Community on foreign sources of oil. Until 1986 the difference in price between Community coal and imported coal on the one side and heavy fuel oil on the other had been large enough in favour of coal to make the transition to a greater use of coal in the industry, as was recommended by the Council, economically feasible. Nevertheless, for several reasons the policy to promote a greater industrial use of coal failed.

4.8.4.3 From 1986-1996

On 16 September 1986 the Council adopted the objectives for the next ten years. Although the Community in that year had been confronted with a fall in oil prices (the Member States of OPEC were engaged in a struggle to obtain a fair market share) as well as with a shocking energy event, namely the accident that on 25 April 1986 took place in reactor number 4 of the nuclear power plant at Chernobyl, Ukraine (see Chapter 3), the Council nevertheless adhered to and confirmed the long-term objectives of the EC's energy policy based on crude oil/oil products substitution and energy efficiency. But it was conceded that in view of the low level of oil prices and the corresponding attractiveness of imported oil, additional measures might be needed in order to continue following a policy of diversification of the EC's external sources of supply. The emphasis was put on the development of the Community's own energy resources under satisfactory economic conditions and geographical diversification of the Community's external sources of supply and diversification between the different forms of energy; and a search for balanced solutions as regards energy and the environment, by making use of the best available and economically justified technologies and by improving energy efficiency, as well as taking account of the desire to limit distortions of competition in the energy markets by a more coordinated approach in environmental affairs in the Community.

Note. With regard to the Community's consumption of oil and natural gas, the following specific objectives were adopted:

(i) with regard to crude oil: to keep net crude oil imports from third countries within reasonable proportions by maintaining a policy of oil substitution and by continuing and, if need be, stepping-up crude oil exploration and production within the Community, particularly in promising areas not yet exploited. Oil consumption should be kept down to around forty per cent of energy consumption and net oil imports thus maintained at less than thirty-three per cent of total energy consumption in the Community in 1995. (At this point it should be mentioned that already in 1986 this objective had been achieved. It did not prove possible to maintain this percentage in the ensuing years); and

(ii) with regard to natural gas: to maintain the share of natural gas in the energy balance on the basis of a policy aimed at ensuring stable and diversified supplies and continuing and, if need be, stepping up natural gas exploration and production in the Community.

4.8.4.4 From 1996-2006

In January 1996 the EC Commission published a White Paper setting out the framework of an energy policy at Community level for the next ten years. Of perhaps only historical interest, the White Paper foresaw a real possibility that, in the planning period, an important share of the domestic production of solid fuels would be fully competitive at world market prices. In parallel, the White Paper recommended that even more efficient combustion technologies should be developed and implemented if solid fuels are to maintain or even increase their share in the long term. These improvements in technology would need to focus on limiting CO_2 emissions since other emissions can already be effectively controlled. The Community's level of coal import dependency was forecast to increase in future, but the characteristics of the world coal market and the large diversity of suppliers meant that the risk of a persistent interruption of supply, even in the long term, is minimal.

Recognizing the contribution that nuclear energy could make to all the Community's energy objectives, the White Paper recommended to keep the nuclear electricity option open but recognized that its future depends to a large extent on its acceptability by society and political leaders. This acceptability problem was seen to derive particularly from concerns on nuclear safety, on transport and disposals of nuclear waste and on nuclear non-proliferation. It was conceded that in reality most Member States either prefer to pursue a non-nuclear energy policy, or had by now decided to reduce dependency on nuclear-based sources of energy.

The Community would support the development of nuclear fusion science. The Community would make every effort to realize the potential of renewables by supporting research projects; by stimulating cooperation in the development and the dissemination of new and competitive technologies, and by establishing a Community framework for national fiscal and other incentives in order to translate technological advances into marketable products. These concerns would be at the centre of the ALTENER II programme.

Note. According to the White Paper the adoption of a CO_2/energy tax would contribute to the goal of sustainable development. However, it is stressed that any policy based on fiscal measures would need to address their possible effects on the Community's global competitiveness, especially for energy-intensive industries. Measures that provide certain tax breaks or incentives as well as other policy changes, such as using the tax revenues to reduce employers' non-wage labour costs and thus improve industrial competitiveness and employment prospects, would be needed to offset these adverse effects. In the White Paper, it is recognized that it may be difficult for the Member States to shift their fiscal systems in this direction if other major industrial competitors, e.g., the US and Japan, are nor prepared to follow suit. The Community, together with the Member States, would need to play a major role to persuade other energy consuming countries to follow the Community's policies in this area.

In the White Paper, research and technological development are identified as important instruments for supporting the Community's energy policy. For this reason the Community would continue to support major programmes for fossil fuels (JOULE/THERMIE) and in nuclear fission and fusion. It also meant establishing close collaboration between those programmes and measures to improve their market application (SAVE, ALTENER).

Note. The JOULE/THERMIE programme aims to promote the efficient use of indigenous resources of energy and reducing environmental emissions, in particular CO_2; it also aims to strengthen industrial competitiveness. Nuclear fission and fusion research activity takes place under the Euratom Treaty and is focused on nuclear security and safety (in order to increase the nuclear fission option's public acceptability) and on controlled fusion by magnetic confinement with the long-term objective of joint construction of safe and environmentally sound prototype reactors and in the end an economically viable power station. Priority was given to establish the engineering design of the first experimental reactor in the context of the International Thermonuclear Experimental Reactor (ITER) programme (see Chapter 3).

4.8.4.5 From 2006 Onwards

In March 2007, the EU leaders agreed to cut the EU's own greenhouse gas emissions by at least twenty per cent by 2020 regardless of what other countries do. The EU plans to achieve this reduction through actions foreseen in the new integrated energy and climate policy together with measures already in place such as the EU emission trading scheme (EU ETS) which was launched in January 2005.

In this context the EU leaders have decided to:

(i) save twenty per cent of energy consumption compared with projections for 2020 by improving energy efficiency;
(ii) increase to twenty per cent by 2020 the share of renewable energies in overall energy consumption, thus almost tripling the current level;
(iii) increase tenfold – to at least ten per cent – the share of biofuels in overall petrol and diesel consumption by 2020, provided that sustainable 'second-generation' biofuels from non-food crops become commercially available;
(iv) develop and promote low- or even zero-emitting technologies, including carbon capture and storage – i.e., preventing CO_2 entering the atmosphere by capturing it and storing it underground in depleted gas fields or old salt mines – so that these can make a major contribution to reducing emissions by 2020.

Note with reference to (iii). The announced target of a share for biofuels of at least ten per cent is still under discussion, in view of warnings about a world food crisis.

As far as nuclear energy is concerned, the Commission emphasized that it is for each Member State to decide whether or not to rely on nuclear electricity. Nuclear energy is mistakenly not considered to be a renewable energy source and can therefore only be of assistance in fulfilling the twenty per cent target with regard to the reduction of CO_2 emissions, for instance by replacing or converting fossil fuel-fired power plants. As mentioned before if use is made of fast breeder reactors, the world usable uranium reserves would increase a hundred-fold providing the feedstock for non-fossil electricity for thousands of years. From this point of view, it makes no sense not to treat and consider nuclear fission energy as a renewable form of energy on a par with solar energy, wind- and hydo-power (see Chapter 3).

Chapter 5
Petroleum Legislation of States

5.1 ORIGIN, SCOPE AND STRUCTURE

The drilling of the first oil well in 1859 did not only mark the beginning of the petroleum industry, but at the same time it signalled the beginning of the development of mining legislation specifically addressing petroleum and petroleum operations, i.e., mining legislation that takes into account and recognizes the special physical characteristics and natural condition of petroleum and the mining methods that have to be applied to extract the same from the pores and fissures of sedimentary rocks in the subsoil. Of necessity, this new mining legislation, hereinafter referred to as petroleum legislation, is different from mining legislation applicable to the mining of coals, metal ores and other hard rock minerals.

Note. In contrast with the manner in which deposits of hard rock minerals e.g., coals, metal ores, usable rocks and clays, are mined i.e., at the surface or in the underground by mechanically separating them from the rock in which or between which they are embedded, accumulations of petroleum are mined by drilling holes from the surface into the reservoir rock in the pores or fissures of which the accumulation occurs.

A State's petroleum legislation comprises all laws, regulations and rules specifically designed for the purpose of regulating petroleum operations in areas falling under the (natural resources) jurisdiction of the State and for awarding authorizations for carrying out said operations. Authorizations are awarded in the form of either licences or contracts of work.

A licence is an administrative authorization issued by the government acting on behalf of the State, and exercising the latter's sovereign powers over natural resources.

A contract of work is a contractual relationship (a contract) entered into by a state entity (the state party) on the one part and a commercial petroleum enterprise from the private or public sector (the contractor) on the other. The state party may be the State itself acting under its sovereign powers over natural resources, or an authority representing the State such as a government ministry or a petroleum directorate established within a ministry. The state party may also be a state oil enterprise i.e., a state owned and controlled oil company, usually referred to as the national oil company specially established for the purpose of managing and developing the national petroleum resources. When entering into a contract, the state oil enterprise is either acting pursuant to a general authority granted by the petroleum law or acting as the holder of a specific exclusive exploration and/or production licence granted in conformity with the terms and conditions of this law.

Both the licence and the contract of work authorize the holder(s) thereof i.e., licensees or contractors, to undertake such petroleum operations (i.e., operations involving the exploration, prospecting and/or development and production of petroleum) and within such area and within such period as are described in their licence or contract of work, subject to fulfilling the obligations and making the payments stipulated therein.

The licence is either exclusive or non-exclusive. The licence is exclusive if it gives the licensee exclusive rights in the sense that as long as the licence is valid no other person than the licensee itself is authorized to exercise the rights conferred by the licence. Exclusive licences are granted for exploration work, including drilling, and for development/production operations. In some instances separate exclusive licences are required for the exploration and development/production phases respectively. But if so, there is a connection between these licences, in that the holder of an exploration licence, which after having made a discovery applies for a production licence, is generally entitled to be granted the applied for licence. In other instances a comprehensive exclusive licence is made available, authorizing its holder to conduct all petroleum operations starting with the search for petroleum and ending, if the search had been successful, with the production (and the disposal) thereof. A non-exclusive licence is made available for general geological and geophysical work that does not involve drilling, and generally covers a large area.

Where (already) a petroleum law exists providing for a licensing regime contracts of work, if that type of petroleum authorization becomes the preferred form of authorization, are integrated into specific exclusive licences. This integration can be accomplished either by granting such a licence to the state party who then in turn concludes a contract of work with another entity (contractor) for the purpose of conducting petroleum operations with respect to the licensed area, or by granting the exclusive licence to the contractor. In both cases the contract area coincides with the licensed area. Furthermore, the duration of the contract, possible renewals or extensions thereof and area relinquishments are all determined by the corresponding conditions of the (underlying) licence.

Contracts of work may be categorized into risk-bearing contracts (risk contracts) and non-risk-bearing contracts (non-risk contracts). Risk contracts are contracts of work under the terms of which the contractor has to invest its own funds in the petroleum operations authorized under the contract. In contrast therewith, a non-risk contract does not require the contractor concerned to supply its own funds; he will be compensated for his efforts regardless of the outcome thereof.

Note. The risk accepted by the risk-contractor is the risk of not being able to find commercially exploitable oil or gas reservoirs or of not being able to dispose of its share of the production in a commercial manner. In both situations the risk-contractor will have lost its investments.

Risk contracts in turn may be divided into:

(i) production sharing contracts; and
(ii) risk service contracts.

Note. Of the two types of risk contracts the production sharing contract is the most widely used by governments of non-Western States. The contracts were first embraced in Indonesia as the legal instrument for permitting foreign oil enterprises to undertake petroleum operations within its territory. The contracts were adopted in replacement of the exclusive licences that had been terminated by virtue of Government Decree No. 44 of 26 October 1960. Although currently the production sharing contract is used and applied in almost all non-Western States no universal model or standard contract exists. Each State that has adopted the concept of production sharing has over the years developed its own variant of the contract, which in some cases will contain elements and features of a preceding or co-existing licensing regime.

From the foregoing it may be concluded that the exclusive licence and the risk contract have much in common. In one important aspect however they differ from one another: the exclusive licence, if comprising the authorization to produce petroleum, i.e., a production licence, confers to the holder thereof ownership of petroleum, if and when produced, whereas a risk contract never grants the holder of the contract such ownership.

Note. As will be discussed hereinafter a contractor operating under the terms of a production sharing contract is entitled to receive, as compensation for the costs incurred and by way of profit, a part of the oil and/or natural gas that is produced under the terms of the contract concerned but this transfer of ownership takes place after the petroleum has been produced and after it has been separated into oil and natural gas.

Petroleum legislation prescribes methods of working and includes all types of working conditions and rules aimed at safeguarding the safety and health of the workers and the general public and at protecting and preserving the environment in which the petroleum operations are carried out. A licensing regime shall include rules and conditions obliging the licensee to pay to the owner of the petroleum in situ, which is usually the State, royalties, also known as production fees, to pay to the State area fees and cash bonuses, and (depending on the regime concerned) to share with the State the net or gross profits realized by the licensee from its production of petroleum.

Petroleum legislation operates in conjunction with any fiscal legislation imposing a tax on income realized by a licensee or contractor from its authorized petroleum operations. The tax concerned may be the tax imposed on income realized by commercial incorporated enterprises generally (hereinafter referred to as 'corporation tax' or simply 'income tax' as the context requires or allows) or, where a licensing regime is concerned, a tax solely levied on income derived by the holder of a production licence from his petroleum operations (hereinafter referred to as 'special petroleum income tax').

Petroleum legislation also operates in conjunction with relevant legislation by which is meant State's legislation that although not specifically directed at petroleum or petroleum operations, nevertheless by the nature of its provisions, also regulates the conduct of petroleum operations. Relevant legislation as described is administered by government ministries and/or by provincial and municipal authorities or administrations.

Note. Examples of relevant legislation are:

(i) generally applicable legislation concerning the conservation and protection of the living and non-living natural resources and of the environment (environmental and conservation legislation); and
(ii) generally applicable legislation concerning the permanent or temporary occupation and use of surface areas for commercial, agricultural, industrial, housing and other purposes and concerning the restoration of these areas after any temporary occupation and use (planning and zoning legislation).

The granting of a petroleum authorization, be it a licence or a contract of work, generally does not imply or guarantee that a corresponding authorization required under relevant legislation concerned shall be granted by the central, provincial or municipal authorities in charge of administering such authorizations. When taking a decision in matters concerning the occupation and use of land, the installation and operation of facilities and the measures that are required to protect the environment said authorities will apply the criteria and yardsticks prescribed in the relevant legislation without being obliged to make an exemption in favour of petroleum operations. If the competent authorities concerned finally decide to refuse to grant to an authorized person the necessary work permits it shall become impossible for such a person to exercise its rights embodied in its (petroleum) authorization. The ensuing conflict must be resolved under administrative law procedures.

Petroleum legislation is structured as follows:

(i) There is a petroleum law i.e., an act or decision of a body (or congress) of people's representatives, dealing with petroleum matters. Where a petroleum law provides for the granting of licences it may be a comprehensive law, containing all the provisions of any importance in regulating petroleum operations and the issue and contents of licences, or an enabling law, hereinafter referred to as a framework law, containing essentially no more than an authorization for the government to issue from time to time the required regulations and licences. Said regulations may provide for a model licence. Where a petroleum law provides for the granting of contracts of work it will be a framework law containing essentially no more than a specific or general authorization for the state entity concerned to enter into contracts of work. In most instances the law provides for a model contract.

(ii) There is subsidiary legislation emanating from the government or from a government authority such as the minister responsible for petroleum affairs, pursuant to its powers under the petroleum law and referred to as petroleum regulations or ministerial decrees.

(iii) Specific ministerial decisions taken by the government minister responsible for petroleum affairs pursuant to the petroleum regulations, which inter alia may concern the grant of the appropriate authorizations, be they licences or contracts of work, or the approval of a work programme, a field development programme or a sales contract.

A petroleum law may be a law specifically designed and meant for petroleum matters, but it may also be a general mining law with a special section reserved for petroleum matters.

In the absence of a petroleum law, constitutional law or the constitution itself may directly authorize the government to regulate mining and/or petroleum matters and issue petroleum authorizations.

The hierarchy of petroleum law, government regulations, ministerial decrees and ministerial decisions determines how easy or difficult it is to make and introduce new rules and/or abolish old rules. Petroleum laws are cumbersome to change, since the legislature will become involved with every proposed change. Government regulations and ministerial decrees, implementing and supplementing the provisions of the law, are easily repealed and replaced by new regulations or decrees; although where government regulations are concerned the possible requirement of parliamentary control might make such changes more difficult. The regulatory freedom of a government (or a minister) depends in final analysis on the character of the petroleum law: a comprehensive petroleum law leaves a government (or a minister) little regulatory freedom and vice versa. Specific ministerial decisions are always related to specific cases (the grant of a licence, the approval of a contract of work, the approval of a development plan or of a sales contracts, etc.) and usually taken within the framework of a ministerial policy, but a change of legislation is not involved. On the other hand ministerial decisions may be subject to appeal under the rules of administrative law.

5.2 PRINCIPAL PROVISIONS OF PETROLEUM LAW

5.2.1 Definition of Petroleum

'Petroleum' is generally defined as all liquid and gaseous hydrocarbons existing in their natural state in the subsoil, as well as other substances produced in association with such hydrocarbons.

Note. This definition is taken from the Norwegian Petroleum Act of 29 November 1996 No. 72. A similar definition is found the UK Petroleum (Production) (Seaward Areas) Regulations. In the latter definition it is added that petroleum does not include coal or bituminous shales or other stratified deposits from which oil can be extracted by destructive distillation.

5.2.2 Ownership of Petroleum In Situ

Almost without exception, a petroleum law (or a constitutional law) vests in the State the ownership of petroleum in situ, i.e., petroleum as occurring under natural conditions on the surface or in the subsoil of the State concerned.

Note. There are however a few, but important exceptions to this rule. Under the territorial jurisdiction of the US ownership of natural resources, e.g., coal, hard rock minerals and petroleum, is vested in the public or private owner of the land on the surface or in the subsoil of which the natural resource is located. In contrast therewith, no ownership of petroleum in situ is acknowledged under the territorial jurisdiction of the German Federal Republic (Federal Mining Law of 20 August 1980). Under this Mining Law petroleum in situ is considered to constitute a *res nullius*. At the same time no person is allowed to acquire ownership by the act of occupation. This means that no person, when discovering petroleum in situ, may claim ownership thereof, unless that person had made the discovery while being in possession of an exploration licence and on account of his discovery has been granted a production licence. As explicitly stated in the German Mining Law, the latter licence entitles its holder to acquire the ownership of petroleum upon production. In this German example ownership of petroleum is not transferred (upon production) from the one person to the other but such ownership is created by the act of production.

In the international context it has been established by the 1958 Convention on the Continental Shelf, that any coastal State has sovereign rights over its part of the continental shelf for the purpose of exploring it and exploiting its natural resources. These rights have been confirmed by the 1982 Convention on the Law of the Sea. On the strengths of these sovereign rights and taking into account that petroleum belongs to the natural resources of the continental shelf coastal States have proclaimed and laid down in their petroleum law to be the owner of the petroleum in the seabed and in subsoil of the continental shelf. The matter of land ownership and associated claim to mineral ownership does not arise in this case because the seabed has no owner.

Apart from a State, a public or private landowner or even nobody, petroleum in situ may have mankind as its owner. Under the 1982 Convention on the Law of the Sea the mineral resources in situ of the deep seabed area (the Area as defined in this Convention) have been declared to be the common heritage of mankind. Mineral resources are defined as to include petroleum.

In a parallel development of international law the issue of state sovereignty over the State's natural resources has been addressed in a more general and principal way, not restricted to the sovereignty of a coastal State over the natural resources of its continental shelf. The first proclamation on this issue can be found in Article 2 of Chapter II of the 1974 UN Charter of Economic Rights and Duties of States. Said article reads:

> Every State has and shall freely exercise full permanent sovereignty, including possession, use and disposal, over all its wealth, natural resources and economic activities.

This position was later confirmed (albeit without qualifying this sovereignty as being 'full' and 'permanent') by Article 18 of the 1994 European Energy Charter Treaty. This article reads:

> The Contracting Parties recognise state sovereignty and sovereign rights over energy resources (defined as to include petroleum). They reaffirm that these must be exercised in accordance with and subject to the rules of international law.

Note. In States where ownership of mineral resources is vested in the public or private owner of the land on the surface or in the subsoil of which the mineral resources are located, there does not have to a conflict with the principle of the State's sovereign rights over these resources, as long as it is accepted that such sovereign rights include the right to determine to whom the ownership should be assigned.

5.2.3 Delegation of Legislative Power to the Government

Any petroleum law provides for a delegation of legislative power to the government. The law describes the subjects, which are left to be provided for by regulations to be issued from time to time by the government, usually the Minister responsible for petroleum affairs. The scope and extent of such delegation, in fact the distribution of subjects to be provided for between law and regulations, depends on the type of law concerned. The more wide-ranging this power to make regulations is the better the petroleum legislation can react to changing technical and economic circumstances and to new insights in safety matters, the protection of the health of workers and of the environment.

5.2.4 Opening Up of Territory

Where the ownership of petroleum in situ is vested in the State, as is generally the case, it is acknowledged in international law that the latter has the absolute discretion to determine whether or not its petroleum resources shall be explored for, developed and produced. In other words no State is obliged to make its territory available for petroleum operations, even if such operations would manifestly be to its own (short term) benefit or to the benefit of the international community.

Note. As a matter of fact this right to choose for development now or later is part and parcel of the State's sovereign right over natural resources but has specifically been enshrined in international or regional Conventions e.g., the 1991 European Energy Charter, the 1994 Energy Charter Treaty and, specifically for petroleum in the EC Directive 94/22/EC of 30 May 1994. As stated in the latter Member States retain the right to determine the areas within their territory to be made available for the exercise of the activities of prospecting, exploring for and producing hydrocarbons. But once a State has decided to open up certain areas for petroleum operations, the agreed rules may oblige Contracting States or Member States, as the case may be, not to discriminate between qualified applicants in allocating authorizations.

5.2.5 Type and Scope of Authorization

Any petroleum authorization is granted for a particular type of petroleum operation e.g., exploration, prospecting, exploration drilling, development and production or for a combination of these activities, with respect to a particular area and for a specific period. The authorized operation, the authorized area and the duration of the different periods comprised in the authorization are all incorporated in the authorization document. In addition the authorized area is depicted on a map attached to the authorization document. The authorization provided for is either a licence (exclusive or non-exclusive), or a contract of work (risk contract or non-risk contract).

5.2.6 Procedures for Applying for and Awarding Exclusive Authorizations

These rules include the laying down of (non-discriminatory) criteria for qualifying as an applicant for an exclusive authorization, the method of choosing between competing qualified applicants and the procedures leading to the granting authorizations: e.g., bidding rounds, auctions or cash tender bids, special invitation, separate negotiations.

Note. A very relevant example of such procedures is offered by the EC Directive 94/22/EC of 30 May 1994.

5.2.7 Size and Shape of the Area for Which an Authorization May Be Granted

Under this heading fall rules about the size and shape of the areas in respect of which an (exclusive) licence or risk contract will be made available. For instance, the area that is made so available may be covered by a geometrical grid, resulting in an area which is divided into blocks.

Note. The EC Directive of 30 May 1994 requires the Member States to ensure that if the geographical areas are not delimited on the basis of a prior geometric division of the territory, the extent of each area is determined in such a way that it does not exceed the area justified by the best possible exercise of the activities from the technical and economic points of view.

5.3 PRINCIPAL PROVISIONS OF EXCLUSIVE LICENCES

5.3.1 STANDARD CONDITIONS

Standard conditions of exclusive licences, i.e., conditions that all exclusive licences have in common, are either laid down in government regulations issued under the authority of the petroleum law or, as the case may be, in a model licence attached to such regulations, or in the petroleum law itself, if the latter has been set up as a comprehensive petroleum law.

The standard conditions of a licence have to be distinguished from its particular conditions: the latter concern the size of licensed area, exploration work commitments for specific periods and obligatory payments but only to the extent that the latter have been left open to competition between qualified applicants, otherwise obligatory payments belong to the standard conditions.

5.3.1.1 Ownership of Petroleum if and when Produced

The holder of an exclusive production licence has the right to acquire the ownership of any petroleum that enters into any well that has been drilled by the former in accordance with all terms and conditions of his licence.

Note. As example of a licensee's ownership of petroleum, if and when produced, may be used the wording used in the UK model offshore production licence (to be found in Schedule 4 of the Petroleum (Production) (Seawards Areas) Regulations 1988 as amended):

> In consideration of the payments and royalties hereinafter provided and the performance and observance by the licensee of all terms and conditions hereof the Minister, in exercise of the powers conferred upon him by the Act of 1934 and the Act of 1964 hereby grants to the licensee EXCLUSIVE LICENCE AND LIBERTY during the continuance of this licence and subject to the provisions hereof to search and bore for, and get, petroleum in the seabed and subsoil under the seaward area comprising an area of . . . square kilometres more particularly described in Schedule 1 to this licence etc.

Another example in this context is provided by the aforementioned German Federal Mining Law of 13 August 1980. First, petroleum is declared to belong to the category of *bergfreie Bodenschatze* (free minerals), which means that petroleum in situ, like any other mineral declared to belong to said category, is a *res nullius*. Then the Law stipulates that any person wishing to search for free minerals needs an *Erlaubnis* (an exploration licence) and that any person wishing to produce a free mineral needs a *Bewilligung* (a production licence). The production licence comprises the right to search for and to produce the free minerals described in the licence within the *Bewilligungsfeld* (licensed area) and the right to acquire the *Eigentum* (the ownership) of the free minerals so produced.

In Chapter 3, Section 3-3 (Production Licence), third paragraph of the Norwegian Petroleum Act of 29 November 1996 No. 72 it is stated:

> A production licence entails an exclusive right to exploration, exploration drilling and production of petroleum deposits in areas covered by the licence. The licensee becomes the owner of the petroleum which is produced.

A similar explicit provision about the ownership of petroleum after its extraction from the subsoil reservoir is found in Article 3.2 of the Netherlands General Mining Law of 31 October 2002. After having established that mineral resources in situ are owned by the State, the article goes on stating that:

> in the process of producing mineral resources in accordance with the terms of the licence their ownership is transferred to the holder of the licence concerned.

Summarizing, a transfer of ownership of petroleum takes place from the owner of the petroleum in situ (i.e., either the State or, as the case may be, the public or private owner of the land overlying the petroleum deposit) to the holder of an exclusive production licence at the moment the petroleum enters this licensee's well. The title to petroleum established upon production is an absolute right that must be assumed to remain applicable even if the petroleum produced originates from outside the boundaries of the licensed area. Last mentioned situation may be encountered if the petroleum is produced from a petroleum reservoir (a straddling reservoir) that extends across the (vertical projection of the) common boundary of two adjacent licensed areas.

Note. This absolute right, i.e., the right to acquire ownership of any petroleum that enters any legitimately drilled well irrespective of or without regard to the origin of the petroleum so produced has been upheld (by courts of law) and respected (by producers) as from the day the first oil well was drilled (1859), preventing a landowner from claiming any loss of ownership of petroleum when 'its' petroleum (i.e., petroleum originally in place beneath his land and of which he had been the owner) was being produced or thought to be produced by the wells drilled on adjacent property. Indeed, it was in those early days accepted practice to drill as many wells as possible on one's property in order to maximize production. By preference the wells (known as offset wells) were placed along the common boundary, in order to prevent petroleum beneath the land to be produced by offset wells drilled in the adjacent block. In Canada (reference: Canadian Oil and Gas Law, § 21; § 32), the licence was classified by the Ontario Courts of law as a *profit a prendre* as early as 1906 (*Mcintosh v. Leckie* (1906), 13 OLR 54). In 1957, the Supreme Court of Canada defined the interest conferred by the usual oil and gas lease (in Canada) as a *profit a prendre* terminable upon the happening of any of the contingencies for which it provides (*Berkheiser v. Berkheiser* (1957) SCR 387, 7 DLR (2nd) 721). The Court of Appeals in Alberta stated the effect of the *Berkheiser Case* to be that the lessee, while having an interest in the land in the nature of a *profit a prendre*, had no title to the oil or gas as long as they remained in the ground and uncaptured (*St Lawrence Petroleum Ltd v. Bailey Selburn Oil and Gas Ltd* (1962) 35 DLR (2nd) 574). There is a clear similarity between the right accorded to the UK licence holder, namely the right 'to get petroleum' beneath the licensed area, and the Canadian description of the licence as a *profit a prendre*. Obviously, when the State is declared to be the owner of petroleum in situ, the matter of the origin of the petroleum is completely irrelevant, because there is no person other than the grantor of the licence itself that may claim loss of ownership, except of course if the boundary of the licence coincides with an international boundary. The position of the person holding a production licence in respect of the adjacent area from which petroleum may flow away and cross the licence boundary needs protection because the more petroleum flows away the less petroleum is left for that person 'to get'. This question is discussed in Chapter 10.

5.3.1.2 Duration, Sub-periods

Generally, there is an initial period and one or more extensions. The total licence period may amount to thirty or forty years, if development and production forms part of the licensed operations. A licence for exploration only shall not extend beyond ten years.

Note. Where the EC Directive 94/22/EC of May 1994 is applicable no standard periods are used, each applied for licence is treated as a special case. In this context Member States must ensure that the duration of an authorization does not exceed the period necessary to carry out the authorized operations. However, the competent authorities may prolong the authorization where the stipulated duration is insufficient to complete the operations in question and where the operations have been performed in accordance with the authorization. Member States must also ensure that operators do not retain exclusive rights in the authorized area for longer than is necessary for the proper performance of the authorized operations.

5.3.1.3 Area Relinquishment Schedule

It is a common condition that at certain intervals, for instance at the end of the initial period, part or parts of the original area of the licence must be relinquished. Part or parts to be relinquished must of such a shape and size that they are suitable to be offered once more for licensing.

5.3.1.4 Obligatory Exploration Work Programme

During the initial period of the licence an obligatory exploration work programme shall have to be carried out. The extension of the licence beyond the initial period shall depend on a satisfactory implementation of this condition. The details of such a work programme belong to the particular conditions of a licence (see hereinafter) and usually form an element of the competitive bidding for a licence. A licence may provide for a sanction on the non-fulfilment of the whole or any part of the obligatory work programme. In case of non-fulfilment the defaulting licensee may have to pay in cash the monetary value of the non-fulfilled programme to the competent authority. Non-fulfilment may also constitute a ground for cancelling the licence, but the problem is usually solved in further negotiations between the competent authority and the licensee.

Note. An obligatory drilling programme to which sanctions are attached may pose a problem for the licensee should the preceding seismic survey not turn up a sufficient number of drillable sub-surface structures as needed to fulfil the said programme. In such eventuality the licensee will be forced to request the competent authority to be relieved of the obligation to complete the agreed drilling programme or surrender its licence. If the competent authority is persuaded that completing the obligatory drilling programme would not serve any useful purpose, at least not when considered in relation to the cost to be incurred by the licensee, it will either grant the licensee the requested relief or propose that the licensee drills a well at another place, perhaps in the area of another licence, or at another time, or accept the surrender of the licence.

5.3.1.5 Development Plan and Production Schedule

The licensee shall be obliged to prepare a development plan and corresponding production schedule with respect to any commercial discovery that has been made and shall submit such plan and schedule for the approval of the competent authority. Only so approved plans may be carried out.

5.3.1.6 Disposal of Production

Under prevailing licensing regimes the licensee is completely free in disposing of its production, except in a case of national emergency.

5.3.1.7 Liability for Damages

The licensee shall have to keep the competent authority indemnified against all claims and demands brought against the authority by any third party in connection with the licence or the licensed operations. Under some petroleum legislation special provisions regarding liabilities for pollution damage caused by the operations are provided. As observed before, where no special rules are provided, liability for damage caused in carrying out the licensed operations is governed by the code of civil law.

5.3.1.8 Methods of Working

The licensee will always be obliged to execute all his operations in accordance with the provisions which prescribe the methods that have to be followed or applied in carrying out the authorized operations, such as rules regarding drilling practices and the installation, use, operation and final abandonment of oil/gas installations and pipelines. The licensee will also be obliged to execute the operations in a safe manner, and without harming the health of the workers employed by him in his operations or causing damage to the environment. All these rules and provisions are to be found in the (model) licence itself and/or in government regulations or ministerial decrees specifically made for the purpose, all depending on how the petroleum legislation has developed over time.

5.3.1.9 Joint Exploitation of Straddling Petroleum Reservoirs

Where a single, continuous petroleum reservoir extends across (the vertical projection of) the boundary of a production licence into the adjacent area such petroleum reservoir is referred to as a straddling petroleum reservoir. If in respect of the adjacent area production licences have been granted each covering a part of the straddling reservoir the joint exploitation of such a reservoir by all persons holding such licenses and by the licensee is usually of benefit to them all and in practice will be carried out on a voluntary basis. But in principle, the licensees involved in a straddling reservoir are free to extract petroleum from their part of the reservoir as long as, and to the extent that, they operate in accordance with the rights conferred and obligations imposed by their respective licences. Within this framework they are free to decide whether or not and in whatever form to cooperate with one another in exploiting the straddling reservoir in question. But governments too expect to benefit from joint exploitation in the form of a more efficient development plan, less waste of capital and a higher ultimate recovery, which are all factors leading to higher tax revenues, and in order to ensure that joint exploitation will actually take place will include in their licences an obligation for the licensee to this effect.

The UK offshore petroleum legislation (Petroleum (Production) (Seaward Areas) Regulations 1988/1995/1996) (see Chapter 6) offers an important example of a statutory requirement to cooperate if and when a straddling reservoir must be developed. The Schedule 4 model Clause 28 'Unit Development' reads as follows:

> 28(1) If at any time in which this licence is in force the Minister shall be satisfied that the strata in the licensed area or any part thereof form part of a single geological petroleum structure or petroleum field (hereinafter referred to as 'an oil field') other parts whereof are formed by strata in areas in respect of which other licences . . . are then in force and the Minister shall consider that it is in the national interest in order to secure the maximum ultimate recovery of petroleum and in order to avoid unnecessary competitive drilling that the oil field should be worked and developed as a unit in cooperation by all persons including the licensee whose licences extend to or include any part thereof the following provisions shall apply: (2) Upon being so required . . . by the Minister the licensee shall cooperate with such other persons (hereinafter referred to as the other licensees) as may be specified in the said notice in

> the preparation of a scheme (hereinafter referred to as 'a development scheme') for the working and development of the oil field as a unit by the licensee and the other licensees in cooperation, and shall, jointly with the other licensees, submit such scheme for the approval of the Minister . . . (4) If a development scheme shall not be submitted to the Minister in time, or if the development scheme so submitted shall not be approved by the Minister, the Minister may himself prepare a development scheme which shall be fair and equitable to the licensee and all other licensees, and the Licensee shall perform and observe all the terms and conditions thereof. (5) If the licensee shall object to any such development scheme prepared by the Minister he may . . . refer the matter to arbitration in the manner provided by clause 43 of this licence.

A second example of mandatory cooperation is offered by the Netherlands Mining Law of 31 October 2002. Article 23 of this Law reads as follows:

> 23(1) A person holding a production licence shall refrain from exploiting a reservoir that is reasonably expected to extend beyond the boundary of the area of his licence as long as he has not entered into an agreement with the person holding a licence in respect of the adjacent area pursuant to which the development and production of the said deposit shall be undertaken in mutual consultation. (2) If during or after the exploitation of a reservoir it appears that the said reservoir extends across the boundary of the area of the licence the licensee concerned is obliged promptly to inform the person holding a licence for the adjacent area and to collaborate with that person in preparing and concluding an agreement as referred to above under (1).

According to this example from the Netherlands it is not required that the straddling reservoir be developed and produced as a unit, as is required under UK legislation as described above. In the Netherlands example, licensees are free to make a choice out of various cooperative forms of exploitation, which may range from outright unitization to separate development with two operators.

A third example is offered by the Norwegian Petroleum Act of 29 November 1996 No. 72. Chapter 4, Section 4 -7 'Joint Petroleum Activities' reads as follows:

> If a petroleum deposit extends over more than one block with different licensees, or onto the continental shelf of another State, efforts shall be made to reach agreement on the most efficient coordination of petroleum activities in connection with the petroleum deposits as well as on the apportionment of the petroleum deposit. This shall apply similarly when, in the case of several petroleum deposits, joint petroleum activities would obviously be more efficient. Agreements on joint exploration drilling shall be submitted to the Ministry. Agreements on joint production, transportation, utilization and cessation of petroleum activities shall be submitted to the Ministry for approval. If consensus on such agreements is not reached within reasonable time, the Ministry may determine how such joint petroleum activities shall be conducted, including the apportionment of the deposit.

The manner in which the licensees concerned intend to cooperate, with regard to the exploitation of a straddling petroleum reservoir, should not be a concern of the government. Generally, the government's interest lies in ensuring that a straddling reservoir as a whole is exploited in an optimum manner, resulting in maximum fiscal and other benefits for the State. From this perspective it would appear that a government would not particularly be interested in how the participating licensees divide the straddling reservoir between themselves. But in a situation where the terms and conditions concerning government take and/or state participation, differ from the one licence to the other the government will be keenly interested in these issues. Such a situation may occur if the respective participating licences have been granted under different licensing rounds involving different conditions.

Note. The existence side by side of two licences, containing different terms and conditions in the matter of government take and/or state participation, is a quite common phenomenon in countries having within their jurisdiction a mature extractive petroleum industry. During the build-up of the petroleum exploration and extraction activities, the conditions, upon which licences were granted, were subject to continuous change in response to changing commercial circumstances (changes in the level of the international price of crude oil, inflation, etc.) and in response to changing operational circumstances (resulting from moving operations to a more demanding and technically more difficult areas, e.g., deeper waters or less accessible terrain). It may then happen that a licence in respect of an area adjacent to the area of an existing licence is awarded on 'new' or 'improved' conditions, depending on the then prevailing circumstances. This adjacent area may in fact be a surrendered area, i.e., an area that originally formed part of the area of a licence granted at an earlier stage. On the North Sea continental shelf whether belonging to the United Kingdom, the Netherlands or Norway the chances are high that due to the practice followed in said countries of organizing successive licensing rounds adjoining licences may happen to contain different conditions depending on the particular licensing round under which the licence concerned was granted.

5.3.1.10 Obligatory Payments

The licensee is or may be obliged to make payments to the State other than payments required to be made under the applicable rules of the corporation tax or any special petroleum income tax. Said payments consist of royalty, annual surface rentals, cash bonuses, net, gross or excess profit shares.

5.3.1.11 Royalty

The royalty is a payment expressed as a percentage of the oil and/or natural gas volumes that are produced by the licensee and is payable in cash or kind. The royalty must be paid to the owner of the petroleum in situ and has only to be paid to the State if the State is such an owner which happens to be the case in almost all States. If exceptionally the landowner is the owner of the petroleum in situ the royalty has to be paid to him. The royalty is the most basic and oldest known petroleum tax. Royalty is a deductible item for the purposes of the income tax. Important elements in the calculation of the royalty are:

(i) the rate expressed at a percentage of the value chosen as the base value;
(ii) the place, where the value is assessed; and
(iii) the base value.

As a rule, a theoretical arm's length market value is taken as the base value. If the wellhead is selected as the place of assessment of the value, all capital and operating costs involved in treating, handling and transporting the oil or natural gas production from the wellhead to the market place may be deducted from the theoretical arm's length market value.

5.3.1.12 Area Rentals

An area rental is a levy that is expressed as an annual amount per square km. Usually the amount increases the longer the licence is kept.

5.3.1.13 Cash Bonuses

Cash bonuses are cash amounts to be paid to the State on certain occasions, e.g., on the day the licence is granted, or the contract is signed (signature bonus) or when the production reaches a certain level (production bonus). The payment of these cash bonuses belong to the particular conditions of the licence.

The cash bonus may also be the (huge) cash amount that the applicant has offered to pay to the State for the licence in the context of a procedure in which licences are put up for competitive bidding.

Most cash bonuses are not allowed as a deductible item for the income tax.

5.3.1.14 State Net Profit Share

The state net profit share is a levy that amounts to a percentage share of the base value. As a rule, a theoretical arm's length market value is taken as the base value. If the wellhead is selected as the place of assessment of the value, all capital and operating costs involved in treating, handling and transporting the oil or natural net profits. The net profit is the profit or income remaining after payment of income tax. The rules of calculation of the state net profit share will be laid down in regulations or in the licence. Similarly as the payment of royalty and area rentals the payment of this share will be allowed as a deductible item for the income tax.

Note. A well known example of a net profit share is the net profit share that may be demanded in the bidding procedure for competitive leases to be granted with respect to the US outer continental shelf under the terms of the US Outer Continental Shelf Lands Act of 7 August 1953.

5.3.1.15 State Gross Profit Share

The state gross profit share is a levy that amounts to a percentage share of the gross profit. The gross profit is the income before payment of income tax, i.e., gross proceeds less operating costs and depreciation charges. The sharing of the gross profit is an alternative for imposing a special petroleum income tax. The rules of calculation of the gross profit and the state's percentage share thereof will be laid down in the regulations or in the licence. The gross profit as defined does not have to be identical to taxable income as defined for the purposes of the income tax. As a matter of fact the sharing of the gross profit allows for flexibility and adjustment of the fiscal burden when needed in the light of changing circumstances, for instance by changing the State's percentage share and/or by introducing special cost allowances.

Note. The concept of the sharing of the gross profits under a licensing regime can be traced back to the 50/50 profit sharing principle that in December 1950 was introduced in the Saudi Arabian petroleum agreement of 25 May 1933. The other Middle East petroleum agreements of the 1930s were similarly revised. Said principle meant that the total sum of income tax and all other payments to the State including royalty should not exceed fifty per cent of the gross profits of the concessionaire concerned. It was subsequently also incorporated in the state participation agreements that for new ventures were concluded in the 1950s and the 1960s in Iran and Egypt.

5.3.1.16 State Share of Excess Profits

The state share of excess profits is a levy that is calculated on a unit basis as a percentage share of the difference between a certain base price, which is indexed for inflation, and the actually realized sales price. The cash payment is in itself a deductible item in calculating and determining taxable income under the common income tax.

Note. A well-known example is the US windfall profit tax, which was in effect from 1980 to 1988.

5.3.1.17 State Participation

Under the applicable petroleum legislation, the government may have been authorized to demand as a condition for the grant of a licence that the State itself or a wholly state owned enterprise designated for the purpose becomes a participant in such a licence and in the venture that is based on the same. Such demand for state participation in the licence and the venture based thereon may be fulfilled by the applicant by accepting the State or the designated state enterprise as a co-licensee and establishing a joint venture on the basis of an agreement of cooperation, also known as a joint operating agreement, or since the State or a state enterprise is participating, a state participation agreement. If the State is directly participating in the licence, as is for instance the case in Norway, a wholly state-owned enterprise may be designated to represent the State in the licence and to manage the State's interest. State or designated state enterprise is hereinafter referred to as the state participant. As far as the holding of the licence is concerned there are two other options: instead of having the licence shared between the state participant and the applicant the licence could be awarded either solely to the state participant or solely to the applicant, everything else (meaning the state participating agreement) remaining the same. If these options are chosen, the consequence will be that the participant (state participant or applicant as the case may be) to whom the licence has been awarded will be solely responsible for exercising the rights and fulfilling the obligations granted or imposed by the licence (e.g., fulfilling the work commitments, timely applying for renewals and extensions and making submissions for getting approvals, etc.).

However this may, the modalities of the state participation which are to be found in the state participation agreement, concern:

(i) the percentage share state participation (whether fixed for the duration of the agreement or made dependent on factors such as production levels, etc.;

(ii) the timing of the state participant's financial participation becoming effective: i.e., a choice has to be made between the state participant's financial contributions (in accordance with its percentage share participation) to start either from the very beginning of the authorized operations, i.e., at the start of the exploration phase, or not before the first commercial discovery has been made.

If the latter option is chosen it has to be settled whether or not the state participant has to reimburse its percentage share participation of the successful exploration expenditure that led to the commercial discovery.

A text of the required agreement may be provided as part of the model licence, if available. Otherwise such agreement is subject to the approval of the competent authority.

Note. Apart from incorporating the modalities of state participation, the contents of a state participation agreement are generally very similar to those of the usual joint operating agreement concluded between oil companies from the private or public sector (see Chapter 11).

The percentage share state participation may be fixed in law or regulation as a standard condition of the licence or state participation agreement or may be negotiated in the context of a bidding round, and so become a particular condition of the licence concerned.

Note 1. The concept of state participation within the framework of a licensing regime originally derived from the Iranian Petroleum Law of 29 July 1957. This Law authorized the government represented in this matter by the National Iranian Oil Company (NIOC) to enter into petroleum agreements with foreign entities with respect to acreage outside the territory of the Consortium. The intended agreement envisaged the establishment of a joint company having NIOC and the foreign entity as shareholders. In such joint company NIOC's shareholding interest should be no less than thirty per cent. A non-incorporated joint venture was also possible. On the basis of this law three offshore agreements were concluded in 1957 and 1958 inter alia one with Agip Mineraria, an affiliate of Italian state-owned oil company ENI (see Chapter 6). In May 1960 the Egyptian government announced that with respect to petroleum operations in the Western Desert it would opt for fifty per cent state participation accompanied by the formation of an Egyptian joint stock company. In 1963 and 1964 the government signed a series of participation agreements, known as the Western Desert participation agreements, in which these principles were incorporated. In the Western Desert agreements the exclusive licence concerned is held by the state participant, the Egyptian General Petroleum Corporation (EGPC), and the foreign participant in equal undivided interest shares (see Chapter 6).

Note 2. In States where the EC Directive of 30 May 1994 (see Chapter 6) is applicable, the influence and power of intervention of the state participant in the management of the operations has been restricted in the sense that any vote by the state participant shall be based exclusively on transparent, objective and non-discriminatory principles, and shall not prevent the management decisions of its co-licensee(s) from being based on normal commercial principles. Furthermore, the state participant shall not be party to information nor exercise any voting rights on decisions regarding sources of procurement for its co-licensee(s).

5.3.1.18 Settlement of Disputes

Since a licence creates a direct relationship between the licensee and the government granting the licence in its capacity as representative of the State, any dispute between the licensee and the government should as a rule be dealt with in accordance with the rules of public administrative law and, in case no amicable resolution of the conflict appears possible, be finally decided by a court of administrative law.

Nonetheless, petroleum legislation of several European States (such as the United Kingdom and Denmark) provide for arbitration in case of a dispute between a licensee and the competent authority regarding the licence in general or regarding a specific aspect of the licence (for instance a refusal to approve a work programme or a development plan). (*Note.* To refer a dispute of this kind to arbitration instead of following the procedures of administrative law and bring the matter before the competent Court or Tribunal has many advantages: because such a dispute usually has a high technical content arbitration allows the direct involvement of experts and leads to faster and more practical results.)

Also the Nigerian Petroleum Act of 1969 (Article 41 of the First Schedule to the Act) provides for arbitration as a means for settling disputes between government (the Minister) and licensee or lessee as follows:

> 41. If any question or dispute arises in connection with any licence or lease to which this Schedule applies between the Minister and the licensee or lessee

> (including a question or dispute as to the payment of any fee, rent or royalty), the question or dispute shall be settled by arbitration unless it relates to a matter expressly excluded from arbitration or expressed to be at the discretion of the Minister.

Note 1. No details are provided on how this arbitration has to proceed. Presumably the rules of procedure of the Nigerian Arbitration Act as amended from time to time have to be followed.

Note 2. An early example of a provision for arbitration in the context of a licensing regime is offered by the petroleum agreements which were made with the governments of the Arab Gulf States in the 1930s. A typical article reads:

> Disputes arising between the parties during the validity of the agreement concerning the interpretation or execution of the agreement or anything contained therein or the rights or liabilities of either party thereunder must, failing agreement to settle such disputes in any other way, be referred to arbitration by two arbitrators and a referee, each of them to be appointed by the British Political Resident in the Persian Gulf, if so needed.

The petroleum agreement concluded with the Sheikh of Kuwait on 23 December 1934 offers in addition the means for settling a dispute in connection with obligatory payments such as royalty and other payments. The article concerned reads:

> If at any time during the validity of the agreement any dispute arises regarding the accuracy of the accounts of the concessionaire in connection with the amounts of the royalty and of other payments due to the Head of State under the agreement, the Head of State has the right to appoint, in consultation with the British government, a registered firm of auditors to examine the books of the company on behalf of the Head of State in Kuwait and or in London as he may consider necessary. All expenditure incurred in connection with such auditing shall be paid by the Head of State. The concessionaire is obliged to provide the registered firm of auditors appointed by the Head of State the necessary facilities to enable them to check the books and registers of the concessionaire and to render every assistance to enable the auditors to thoroughly examine such accounts and must in every way assist them to safeguard the interests of the Head of State.

5.3.1.19 Stability of Terms

In the ME petroleum agreements of the 1930s provision was made for guaranteeing the stability of terms of the agreement. It was agreed that no alteration shall be made to its terms by either the Head of State or the concessionaire except in the event of the Head of State and the concessionaire jointly agreeing that it is desirable in the interests of both parties to make certain alterations, deletions or additions to the agreement.

Note 1. A classic example of the care taken to guarantee stability of terms is offered by the Libyan Petroleum Law of 1955. This petroleum law, which was the first example of a modern petroleum law appearing in the Northern Africa/Middle East region, laid down the standard conditions for concessions. Clause 16 of the standard conditions (after the amendment of the Petroleum Law in 1966) reads as follows:

> The government of Libya will take all steps necessary to ensure that the concessionaire enjoys all the rights conferred by this concession. The contractually rights expressly created by this concession shall not be altered except by mutual consent of the parties. This concession shall throughout the period of its validity be construed in accordance with the Petroleum Law and the Regulations in force on the date of execution of the agreement of amendment by which this paragraph (2) was incorporated into this concession agreement. Any amendment to or repeal of such Regulations shall not effect the contractual rights of the concessionaire without its consent.

This Clause 16 was supported and strengthened by the standard Clause 28 which, after the amendment of the Petroleum Law in 1966, reads as follows:

> This concession shall be governed by and interpreted in accordance with the principles of the law in Libya common to the principles of international law and, in the absence of such common principles, then by and in accordance with the general principles of law, including such of those principles as may have been applied by international tribunals.

Note 2. After the founding of the Organization of Petroleum Exporting Countries (OPEC) on 1 September 1960, the Member States and oil exporting non-Member States, such as Oman, Malaysia and Brunei, introduced in the concessions under their jurisdiction a clause which became known as the 'most-favoured-nation' clause (see Chapter 6). By virtue of this clause which obliged the concessionaire to consult with his government about altering the terms of his concession, if an increase in benefits to governments in the Middle East should generally to be received by them, all concession containing such a clause were linked together as far as the level of the government take was concerned. Obviously, the presence of such a clause in a concession is only relevant in as much as such concession comprise the conditions that determine 'the government take'. If the government take is to a large extent determined by fiscal legislation outside the control of the petroleum legislation or the concession based on this legislation, the incorporation of a 'most-favoured-nation' clause would serve no purpose.

Note 3. A more recent example of a so-called economic stabilization guarantee can be found in the Model Joint Operation Agreement that in October 1993 was made available by the Republic of Bulgaria for promoting the second licensing round offshore Black Sea.

The relevant articles of this Model Agreement are Articles 34.5, 34.6, 34.7 and 34.8.

Article 34.5 reads:

> It is hereby agreed that if the promulgated laws of Bulgaria were, subsequent to the signing of this agreement, to be added to or changed in such a way as to materially diminish or expand the licensee's rights or diminish or expand the licensee's obligations hereunder, the parties shall be bound to readjust the

> licence or this agreement so as to restore the licensee's rights and obligations as existing on the effective date.

Article 34.6 reads:

> In the event (I) that any new laws, rules or regulations are promulgated in Bulgaria or (ii) that any change to the promulgated laws, rules or regulations of Bulgaria, occurs subsequent to the effective date (which laws, rules or regulations may relate to the conduct of petroleum activities in Bulgaria, to the environment or to ancillary matters) which materially restricts, divests, limits or expands any rights or benefits accruing to the licensee or which materially increases or decreases the licensee's obligations under this agreement or under any laws, rules or regulations of Bulgaria as in existence on the effective date hereof, the licensee or the Bulgarian Committee of Geology and Mineral Resources (CGMR) may, at any time thereafter, so notify the other in writing.

Article 34.7 reads:

> Within 30 days from receipt of such notice, CGMR and the licensee shall meet to negotiate in good faith and agree upon the modification which need to be made to the terms of this agreement or any agreement related hereto, to restore the licensee's economic rights and benefits hereunder to close to the maximum level to what they would have been had such change not occurred, or upon such other remedy as they agree may be appropriate. In the event that the parties are unable to agree within 120 days after the licensee's notice to CGMR (I) whether an issuance or a material change to the promulgated laws, rules or regulations of Bulgaria, has occurred subsequent to the effective date which materially restricts, divests, limits or expands any rights or benefits accruing to the licensee or which materially increases or decreases the licensee's obligations as aforesaid, or (ii) upon the modifications which may be needed to the Agreement or any such related agreement or upon such other remedy as may be required, then either party may refer the matter or matters in dispute for arbitration under Article 31 (in accordance to which article the matter shall be submitted for final determination to an internationally recognised independent expert appointed by the parties or if the latter fail to agree on this by the International Chamber Of Commerce Center for Technical Expertise, Paris, France).

Article 34.8 reads:

> The licensee shall not invoke the provisions of articles 34.5, 34.6 and 34.7 in instances where the new or changed laws, rule or regulation (I) merely require the licensee to do what the licensee, at the time of such invocation, was doing or ought to have been doing in accordance with good international petroleum industry practice, or (ii) taken as a whole, and after considering any offsetting benefits, may reasonably be expected to have a material adverse impact upon

> licensee's financial position in Bulgaria that is less than 5 per cent annually of net earnings.

The aforesaid Article 34.6 is also applicable for the purpose of stabilizing taxation. As stated in Article 10.2 of thc Model Agreement taxes shall be paid by the licensee based on the licensee's net profit. In the second paragraph of the preceding Article 10.1 it is stipulated that:

> should any Bulgarian laws be introduced, modified or repealed after the effective date of this agreement which materially improves or damages the economic position of the parties, Article 34.6 shall be applied so as to restore the economic position of all parties.

Under petroleum legislation adopted in Western States the status of licences granted in accordance with a framework petroleum law and regulations issued under its authority is 'saved' in the event new regulations are made which amend or even completely replace the previous regulations including the model licence attached thereto, if any. In principle, the new conditions only apply to licences which are granted after the effective date of the new regulations except for rules, conditions and permits pertaining to working methods or to safety, health or environmental matters. The latter will be made applicable subject to the proper transitional provisions to already existing or new licences alike. If the petroleum law itself is replaced existing licences will be upheld to the extent the transitional articles of the new provide therefore. Anyway the particular conditions of the licence are always upheld.

Note. A modern example of protection against any unilateral change of established conditions is provided by the German Federal Mining Law of 13 August 1980. Article 16 of this Law, covering the main aspects of exploration and production licences, states in its third paragraph that to change or modify any imposed condition after the licence had been granted is permissible, provided such change or modification is commercially acceptable to the licence holder, technically feasible and is serving the purposes of the licence.

5.3.2 Particular Conditions

Any licence document will contain conditions that are particular to the licence concerned. Following conditions are involved:

5.3.2.1 Description of the Licensed Area

The description is usually scheduled to the licence. Where a geometrical grid is applicable the area of the licence is described in terms of grid units, otherwise in terms of coordinates and accompanied by a map.

5.3.2.2 Cash Bonuses to Be Paid on Certain Occasions

The licensee may be obliged to pay a cash bonus on certain occasion as described in the licence document: e.g., a cash bonus on the occasion of the granting of the licence (signature bonus) and/or when the production has reached a certain level (production bonus).

5.3.2.3 Duration of the Licence

The duration of the licence (including extensions) is dependent on the type of licence involved and determined on the basis of the particular circumstances affecting the licence.

5.3.2.4 Obligatory Exploration Work Programme

No licence is granted without imposing on the licensee an obligation to carry out some exploration work in the form of geophysical surveys or drilling holes. The scope, contents and timing of the work programme depends entirely on the circumstances of the case.

5.4 PRINCIPAL PROVISIONS OF THE PRODUCTION SHARING CONTRACT

5.4.1 STANDARD CONDITIONS

Preliminary Note. The following description of the standard conditions of the production sharing contract is based on a review of several recently published Model Production Sharing Contracts.

5.4.1.1 The Parties

Production sharing contracts are made and entered into on the one part by a state entity, hereinafter 'state party' and on the other by an entity consisting of one or more, mostly foreign oil companies from the private or public sector, hereinafter 'contractor'. This state party may be the State itself, a state authority representing the State, such as a ministry of petroleum or a petroleum directorate established within such a ministry, or, as is more commonly the case, a duly authorized state owned and state controlled oil enterprise.

Note. In some host States (e.g., Egypt, Syria), where the contracts are made with a duly authorized state oil enterprise, the government itself is also a party to the contract and will co-sign the same. As a co-signatory of the contract the government will decide typical governmental matters such as the approval of assignments of interest in the contract and the termination or cancellation of the contract.

5.4.1.2 Integration into a Specific Exclusive Licence

Where (already) a licensing regime is in existence, production sharing contracts, if so desired by the government, may be integrated into specific exclusive licences relating to the intended petroleum operations and area. This integration can be accomplished either by granting the required licence to the state party who then in turn concludes a production sharing contract with a contractor for the purpose of conducting the authorized petroleum operations (exploration and/or development and production) within and with respect to the licensed area, or by granting upon concluding the contract the required exclusive licence to the contractor. In both cases the contract area coincides with the licensed area. Furthermore, the duration of the contract, possible renewals or extensions thereof and area relinquishments are all determined by the corresponding conditions of the (underlying) licence and in case the licensing regime involved prescribes separate licences for the exploration and production stage, the state party or contractor, as the case may be, will have to apply for and need to be granted a production licence in the event a discovery of petroleum is declared commercial in order to be able to develop and produce this discovery.

5.4.1.3 Contract Area

The area of the production sharing contract is described in the contract. In practice it is described in terms of blocks with reference to a geographical grid system. The area of a contract may comprise one or more of such blocks.

Alternatively, if the production sharing contract is integrated into a specific exclusive licence, the area of the contract coincides with the area of the underlying licence, which in turn is also described in terms of blocks with reference to a geographical grid system.

5.4.1.4 Duration of the Contract and Area Relinquishment

Distinction is made between an exploration phase and a development/production phase. The exploration phase is usually divided in sub-periods of two or three years. Parts of the original contract area may have to be surrendered at the end of each sub-period. The contract will be terminated if within the duration of the exploration phase no declaration of commercial discovery is made. At the end of the total exploration period the area not converted into the areas containing commercial discoveries and earmarked and demarcated for development and production must be surrendered. Development work and production operations in converted areas are bound to specific development and production periods and extensions thereof.

The overall duration of a contract may be fixed at a specified number of years (e.g., thirty years) but should comprise the successive holding/development and/or production periods (plus possible extensions) and, if so needed, be extended accordingly.

Alternatively, if the production sharing contract is integrated into an exclusive licence that is held either by the state party or by the contractor, as the case may be, the conditions of the licence regarding duration and renewal or extension of periods, area relinquishment and conversion from exploration licence into a production licence, if applicable, apply. The holder of the licence is responsible for taking action in these matters when required in accordance with the conditions of the licence in order to prevent that rights are lost or obligations remain unfulfilled with all the consequences thereof for the status of the licence.

5.4.1.5 Obligatory Exploration Work

With respect to each sub-period of the exploration phase the contractor will be obliged to carry out a minimum exploration work programme and/or to spend a specified minimum amount of US dollars on exploration work.

5.4.1.6 Declaration of Commercial Discovery (General)

After making a discovery, the contractor is obliged to carry out within a specified period a programme of evaluation work in order to decide whether the discovery constitutes a commercial discovery. Only discoveries declared by the contractor to be commercial and as such approved by the state party may be developed and produced in accordance with the rules of the contract, in which case the whole area capable of production from the discovery is turned into a development area. Any such area is considered automatically terminated if within a specified number of years from the date of commercial discovery in said area no oil shall have been lifted as part of a regular program of lifting in accordance with the lifting schedule.

Note. If the production sharing contract is integrated into a licensing regime and if this regime prescribes separate licences for the exploration and production stage, the state party or contractor, as the case may be, will need to be granted a production licence in order to be able to develop and produce the discovery declared to be commercial.

5.4.1.7 Declaration of Commercial Discovery (Non-associated Gas)

Where a discovery of non-associated gas is made and difficulties are foreseen with the availability of markets and/or with the long distance transportation of the natural gas production to available markets contracts may provide for a postponement of a declaration of commercial discovery and allow the contractor a holding period of say five years with possible extensions. During such a holding period the contractor is not obliged to undertake development work with respect to the particular gas discovery, but is expected to search for other non-associated gas deposits in the vicinity in order to aggregate the discovery with possible other discoveries and to investigate, jointly with the state party, the commercial options of development and transportation.

Note. Such joint investigation into the commercial possibilities of development of a non-associated gas discovery may lead to the conclusion that the discovery would become a commercial discovery if terms and conditions would be applicable more favourable to that effect than the ones stipulated in the contract. Interestingly, such possibility is envisaged in the 1997 Model Production Sharing Agreement made available Turkmenistan. Article 19.5 of that Model Agreement reads:

> In the event that a discovery of non-associated gas may become a commercial discovery under terms and conditions more favourable, to that effect, than the ones stipulated herein (including such terms as the cost recovery petroleum and the production sharing provisions under Article 13 above), at the initiative of either party, the Competent Body [i.e., the State Body to which the Cabinet of Ministers delegate powers to issue licences and enter into contracts with contractors] and contractor shall meet in order to discuss and agree on revised terms, such as would allow contractor to develop commercially such non-associated gas discovery and to achieve a net economic return thereon equivalent to the one which would have derived from a comparable crude oil accumulation. If agreement is reached on such revised terms and conditions within six (6) months from start up of discussions, then such terms will be incorporated into a supplemental agreement which shall be an integral part hereof. Failing agreement within the aforesaid time limit, the non-associated gas discovery in question shall be treated as a non-commercial discovery for all purposes herein.

By contrast, in the 2006 Model Production Sharing Agreement made available in Angola a more simple approach is adopted, whereby the state party is given the option to keep the gas discovery for itself. Article 29.3 of that Model reads:

> If non-associated natural gas is discovered within the contract area then Sonangol will have the exclusive right to appraise, develop and produce said gas for its own account and risk. If Sonangol so determines and if agreed with contractor within a term determined by Sonangol, the discovery of non-associated natural gas shall be developed jointly by Sonangol or one of its affiliates and contractor.

5.4.1.8 Operatorship, Programmes and Budgets

In principle, the contractor acts as the operator and as such is responsible for and in charge of carrying out the operations in accordance with the programmes and budgets as approved under the rules of the contract. If the contractor is a consortium of companies, one of them shall be appointed as operator. During the exploration phase of the contract the contractor is responsible for fulfilling the minimum exploration work programme and to spend the minimum expenditure commitments.

Note. Departures from this basic pattern occur. According to contracts applicable in Egypt and Syria a joint non-profit, non-asset owning operating company (having the contractor and the state party as its shareholders) is established as soon as for the first time a commercial discovery has been declared. This joint operating company is responsible for carrying out the development and production operations and any further exploration work and is funded by the contractor. It can also be organized differently: the state party may have been given the option to become operator with respect to any individual discovery or field, leaving the contractor in charge of the operations until the state party exercises this option. This approach is followed in contracts applicable in the People's Republic of China. The state party (in this case the Chinese National Offshore Oil Corporation (CNOOC)) has the option to take over the role of operator with respect to any field that has been brought into production. Such an option may be exercised as from the moment all development costs incurred with respect to the field concerned have been recovered from cost oil generated by the field in question.

The operator, be it a jointly owned operating company or the contractor itself, is responsible for preparing the annual work programmes and corresponding budgets and any specific plan for evaluation of a discovery and development thereof if commercially warranted. After the proposed programmes and budgets have been approved by the supervisory body, the operator is responsible for the execution of these programmes and plans within the agreed budgets and in accordance with the rules of the contract and any applicable petroleum legislation and relevant legislation. In absence of relevant legislation the necessary operational rules shall be provided by the contract. If so, these regulatory provisions are formulated as duties of the operator. Important duties include the operator's obligation to carry out petroleum operations in a proper and workmanlike manner and in accordance with good oil field practice, to prevent causing damage to the environment, pay for the costs associated with clean-up of any pollution caused, to take appropriate abandonment measures upon termination of the contract, and to fulfil any of the national submitting to the state party copies of all relevant data and information obtained in the course of the operations, such as geological, geophysical and any type of well data, providing all personnel, financial and technical resources required for a proper performance of the operations. An important task of the operator consists in organizing the procedures to be followed in the awarding of major contracts.

Note. The operator shall be required to award contracts exceeding certain financial limits exclusively by international tender. Exceptions may have to be made with respect to the acquisition of local services and locally manufactured goods.

5.4.1.9 Supervisory Body

Operations are supervised by the state party directly or exceptionally, by a management committee, consisting of representatives of state party and contractor.

The task of the supervisory body (the state party or a management committee) consists in reviewing and approving, in the form as prepared and submitted by the operator:

(i) any declaration of commercial discovery and any plans to develop such a discovery;

(ii) annual programmes of work and corresponding budgets; and
(iii) the award of major contracts.

The supervisory body has the right to propose changes to the submitted work programmes.

Note. The requirement that work programmes, budgets, a declaration of commercial discovery, the corresponding development plan and the award of major contracts shall obtain the approval of the supervisory body is an essential condition of a contract because, as stipulated therein, the operator is only allowed to execute work programmes and development plans and incur costs in connection therewith if these programmes and plans and corresponding budgets and contracts have been so approved. To reinforce this requirement, it is generally expressly stated that no compensation for or reimbursement of costs are given if these costs have been incurred in doing work that was not part of an approved work programme or have been incurred on the basis of a non-approved contract.

Within limits the contractor has the right to change an approved programme in order to adapt the latter to changed circumstances. Similarly the supervisory body may, within pre-determined limits, change an approved programme.

5.4.1.10 Cooperation between State Party and Contractor

The cooperation between state party and its contractor may be formalized by setting up either a jointly owned operating company, in which the parties become shareholder, or a management committee, in which both parties are represented. Such a management committee may also act as a supervisory body (see above). When a management committee is meant to be an organizational instrument for consultation its task consists in discussing and reviewing any proposal prepared by the operator before they are submitted to the supervisory body, in this case the state party, for approval.

5.4.1.11 Operations for the Sole Account and Risk of the State Party

A contract may provide for the state party having an option to conduct for its own risk and account certain exploration, appraisal or development drilling operations which are deemed by the contractor not to be economically justified when applying his own economic criteria and yardsticks including the development of any discovery finally not declared to be commercial. If the state party would be interested in a sole risk and account operation, it may have to use the services and facilities of the contractor for which the latter will be compensated. To provide for such operations the contract will adopt the corresponding rules incorporated in the standard joint operating agreements made between commercial oil enterprises (see Chapter 11).

5.4.1.12 Bearing the Costs and Risks of the Operations

In principle, the costs and expenditures incurred in the authorized operations as well as any losses and risks derived therefrom shall be borne by the contractor and the state party shall not be responsible to bear or repay any of the aforesaid costs and expenditures. But contracts may provide for an option on the part of the state party to share in the funding of the development phase of any particular field as soon as the decision is taken to develop such field. Without exception, however, bearing the costs and risks of exploration work is the sole and exclusive responsibility of the contractor.

Note. In countries with a weak, not readily convertible currency the foreign contractor is generally obliged to provide the funds for the operations in convertible foreign currency (usually US dollars are stipulated). To alleviate the burden of this currency obligation a foreign contractor is allowed to keep the proceeds of its export sales abroad, at least to the extent that any local payment obligations, such as payment of local taxes and locally incurred costs, have been fulfilled. In this way the foreign contractor is exempted from the generally applicable obligation (imposed on domestic exporters) to surrender foreign currency proceeds to the central bank of the host State (in exchange for local currency). Furthermore, the contractor will be allowed to hold bank accounts in convertible foreign currency, to repatriate any surplus foreign exchange and to keep the books of account in US dollars.

5.4.1.13 Allocation of Oil and/or Natural Gas Production

Available oil and/or natural gas production is divided between the State, the state party and the contractor in accordance with the rules of the individual contract. Depending on the contract and the circumstance of the case rules are provided either for the allocation of the oil production alone whereby the allocation of non-associated natural gas production is referred to future agreement between state party and contractor, or for the allocation of both oil and natural gas production. In the latter case oil and non-associated natural gas are allocated in accordance with the same rules and methods but the absolute numbers and percentage figures that will be applied in each case will be different.

Actual figures i.e., the figures to be put in the Model (for both oil and natural gas) are arrived at after negotiations with applicants and depend on competition, the particular circumstances of the contract and the oil and/or non-associated gas prospectivity of the contract area. The expectations regarding future oil prices play a lesser role because of the built-in flexibility in the sharing of the benefits of the contract, which should ensure that the state party and its government will be the main beneficiaries of a high oil price level.

Preliminary Note. In what follows hereinafter only the rules applicable to the allocation of oil production will be described, bearing in mind that those meant for non-associated gas will *mutatis mutandis* be the same or be agreed at a later date. Nonetheless, in contract areas where oil as well as non-associated gas are developed and produced the allocation systems may be linked together as will be explained below.

The allocation procedure starts with setting aside a portion destined for the State by way of payment in kind of a royalty. (*Note*. The required payment of a royalty is really a feature of the licensing system.) This portion is referred to as royalty oil. Applicable percentages vary between fifteen and ten per cent, usually made dependent on the level of daily production. When made dependent on production volumes a series of successive production slices or tranches may be specified where a higher production tax percentage is applicable to each tranche higher in the series. The portion remaining after deduction of royalty oil is put at the disposal of the contractor for recovery of the costs and expenditures incurred by him in carrying out the petroleum operations. This portion is referred to as cost recovery oil.

Production remaining after royalty oil and after cost recovery oil is referred to as profit oil. This portion is shared in complex proportions between the state party and the contractor.

5.4.1.14 Determination of Cost Recovery Oil

Cost recovery oil is bound to a maximum of X per cent of all oil produced and saved in the contract area after deduction of royalty oil. But instead of using a single percentage figure X, the cost recovery process can be made more complex by dividing production into successive tranches with each tranche higher in the series being assigned a lower cost recovery percentage. In contrast with this, some contracts do not recognize or impose any volume restriction on the production available for cost recovery, in other words total production, after deduction of royalty oil, is put at the disposal of the contractor for cost recovery (as further specified).

If in any given year costs or expenditures are less than the maximum value of the cost recovery oil the difference shall become part of and included in the contract area profit oil. In the event that in any given year costs or expenditures exceed the value of cost recovery crude oil for such year the excess shall be carried forward for recovery in the next succeeding year or years, but in no case after the termination of the contract.

In contract area where oil as well as non-associated gas is being developed and/or produced the allocation systems may be linked in as much as costs incurred with respect to natural gas development operations are allowed to be recovered from oil production and vice versa. This is a powerful incentive to develop non-associated gas reservoirs.

A crucial element of the cost recovery process is the determination of the monetary value of the oil (or the natural gas as the case may be) that is put at the disposal of the contractor for cost recovery. If the price used for the determination of this value is higher than the objective market value, the cost recovery remains incomplete to the disadvantage of the contractor. To meet this problem a contract will include a description of the market value of the oil or natural gas that is expected to be produced and stipulate to use this value for the calculation of the monetary value of the available cost production.

Not all costs incurred qualify for recovery out of production. Distinction is made between 'recoverable costs' and 'non-recoverable costs'.

Note. Non-recoverable costs are costs made by the contractor in deviation from the rules of the contract. In particular those costs are excluded from recovery which have been incurred in non-authorized work, are the result of gross negligence or wilful misconduct, have been made on the basis of a contract that has been awarded in contravention of the prescribed tender procedures, or which do not appear in an approved budget. Most contracts contain an extensive listing of cost categories that do not qualify for cost recovery and/or contain criteria on the basis of which it shall be judged whether or not costs are acceptable for cost recovery.

Recoverable costs are categorized as exploration costs, capital expenditure, operating costs and abandonment costs. Each cost category, except abandonment costs, may be recovered in accordance with the rules of amortization and depreciation as applicable to such category. Abandonment costs arise at the end of the contract. These costs must be estimated and a quarterly provision should be allowed as part of the recoverable costs.

Cost recovery operates on a quarterly basis, i.e., recognized costs incurred in any quarter are recovered from the cost production available in any such quarter. In the case of under-recovery recognized costs are carried forward to the next quarter.

In order to provide incentives meant to help contractors faced with high cost operations, special cost related deductions (investment credits) may be granted. Such special deductions or credits are expressed as a percentage of the capital expenditure incurred and form part of the recoverable costs.

5.4.1.15 The Sharing of Profit Oil

Profit oil is divided between the state party and the contractor in many different ways, the one more complex than the other. A common method uses a series of successive daily production tranches and assigns to each tranche higher in the series a sharing proportion that is more favourable for the state party. The sharing itself concerns the profit oil part of each individual tranche, i.e., each tranche has first to be divided in royalty oil, cost recovery oil and profit oil before the sharing percentages can be applied. A closely related method consists in dividing cumulative oil production in successive tranches and assigning to each tranche a different profit oil split. Over time, as cumulative production increases, the sharing of the profit oil will become more favourable to the state party.

In the 1997 Model Production Sharing Agreement (Turkmenistan) the sharing proportions applied to profit petroleum (i.e., oil and natural gas) produced in any particular year are determined on the basis of the ratio of the cumulative contract revenues earned by contractor from cost recovery petroleum and his share of profit petroleum from inception of such earnings up to the end of the year preceding the year in question to the cumulative contract expenses incurred by contractor from inception thereof to the end of the year preceding the year in question (which ratio is referred to as the 'R factor'). The sharing of profit petroleum between the state party and the contractor is made a function of a stepwise increasing value of the R factor in the sense that the higher the value of the R factor the more the sharing proportions change in favour of the state party. The numbers to be filled in are subject to negotiation.

A closely related method of profit production sharing is contained in the 2006 Model Production Sharing Agreement (Angola) and envisages the sharing of profit oil in the current quarter according to the internal rate of return of the income tax paid cash flow (cash flow after income tax) realized by the contractor at the end of the preceding quarter (see Chapter 7).

5.4.1.16 Petroleum Income Tax

Generally, the contractor is subject to payment of the tax imposed on the income realized by him under the contract. The tax to be paid is a percentage of contractor's taxable income, which consists of the revenues realized by the sale of the cost recovery oil and of contractor's share of the profit oil less operating costs, depreciation charges and possible tax incentives. Sometimes the state party has also to pay income tax, sometimes the state party is exempted therefrom.

Note. In Indonesia, where the concept of production sharing had originated, the production sharing mechanism was initially meant to replace or absorb the payment of income tax and to exhaust the methods to determine government take. This brought US and European oil companies whose affiliates had entered into such non-income tax paying production sharing contracts in a difficult position in relation to the fiscal authorities in their home countries. Such affiliates were supposed to have been subject to or even to have actually paid income tax in the country where they were operating. The Indonesian government and governments of other host countries that were confronted with these fiscal problems arising in the home countries of the parent companies of their contractors have reacted positively by changing the contracts to the effect that the contractors were made subject to and had to tax on the income generated under the contract, while at the same time integrating this tax into the production sharing mechanism in such a way that the overall government take was not affected (see Chapters 7 and 8).

5.4.1.17 Lifting and Disposition of Oil

It is the right and obligation of each of the Parties to separately take at the delivery point (in accordance with a lifting schedule) its respective oil entitlements as determined under the contract. For the contractor this means that he must lift the cost recovery oil and his share of the profit oil.

5.4.1.18 Disposition of Natural Gas

Only after the possibilities of development and marketing, either locally or abroad, have been studied and evaluated arrangements for the disposition of natural gas, whether it concerns associated or non-associated gas, can be made. For non-associated natural gas export markets have to be sought and developed to which the gas will be transported (by long-distance pipeline or LNG tanker) and delivered under the terms of long-term gas sales agreements. For these purposes the state party and the contractor will each dedicate its contractual share of the gas production. For associated natural gas joint schemes may be developed such as a scheme for selling the gas produced on the domestic gas market, or a scheme for building near the oil field where the gas is coming from a gas plant to extract liquid petroleum gases (LPG, a mixture of propane and butane), or energy intensive industrial plant.

Note. In recent Egyptian production sharing contracts the disposal of natural gas production has received more attention. Priority shall be given to meeting the requirements of the domestic gas market. If gas is to be disposed of in this market, the state oil company, viz. the Egyptian General Petroleum Corporation (EGPC), shall be the buyer under a long-term gas sales agreement. The parties shall consult together whether to build a gas plant to extract liquid petroleum gases (LPG, defined in the contract as a mixture of propane and butane) from gas produced under the contract. The costs of such a gas plant are recoverable from cost recovery petroleum, unless the Minister agrees to accelerated recovery. EGPC has the option to pay for the gas purchased under any long-term gas sales agreement and for the LPG bought from the gas plant either in cash or in crude oil. If in the absence of a long-term gas sales agreement, natural gas or LPG is exported the contractor may retain abroad the proceeds from exporting the contractor's share thereof.

5.4.1.19 Purchase of State Party's Share of the Profit Oil

Where it is foreseen that the state party faces problems with disposing of its share of the profit oil, the contract concerned may require the contractor to purchase any part thereof under normal commercial terms and conditions in the international oil industry and at the contractual market price in force at the time the oil is lifted. Alternatively, the state party concerned could have been given the right to require its contractor to use his best efforts to arrange the sale of all or part of the state party's share of production on terms and conditions to be agreed between the parties and subject to contractor's right to recover all reasonable costs associated with such marketing and disposal.

5.4.1.20 Domestic Supply Obligation

Generally, the contractor is allowed freely to export his cost recovery oil and his share of the profit oil, albeit this freedom under many contracts is restricted by the imposition of an obligation in one form or another to supply the domestic market (other than in case of national emergency). Usually the domestic supply obligation is translated into a right of the state party to purchase at market price (as defined in the contract) up to fifty per cent of contractor's oil entitlement. This right is qualified by stipulating that it will be exercised only after the state party has fully utilized royalty oil and its share of profit oil and then pro rata among all producers in the country.

Note. Under the terms of the 2006 Model Production Sharing Agreement (Angola) the state party, i.e., Sonangol shall have the right to buy from the contractor oil from the contract area equivalent to the value of the petroleum income tax due by the contractor to the Ministry of Finance, i.e., fifty per cent of contractor's share of profit oil.

5.4.1.21 Joint Exploitation of Straddling Petroleum Reservoirs

Production sharing contracts usually contain provisions in which attention is paid to compulsory joint exploitation of straddling petroleum accumulations, although strictly speaking on both sides of the common boundary only the state party, in its capacity as licensee, is involved and as such will dictate the arrangement to be made for the development of the straddling reservoir. The contractors on either side of the boundary, not being licensees, are not directly involved in this matter but have to follow the instructions given by the state party.

This situation is illustrated by Article 26 'Unitization' of the 1997 Model Production Sharing Agreement (Turkmenistan) reading as follows:

> 26.1 The Competent Body may require that the development and production of any field within the contract area be carried out jointly by Contractor with another entity designated by the Competent Body if:
>
> > (a) such field extends into an area which is subject to a separate petroleum exploration and production agreement with such other entity;
> > (b) such field is not a commercial discovery unless the development and production of such field is combined with the development and production of another field lying within an area which is subject to a separate petroleum and exploration and production agreement with such other entity;
> > (c) another field lying within an area which is subject to a separate petroleum exploration and production agreement with such other entity is not a commercial discovery unless development and production of such other field is combined with the development and production of a field within the contract area.
>
> 26.2 If the Competent Body requires the joint development and production of a field pursuant to paragraph 26.1 the Contractor shall collaborate with the other entities identified by the Competent Body in preparing a joint operating program for the development and production of the field(s) involved, and promptly submit such program to the Competent body. Lack of acting by the Contractor and the other entities in this respect may cause the Competent Body to prepare or cause to be prepared a joint operating program to achieve the efficient development of the field(s) involved.

Note. Competent Body is described in the Petroleum Law as the State Body to which the Cabinet of Ministers delegates powers to issue licences and enter into contracts with contractors.

5.4.1.22 Stability of the Contractor's Economic Position

Apart from the efficiency of the cost recovery mechanism the contractor's economic position under a production sharing contract is determined by the percentage and modalities of the petroleum income tax, the percentage of royalty oil, and the state party's percentage share of profit production. The question arises as to what guarantees are available with respect to the stability of these conditions. The volume of royalty oil and the state party's share of profit production are contractually fixed contributions that have been agreed and accepted against the background of the payments due under the income tax law(s) as prevailing at the time the contract was negotiated and entered into. These contractual payments are assumed to remain unchanged during the validity of the contract. The petroleum income tax paying contractor has however no guarantee that the petroleum income tax will similarly remain unchanged during that period.

In the context of a production sharing contract, there are two methods available to achieve the desired stability of the government take and safeguard the economic position of the contractor:

(i) to freeze the rate of the applicable petroleum income tax for the duration of the contract; or
(ii) to adjust the contractual payments, such as the sharing of profit oil, in order to counter-balance any negative effects on contractor's economic position caused by a future adverse change of the applicable petroleum income tax.

If the first method is adopted, the contract will provide that petroleum income tax will be levied at the rate existing on the effective date of the contract. Future changes to the rate of the income tax shall then not be applicable to income derived from the contract. Such protection against future changes of the tax rate can only be offered by a contract that is a law in itself (e.g., contracts as applicable in Egypt and Syria) or by a contract made with the State itself.

Note. An example of the method of contractual adjustment is provided by production sharing contracts adopted in Malaysia. Contractors there are subject to the Petroleum (Income Tax) Act 1967, as amended with effect from 1 April 1975. It is stated in the contracts that if income tax were changed or new taxes specially meant for the petroleum industry were introduced in such a manner that the burden of taxation were significantly increased or decreased, Petronas (the national oil company acting as the state party) and contractor shall mutually agree on an arrangement which is directed at restoring the original economic position of the Parties. A stability provision in this form works both ways: if the burden of taxation were decreased, e.g., if a lower rate for the income tax were introduced, the contractor is, in principle, not supposed to benefit from such relief. Obviously, where the contractor is exempted from paying income tax, or where the income tax, for which he is liable, is paid on his behalf by the state party concerned, the rate of the tax or any change thereof in the future has no consequences for the economic position of such contractor. Another example of a stability-of-terms provision is offered by Article 17 of the Russian Federal Law No. 225 'On Production Sharing Agreements' of 31 December 1995 (PSA Law). This article (in translated form) reads as follows:

> The terms and conditions of the production sharing contract shall remain in force during the whole period of the contract. Amendments are not only permitted with the agreement of the parties but also at the request of only one of the parties in cases foreseen under civil law. Amendments by agreement between the parties shall become effective under the same procedure as the initial contract. If during the period of the contract under the legislation of the Russian Federation, or of the regional administrations or under statutory acts of local self-government bodies normative acts are established resulting in a worsening of the investor's commercial results under the contract, the contract can be amended to ensure that the investor obtains the commercial results which he could have obtained if the legislation existing at the time of signing the contract had applied. (*Note*. Normative acts mean regulations, investor means contractor.)

5.4.1.23 State Participation

Although production sharing contracts may be considered to be an alternative to obligatory state participation under a licensing regime, elements of state participation have nevertheless been introduced in the contracts. As a matter of fact, state participation has manifested itself in three different forms, namely in the form of an option (or even commitment) for the state party to contribute to and share the development expenditure (as exemplified by contracts applicable in Tanzania, Libya and the People's Republic of China); in the form of establishing a jointly owned operating company (as exemplified by contracts applicable in Egypt and Syria); or by the sharing on a proportional basis of the rights and obligations of the contractor, the state party (or a subsidiary of the state party or a national enterprise designated by the state party), assuming in this manner the position of a co-contractor (as exemplified by contracts applicable in Indonesia and Malaysia).

5.4.1.24 Recruitment and Training of Personnel

Contracts will generally contain clauses requiring contractors to employ by preference nationals of the host country in all categories and functions. Apart from that contractor will be obliged to train all its national personnel for the purpose of improving their knowledge and professional qualification measured against the level of knowledge and professional qualification held by Contractor's foreign personnel. The required training of personnel may also extend to and involve the training of the state party's personnel and may take place within the host country or by preference even abroad using the facilities of any one of contractor's foreign affiliated companies. The cost of the training may or may not be partially or wholly recoverable from cost oil. If not, it ranks as a payment of a bonus by contractor.

5.4.1.25 Transfer of Technology

The required transfer of technology may be part of the training of nationals employed by the contractor or by the state party but is more specifically meant for the benefit of the state party. It is expected that transfer of technology will enable the state party to perform more efficiently the functions assigned to it under the petroleum law as well as under the contract, such as acting as the supervisory body, participating in the activities of a management committee or the board of directors of a joint operating company.

Note. The contracts applicable in the People's Republic of China are characterized by the emphasis put on transfer of technology. In these contracts transfer of technology is understood to mean and involve the application and use of the contractor's or rather the contractor's affiliated companies' technology to and for the benefit of the authorized petroleum operations. The use and application of the contractor's technology includes giving training to personnel in handling such technology.

5.4.1.26 Use of National Products and Services

The contract will contain clauses that impose on the contractor obligations and commitments concerning the use of locally manufactured goods, use of local services and local sub-contractors

Note. An example for such type of clause may serve Article 27 of the Angolan Law No. 10 on Petroleum Activities of 12 November 2004. This Law and its Article 27 is applicable inter alia to production sharing contracts entered into by Sonangol. Said Article 27 reads as follows:

> 27.1 Licensees, the National Concessionaire (i.e., Sonangol) and its associates, and any other entities which cooperate with them in carrying out petroleum operations shall:
>
> > (a) acquire materials, equipment, machinery and consumer goods of national production, of the same or approximately the same quality and which are available for sale and delivery in due time, at prices which are no more than 10 per cent higher than the imported items including transportation and insurance costs and custom charges due;
> > (b) contract local services providers, to the extent to which the services they provide are similar to those available on the international market and their prices, when subject to the same tax charges, are no more than 10 percent higher than the prices charged by foreign contractors for similar services.

5.4.1.27 Financial Guarantees

Host governments will insist on being given financial guarantees (say bank guarantees expressed in US dollars) with respect to the fulfilment of the obligatory exploration work programme and/or the exploration expenditure commitment. Sometimes host governments insist on a general performance guarantee to be issued by the parent company of the contractor guaranteeing the fulfilment by contractor of its obligations under the contract.

5.4.1.28 Liabilities

The contractor will be required to indemnify and keep the State and state party harmless against all claims from third parties for loss or damage to property or injury to persons caused by or resulting from the contractor's operations. Exceptions are loss, damage and injury caused by any action of the personnel of the state party or the State (or, alternatively, caused by the proven negligence and wilful default of the state party). In some contracts it is additionally provided that the state party concerned shall hold the contractor harmless from any claim brought against the contractor by the employees of the state party for their personal injury or damage to personal property due to the fulfilment or non-fulfilment of the contract.

If the contractor consists of more than one oil company each such contractor-company is jointly and severally liable for the performance of all obligations of contractor.

5.4.1.29 Assignment of Interests

Distinction must be made between external assignments and internal assignments. If the contractor consists of more than one oil company and anyone of the contractor-companies wishes to assign its interest in the contract to the others such an assignment qualifies as an internal assignment. If the contractor or anyone contractor-company wishes to assign its interest in the contract to a third party including an affiliate, such an assignment is an external assignment. Generally, any assignment to third party, that is not an affiliate, needs the prior written consent of the state party or of the government in case the government is a co-signatory of the contract or if the contract had been subject to the latter's approval.

5.4.1.30 Title to Assets

In principle, all movable and fixed assets acquired by the contractor for the authorized operations become the property of the state party. The contractor retains the right to the full and exclusive use of the assets for the operations and shall not have pay for such use. Apart from this, the contractor shall of course have the right to recover the cost of their acquisition out of cost recovery oil in accordance with the rules of the contract. It is generally understood that contractor, so long as such assets are used exclusively for petroleum operations, shall be liable to keep the same in good repair and working order, normal wear and tear excepted. Contracts may differ in the matter of stipulating the exact moment when the assets are acquired by the state party, but in all contracts the ultimate moment of transfer of ownership comes when the contract is terminated.

In the light of this kind of provisions, contractors strive to rent land and buildings and to make use of subcontractors for drilling and other operations, and, where possible, to lease rather than to purchase equipment and installations.

Note. Under the terms of the contracts used in Egypt land becomes the property of EGPC as soon as purchased. Title to fixed and movable assets is transferred from contractor to EGPC gradually as the costs thereof are being recovered by the contractor out of cost production, provided that a complete transfer takes place at the termination of the contract. Other contracts provide for an immediate as opposed to a gradual transfer of ownership. The moment chosen for such transfer may be the moment when the total costs of the assets have been recovered by the contractor or when such assets have been fully depreciated for petroleum income tax purposes (as stipulated in the 1997 Turkmenistan Model Agreement) or the moment when purchased in the host country or, if purchased abroad, when landed in the host country (as envisaged in the 2006 Angolan Model Agreement).

5.4.1.31 Confidentiality

The contractor is required to keep data and information obtained pursuant to the contract (and which are therefore the property of the state party) confidential and is not allowed to disclose such data and information to third parties that are not affiliates, without the prior written consent of the state party, except:

(i) when disclosure is needed to permit such a third party to render services under the contract;
(ii) in connection with arranging financing;
(iii) in connection with an assignment of interest; or
(iv) when required under applicable legislation.

5.4.1.32 Settlement of Disputes

By preference, disputes arising between the state party and its contractor that can not be resolved by agreement between the two parties are settled by international commercial arbitration in accordance with the rules of the chosen arbitration procedure. The arbitral award is declared binding on the parties. Applicable law (for the purposes of the arbitration) is specified. Usually, English law is chosen, but exceptionally the law of the host State has to be accepted. The contract may contain its own arbitration procedure, prescribing the number of arbitrators (usually three); the way in which these persons shall be selected (their qualifications and nationalities) and shall be appointed (requiring the designation of an international and neutral appointing authority, e.g., the Secretary General of the Permanent Court of Arbitration, for the appointment of a single arbitrator or for that of the third and presiding arbitrator); their task and duties, etc.

Most contracts however refer to established international arbitration procedures, such as the rules of arbitration of respectively the International Chamber of Commerce, the UN Commission on International Trade Law (UNCITRAL) or the

International Centre for Settlement of Investment Disputes (ICSID), the latter having been established under the Convention on the Settlement of Investment Disputes between States and nationals of other States. The use of the services of the ICSID is very appropriate in the context of production sharing contracts, because the latter may be considered as a foreign investment agreement and in this manner to fall entirely within the scope and purposes of the aforesaid Convention. Nonetheless, arbitration pursuant to UNCITRAL Rules has become more and more popular.

In some host States the prevailing production sharing contract recognizes situations in which there may arise a dispute between the Parties (viz. the state party and its contractor) on the one side and the government on the other. Such disputes must be submitted to the regular courts of law of the host State concerned.

Note. The recognition that such a dispute may arise introduces an additional complication for the contractor in a situation in which the latter has a complaint or grievance in connection with the behaviour of the state party (e.g., the state party refuses to give its approval in a case where its approval is required under the terms of the contract and the contractor cannot proceed without it). If the contractor brings its complaint to the attention of the state party, the latter may inform its contractor that in acting the way it did and about which the contractor complained it has simply followed orders from the government. If the state party refers to the government it will be impossible for the contractor to submit its dispute with the state party to international arbitration, the only available course of action would be to submit the complaint for final settlement to a court of administrative law which brings the case of the contractor within the orbit of the law of the host State.

5.4.1.33 Amendment

It will be stipulated in the contract that the latter can only be amended by mutual agreement of the Parties but where the original contract had been subject to the approval of the government any agreed amendment is likewise subject to the approval of the government. If a contract has obtained the status of a law, as is the case when a law is required to authorize the government (or the competent minister) to sign the contract, any amendment needs to be approved by law and will involve (again) the legislature.

5.4.1.34 Termination

The contract will be terminated in accordance with its rules. These rules envisage:

(i) voluntary termination, i.e., the right of the contractor to terminate the contract at any time (without being entitled to any compensation and subject to take the appropriate abandonment measures);
(ii) obligatory termination if no commercial discovery of petroleum has been made during the exploration phase of the contract; and
(iii) termination by the government if the contractor is in breach of any material provision of the contract or if the contractor is declared to be bankrupt.

5.4.1.35 Applicable Law

Without exceptions, the contract is governed and construed in accordance with the laws of the host State. To these laws should also belong any relevant provision of an international treaty or agreement to which the host State is a party (e.g., bilateral agreements for the protection of reciprocal investments).

Note. Where some doubts exist regarding the stage of development of the legal system of the host country, the following statement is made:

> This Agreement shall be governed by, interpreted and construed in accordance with the laws of Turkmenistan and, as applicable, the principles of international law and the decisions of international tribunals and international treaties to which Turkmenistan is a Party (see 1997 Model Production Sharing Agreement for Petroleum Exploration and Production in Turkmenistan).

5.4.1.36 Approval and Ratification

With a few exceptions, production sharing contracts entered into by the state oil enterprise need the approval of the host government before they become effective. In some instances (e.g., Egypt and Syria), where the government itself is a Party to the contract in conjunction with the state party, a special law has to be passed authorizing the minister responsible for petroleum affairs to sign the contract, thereby approving and ratifying the contract and giving it full force and effect of law.

5.4.2 Particular Conditions

The particular conditions of the contract comprise: the description of the contract area; signature and production bonuses to be paid, exploration work commitments and the actual (percentage) rates and factors to be used in the production sharing mechanism.

5.5 CONTENTS OF RISK SERVICE CONTRACTS

A risk service contract is usually offered to the petroleum industry by governments that are looking for one reason or another for an alternative to the production sharing contract. Like the latter the risk service contract is made with the state entity that is authorized by law to carry out petroleum operations in the country and to enter into contract with other entities in order to be able to perform its task efficiently.

The main features of a risk service contract can be described as follows: the contractor is responsible for funding all capital expenditure and operating costs and for the management of the authorized operations. In case a commercial discovery is made and subsequently developed leading to the production of petroleum the contractor is entitled to recover (with interest) its funding and to receive in addition thereto compensation in cash (not in production). Compensation may be a single US dollar amount per unit of production (a fee per barrel) plus all kinds of other cash amounts by way of incentives. The compensation may be also expressed as the market value of a certain portion of the production to be received during a fixed number of years. In the latter case the contract could give the contractor the option to convert the amount of the compensation during the period of its validity into the oil and/or gas that is produced under the contract. For this purpose the oil and/or gas to be so purchased by the contractor is priced at market value. Obviously, whether or not the contractor will make use of such production-for-cash opportunity depends entirely on the terms agreed for the purchase/sale agreement. However if the risk-contractor is making use of this option and decides to convert its cash reward into production, there is not much difference with a production sharing contract.

Risk service contracts include:

(i) contracts covering special petroleum development projects, where no preceding exploration and/or appraisal work is involved: for instance a contract for the development of a discovery or the rehabilitation of a mismanaged oil or gas field; a contract to apply secondary or tertiary production methods in order to boost production in an otherwise depleted field; etc. Each such contract is unique in the sense that it is tailor-made for a specific project;

(ii) association contracts as applicable in Colombia. These contracts, which are made between a foreign oil company and Ecopetrol, the state oil enterprise and sole licence holder, are in essence and substance very similar to a state participation agreement, in the sense that investments and production are proportionally shared between contractor and Ecopetrol; and

(iii) risk service contracts as practised in several other Latin-American countries, such as Ecuador, Peru, Brazil and Argentina.

The origin of the concept of the risk service contract can be found in the Iranian Petroleum Act of 1974. This Act entered into force a few years before the outbreak of the revolution in December 1978. The Act was meant to regulate petroleum operations outside the Consortium area and followed the line set out before by the risk service contract made in 1967 between NIOC and an affiliate of ERAP, the French state-owned oil company. The Act authorized NIOC to conclude risk service contracts with foreign oil companies. The total area for a single contract was maximum eight thousand sq. km.

Note. The main features of the Iranian risk service contract can be described as follows. The duration of the exploration phase should not exceed five years. The development phase starts when a commercial discovery is made and ends when commercial production starts. At that moment the field operations are taken over by NIOC and the contract is terminated with respect to the field concerned. The contractor is responsible for supplying all the funds required for exploration and also, if so required by NIOC, the funds for development. In respect of each developed field NIOC must conclude with the contractor a sale agreement not exceeding fifteen years and with respect of no more than fifty per cent of field production. For the oil sold under such sale agreement the contractor has to pay the prevailing FOB price less in each year one tenth of the funds supplies by the contractor for the exploration and possibly (i.e., if so requested by NIOC) development during the first ten years of production; and less a discount of maximum five per cent for as long the sale agreement endures. If the contractor supplies funds for development he is entitled to receive interest on the amounts supplied, the maximum to be settled in the contract. Actual contracts (all for onshore exploration acreage and entered into by (affiliates of) US and West-European oil companies) provided for the following conditions: the contractor had to pay for all development expenditure bearing interest not exceeding nine per cent. The discounts on the market prices varied between three per cent and five per cent. Sales quantities varied from thirty per cent to fifty per cent of field production. In all instances the duration of the sale agreement was fifteen years.

The principles of the Iranian style risk service contract were adopted and practiced in Brazil, albeit in a more sophisticated form.

Note 1. Brazil is one more example of a developing host country that, because of lack of financial and technical capabilities, decided to switch from state control to risk service contracts in an attempt to engage foreign oil companies on politically acceptable terms. In October 1975 the Brazilian government formally announced that involvement of foreign oil companies in the exploration and exploitation of petroleum within the country would be welcome. In 1979 a model contract was published, which was designed to be concluded between Petroleos Brasileiro SA (Petrobras) in the latter's capacity as sole entity authorized to undertake petroleum operations in the country and sole owner of petroleum in situ and a foreign oil company acting as contractor.

Note 2. The main features of the Brazilian contract are as follows. The contract is terminated at the end of the exploration period (three plus one year) if, within this period, no commercial petroleum discovery has been made, or at the moment Petrobras with respect to any field brought into production has fulfilled its payment obligations in relation to the contractor. The contractor is responsible for funding all exploration and development work. A minimum exploration work programme, on which a minimum amount of US dollars should be spent, must be carried out. The development period starts as soon as the parties reach agreement on a development plan prepared by the contractor with regard to any discovery considered to be commercial. Petrobras has the right of developing a non-commercial discovery for its own account and risk. The development of a field ends at the moment all production installations are handed over to Petrobras. As from that moment all production operations are taken over by Petrobras. As from the same moment Petrobras is responsible for paying all costs and must start with the reimbursement of past exploration and development expenditure as paid by the contractor with respect to the field concerned, the payment of a remuneration and the payment of interest on outstanding development expenditure. All these payments are quarterly payments calculated per field. The sum thereof may not exceed the difference between revenues received by Petrobras from the sale of field production and field production costs. Reimbursement of costs takes place in equal quarterly instalments (e.g., twenty instalments). The remuneration consists in quarterly cash payments in US dollars representing a complex fraction (dependent on field production performance) of the market value of the quarterly oil or gas production.

The remuneration is payable over a certain number of years to be specified in the contract (e.g., fifteen years). The contractor is entitled, during the period that the remuneration is payable, to buy crude oil from Petrobras at a price equal to the market value, provided the value of the oil so bought does not exceed the monetary amount of the remuneration. Before commercial production commences, the parties must agree on a production profile for the field concerned. If new investments have to be made, the contractor is entitled, but not obliged, to take these new investments for its account. The same rules regarding reimbursement, remuneration and interest apply, except that the remuneration period is not extended. If the contractor is not prepared to fund the new expenditure, and if Petrobras should decide to take the expenditure for its account, the remuneration due to the contractor will be fixed on the production profile as determined on the day preceding the start of the use of any new production facilities. The contractor is subject to an income withholding tax with respect to its remuneration and the interests receipts. This imposition is accompanied by a stability guarantee from the side of Petrobras in the sense that any increase in taxation will be for the account of Petrobras if such increase is not generally applicable to all legal persons domiciled in the home country of (the parent company of) the contractor and generating income in Brazil.

Chapter 6

Past and Current Licensing Regimes

6.1 US

6.1.1 Introduction

In the year 1859, the first oil well in history was drilled near the town of Titusville, Pennsylvania in the US. This event marked the establishment of oil production as an industrial undertaking. It marked also the beginning of the development of a legal system that would be suitable for regulating this new mining industry.

Under the US legal system for mining mineral resources as it existed at the time of the first drilling for oil and as it exists still today these mineral resources to which oil was reckoned to belong are the property of the public or private owner of the land on which or beneath which the mineral resource is located. In essence this legal system was no more than property law extending the ownership of land (downwards) to the mineral resources on or beneath the surface. On account of this property right mineral acquisition rights could be obtained from the public or private owner of the land concerned. Any person interested to acquire and produce or mine a desired mineral resource could apply for and obtain a right (lease) permitting such acquisition. Usually the only obligation imposed by the lessor on the lessee was to pay a royalty in cash to the former.

The most important public owner of land and the minerals belonging thereto was and is the federal government. In accordance with the Constitution federal lands (and their mineral resources) are put under the legislative power of Congress.

Note. Congress was assigned in the Constitution the power to 'to dispose of and make all needful Rules and Regulations respecting the Territory or other Property of the US'.

In pursuance of this constitutional legislative power the Congress has enacted inter alia acts dealing with the federal mineral property. In fact, the Constitution must be considered to be a rather elementary mining law forming the fundament of all mining and petroleum legislation for federal lands, including (much in the future) the continental shelf.

The first drilling for oil in 1859 introduced a completely different method of mining, because in this case the mineral resource to be mined consists of a mixture of liquid and gaseous substances which is stored in the pores and fissures of sedimentary rock and which has to be extracted therefrom by drilling holes from the surface into the rock.

Lawmakers, regulators, legal theorists, landowners and oil producers in the US (and Canada) were forced to reconcile this new physical reality in mining operations with the prevailing principles of mineral property rights. Landowners or the persons to which the former had granted a lease, took the view that on their (leased) property they were entitled to drill holes into the petroleum-bearing reservoirs and to take possession of and to sell (to the refining companies) any petroleum that through the borehole reached the well heads located on their property.

Note. As soon as the commercial importance of oil (as feedstock for the distillation of kerosene) was understood and the success of the drilling process was recognized, landowners or their lessees started to drill as many wells on their own or leased land as possible in order to achieve the greatest production in the shortest time possible. As discussed in Chapter 1 the idea of drilling holes to produce oil (and associated gas) came from looking at the way in which at the time salt water was produced. This was done by drilling wells into salt water bearing rock layers (aquifers) of almost limitless extension, providing an inexhaustible supply of salt water. Initially, oil producers expected the same results with respect to oil and were of the opinion that the quantity that could be produced was only dependent on the number of wells that could be accommodated within their plot of land. By preference wells were drilled along the boundaries of the land, offsetting any wells drilled on the adjacent plot. Soon however it became clear that there was a significant difference between the production of oil and that of salt water. Whereas there was indeed an unlimited supply of saltwater, putting no restriction on the number of saltwater producing wells this appeared not to be the case with oil. The unrestrained drilling of wells into a reservoir and the resulting unrestrained extraction of oil therefrom confronted the producers with wells that suddenly started to produce water instead of oil. What happened was that, instead of pushing the oil forwards in the direction of the wells, formation water bypassed the oil and flowed directly into the wells, leaving the oil behind in the pores of the reservoir. Since the early oil producers were operating on and from small-sized plots of land, they were extracting with their wells oil from a reservoir from which also other producers operating from the adjacent land were extracting oil. Such reservoir was known as a 'common pool'. Remarkably, although every owner realized that the oil in the subsoil was always moving to the nearest well, no claims were made with respect to oil that originally, i.e., before the extraction from the common pool had started, resided beneath the owner's land but due to its migration surfaced in wells drilled on adjacent property. The problem was addressed in a practical manner. Each owner of land that covered part of a common pool took measures to protect its interest in the common pool by drilling as many wells as could physically be accommodated within the confines of his area. Only wells drilled vertically were considered appropriate. Wells that would deviate and end up beneath the adjacent territory were not allowed.

From the point of legal theory the migratory character of petroleum was correctly seen as posing a problem causing the development of theories on the nature of the ownership of petroleum. To define the problem, distinction was made between the ownership of petroleum in situ and the ownership of petroleum as it emerged at the wellhead at the surface. Initially, there were two views. According to the most radical view petroleum could not be owned by the owner of the land overlying the petroleum deposit concerned similarly as solid minerals were owned precisely because of the migratory nature of such petroleum which sets it apart from the latter: in other words the normal rules of land-related ownership applicable to solid minerals cannot be applied to petroleum. In this view, ownership of petroleum can only be obtained by and when taking possession of petroleum, i.e., only after succeeding in extracting petroleum and conveying the same to the wellhead at the surface. The other view is that as long as petroleum remains in situ there is no difference with solid minerals, i.e., petroleum in situ belongs to the owner of the land overlying the petroleum deposit but that such owner will lose his ownership as soon as the petroleum moving in the direction of wells that were drilled on the adjacent plot actually leaves (the subsoil of) his plot. It is the latter view that apparently has prevailed. (Source: Canadian Oil and Gas Law, Part II, § 31 Theories of Ownership in the US.)

US petroleum legislation has developed along two lines: first, there is legislation emanating from the federal government and made with respect to petroleum operations on land owned by the Federation or on land in respect of which the federal government has retained (after the transfer of ownership of the land) the title to minerals. Secondly, there is legislation emanating from the respective Union States. This state legislation regulates petroleum operations carried out on land owned by the State concerned and those carried out on privately-owned lands which have retained their mineral property rights.

The federally owned territory consists for the largest part of the public domain which is made up of areas that were transferred by the thirteen original Union States to the federal government or were ceded to the federal government by foreign governments. The remainder of the federally owned territory is composed of 'acquired lands', which are lands acquired by the US by private contract; 'Indian lands'; 'reservations', such as the Naval Petroleum Reserves and national parks and other protected environmentally sensitive areas; and the 'outer continental shelf lands', which are areas seaward of the boundaries of the respective coastal Union States and extending to the limits established by the 1982 Convention on the Law of the Sea.

Note. The 1982 Convention on the Law of the Sea which became effective on 16 November 1994 has still not been ratified by the US. Nonetheless as far as limits in the sea are concerned the provisions of the Convention apply (see further Chapter 8).

6.1.2 Federal Petroleum Legislation for the Public Domain

The first federal legislation specifically meant for petroleum operations to be carried out in federally-owned land belonging to the public domain was the Mineral Leasing Act of 1920 as amended from time to time. The Act introduced and provides for a system of leasing: any person, provided he is a citizen of the US or an association of such citizens or a corporation organized under the laws of the US or under those of a Union State, wishing to explore for petroleum involving the drilling of exploratory wells or wishing to exploit a petroleum deposit needs to apply for and acquire a lease. Leases are issued by the Department of the Interior (Bureau of Land Management [BLM]). Lands within the known geological structure of a producing oil or gas field (KGS-lands) are leased on a competitive basis (competitive leases). Other lands are leased on a non-competitive basis (non-competitive leases).

Competitive leases are granted through and by means of an auction. KGS-lands are divided into leasing blocks not exceeding 640 acres each and are offered for lease at a royalty and rental specified in the notice in which the sale of leases is announced to the qualified applicant offering the highest cash bonus in the auction. Competitive leases have an initial term of five years, but will be extended as long as commercial production takes place.

Lands lying outside a KGS are subject to non-competitive leases which remain valid for an initial period often years and thereafter for as long there is commercial production. The size of the leased area may not exceed 2,560 acres or be less than 640 acres. If a non-KGS land has never before been leased, the lease is granted to the first qualified applicant. Otherwise such lands are leased on the basis of a kind of lottery organized by the BLM between and among qualified applicants.

As long as the petroleum is not extracted from the reservoir it is and remains owned by the federal government. Once produced however the holder of the lease becomes the owner.

The Act authorizes lessees to agree with others to develop and operate a common pool under a cooperative or unit plan of development. Since the amendment of the Act in 1960 compulsory unitization may be imposed by the Secretary of the Interior. The said Secretary is authorized to provide that leases contain a provision requiring operation under a reasonable unit plan and the Secretary may prescribe such a plan under which the lessee shall operate which shall adequately protect the rights of all lessees involved in the common pool and those of the federal government.

6.1.3 Federal Petroleum Legislation for the Outer Continental Shelf Lands

6.1.3.1 Introduction

In 1945 US President Harry S. Truman proclaimed that the Government of the US regarded the natural resources of the subsoil and the seabed of the continental shelf beneath the high seas, but contiguous to the coasts of the US, as appertaining to the US, subject to its jurisdiction and control. In the accompanying press release the continental shelf is described as submerged land which is contiguous to the continent and which is covered by no more than one hundred fathoms (six hundred feet) of water.

Note. The Truman Proclamation and its consequences for the development of the international law of the sea are discussed in Chapter 8.

On the basis of the presidential Proclamation the Act of 7 August 1953 was enacted, known as the Outer Continental Shelf Lands Act. The Act has substantially been modernized by means of amendment by the Act of 18 September 1978. On that occasion the rule was introduced that development and production shall only be undertaken in accordance with a plan to be approved by the Secretary.

6.1.3.2 Congressional Declaration

Insofar as relevant for the interpretation of the Act and the understanding of the underlying government petroleum policy, the following elements of the accompanying Congressional declaration of policy; jurisdiction; construction (after its amendment in 1978) should be mentioned.

It is inter alia declared to be the policy of the US that:

(i) the subsoil and seabed of the outer continental shelf appertain to the US and are subject to its jurisdiction, control, and power of disposition as provided in the Act;
(ii) the Act shall be construed in such a manner that the character of the waters above the outer continental shelf as high seas and the right to navigation and fishing therein shall not be affected;
(iii) the outer continental shelf is a vital national resource reserve held by the federal government for the public, which should be made available for expeditious and orderly development, subject to environmental safeguards, in a manner which is consistent with the maintenance of competition and other national needs;
(iv) operations in the outer continental shelf shall be conducted in a safe manner by well-trained personnel using technology, precautions and techniques sufficient to prevent or minimize the likelihood of blowouts, loss of well control, fires, spillages, physical obstruction to other users of the waters or subsoil and seabed, or other occurrences which may cause damage to the environment or to property, or endanger life or health.

In the Act the term 'outer continental shelf is defined as all submerged lands lying seaward and outside of the area of lands that have been assigned to the coastal Union States and of which the subsoil and seabed appertain to the US and are subject to its jurisdiction and control.

Note. Although no specific reference is made to international law, the seaward boundary of the outer continental shelf lands must be considered to be determined by the seaward limits agreed in the context of the 1982 Convention of the Law of the Sea (effective date 16 November 1994). Said limits apply even if a particular coastal (federal) State has not (yet) ratified the Convention as is the case with the US. On the other hand the predecessor of the 1982 Convention, namely the 1958 Convention on the Continental Shelf, became in 1964 effective as law in the US and automatically superseded as from that moment any incompatible terminology in the Act.

6.1.3.3 The System of Leasing and its Administration

The Act provides for a system of leasing. A 'lease' is defined as any form of authorization which is issued by the Secretary under the applicable provisions of the Act and which authorizes exploration for, and development and production of, minerals. The term 'minerals' includes oil, gas, sulphur, geopressured-geothermal and associated resources, and all other minerals, which are authorized by an Act of Congress to be produced from public lands. The term 'Secretary' means the Secretary of the Interior, except that with respect to the administration and grant of leases the term 'Secretary' means the Secretary of Energy.

Note. The functions and regulatory powers with respect to the administration and grant of leases were transferred to the Secretary of Energy by the Act of 4 August 1977, whereby the Congress created the Department of Energy and set up the Federal Energy Regulatory Commission (FERC).

The Act assigns to the Secretary the task and responsibility to administer the leasing system. The Secretary is empowered to prescribe such rules and regulations as may be necessary to carry out the said provisions. The Secretary may at any time prescribe and amend such rules and regulations as he determines to be necessary in order to provide for the prevention of waste and conservation of the natural resources of the outer continental shelf and the protection of correlative rights therein and such rules and regulations shall as of their effective date apply to all operations conducted under a lease. The regulations prescribed by the Secretary may include provisions:

(i) for the suspension or temporary prohibition of any operation, including production, pursuant to any lease (or permit):

(a) at the request of a lessee, in the national interest, to facilitate proper development of a lease or to allow for the construction or negotiation for use of transportation facilities; or
(b) if there is a threat of serious, irreparable, or immediate harm or damage to life (including fish and other aquatic life), to property, to any mineral deposits (in areas leased or not leased), or to the marine, coastal, or human environment, and for the extension of any lease (or permit) affected by suspension or prohibition under (a) or (b) by a period equivalent to the period of such suspension or

prohibition, except that no lease (or permit) shall be so extended when such suspension or prohibition is the result of gross negligence or wilful violation of such lease (or permit), or of the regulations issued with respect to such lease (or permit);

(ii) with respect to cancellation of any lease or permit;
(iii) for the assignment or relinquishment of a lease;
(iv) for the unitization, pooling, and drilling agreements;
(v) for the subsurface storage of oil and natural gas;
(vi) for drilling or easements necessary for exploration, development, and production;
(vii) for the prompt and efficient exploration and development of a lease area; and
(viii) for the compliance with the national ambient air quality standards pursuant to the Clean Air Act.

Whenever the owner of a non-producing lease fails to comply with any of the provisions of the Act, or of the lease, or of the regulations issued under the Act, the lease may be cancelled by the Secretary, subject to the right of judicial review, if the default continues for the period of thirty days after mailing of notice to the lease owner.

Whenever the owner of any producing lease fails to comply with any of the provisions of this Act, of the lease, or of the regulations issued under the Act, the lease may be forfeited and cancelled by an appropriate proceeding in any US district court having jurisdiction under the Act.

Except for field pipelines, any authority for the transportation by pipeline shall require that the pipeline concerned be operated in accordance with competitive principles, such as the principle that the pipeline must provide open and non-discriminatory access to both owner and non-owner shippers and that under circumstances the Federal Energy Regulatory Commission may order an expansion of throughput capacity of a particular pipeline.

The lessee shall produce any oil or natural gas, or both, obtained pursuant to an approved development and production plan at rates consistent with any rule or order issued by the President in accordance with any provision of law. If no such rule or order has been issued, the lessee shall produce such oil or gas, or both, at rates consistent with any regulation promulgated by the Secretary of Energy which is to assure the maximum rate of production which may be sustained without loss of ultimate recovery of oil or gas, or both, under sound engineering and economic principles, and which is safe for the duration of the activity covered by the approved plan.

The Secretary may permit the lessee to vary such rates if he finds that such a variance is necessary.

No lessee shall be permitted to flare natural gas from any well unless the Secretary finds that there is no practicable way to complete production of such gas, or that such flaring is necessary to alleviate a temporary emergency situation or to conduct testing or work-over operations.

6.1.3.4 Grant of Leases by Secretary

The Secretary is authorized to grant to the highest responsible qualified bidder or bidders by competitive bidding, under regulations promulgated in advance, any oil and gas lease on submerged lands of the outer continental shelf which are not covered by prior leases that have been validated and maintained. The bidding shall be by sealed bid and, at the discretion of the Secretary, on the basis of:

(i) cash bonus bid with a royalty at not less than 12.5 per cent fixed by the Secretary in amount or value of the production saved, removed or sold;
(ii) variable royalty bid based on a per cent in amount or value of production saved, removed or sold, with either a fixed work commitment based on dollar amount for exploration or a fixed cash bonus as determined by the Secretary, or both;
(iii) cash bonus bid, or work commitment bid based on a dollar amount for exploration with a fixed cash bonus, and a diminishing or sliding royalty based on such formulae as the Secretary shall determine as equitable to encourage continued production from the lease area as resources diminish, but not less than 12.5 per cent at the beginning of the lease period in amount or value of the production saved, removed or sold;
(iv) cash bonus bid with a fixed share of the net profits of no less than thirty per cent to be derived from the production of oil and gas from the lease area;
(v) fixed cash bonus with the net profit share reserved as the bid variable;
(vi) cash bonus bid with a royalty at no less than 12.5 per cent fixed by the Secretary in amount or value of the production saved, removed or sold and a fixed per cent share of net profits of no less than thirty per cent to be derived from the production of oil and gas from the lease area;
(vii) work commitment bid based on a dollar amount for exploration with a fixed cash bonus and a fixed royalty in amount or value of the production saved, removed or sold; or
(viii) any modifications of the above bidding systems or any other system of bid variables, terms, and conditions which the Secretary determines to be useful to accomplish the purposes and policies of the Act, except that no such bidding system or modification shall have more than one bid variable.

The Secretary may, in order to promote increased production on the lease area, through direct, secondary, or tertiary recovery means, reduce or eliminate any royalty or net profit share set forth in the lease for such area.

The Secretary shall submit any modification or alternative bidding system as referred to in (viii) above to the Senate and the House of Representatives. The Secretary may institute this bidding system unless either the Senate or the House of Representatives passes a resolution of disapproval within thirty days after receipt of the bidding system.

Not later than thirty days before any lease sale the Secretary shall submit to the Congress and publish in the federal register a notice:

(a) identifying any bidding system which will be utilized for such lease sale and the reasons for the utilization of such bidding system; and
(b) designating the lease tracts selected which are to be offered in such sale under the bidding system mentioned in (i) above and the lease tracts selected which are to be offered under any one or more of the bidding systems mentioned in (ii) through (viii) above and the reasons such lease tracts are to be offered under a particular bidding system.

Within six months after the end of each fiscal year, the Secretary of Energy, in consultation with the Secretary of the Interior, shall report to the Congress with respect to the use of various bidding options provided for in the Act.

Following each notice of a proposed lease sale and before acceptance of bids and the issuance of leases based on such bids, the Secretary shall allow the Attorney General, in consultation with the Federal Trade Commission, thirty days to review the results of such lease sale. The Attorney General may, in consultation with the Federal Trade Commission (FTC), conduct such antitrust review on the likely effects the issuance of such leases would have on competition as the Attorney General, in consultation with the FTC, deems appropriate and shall advise the Secretary with respect to such review. The Attorney General, in consultation with the FTC, may make such recommendation to the Secretary, including the non-acceptance of any bid, as may be appropriate to prevent any situation inconsistent with the antitrust laws.

Notice of sale of leases, and the terms of bidding shall be published at least thirty days before the date of sale.

No bid for a lease may be submitted if the Secretary finds, after notice and hearing that the bidder is not meeting due diligence requirements on other leases.

The Secretary may cancel any lease obtained by fraud or misrepresentation.

No lease issued under this Act may be sold, exchanged, assigned, or otherwise transferred, except with the approval of the Secretary. Prior to any such approval, the Secretary shall consult with and give due consideration to the views of the Attorney-General.

6.1.3.5 Terms and Provisions of Oil and Gas Leases

An oil and gas lease issued pursuant to a lease sale shall:

(i) be for a tract consisting of a compact area not exceeding 5,760 acres, as the Secretary may determine, unless the Secretary finds that a larger area is necessary to comprise a reasonable economic production unit;
(ii) be for an initial period of five years or not to exceed ten years where the Secretary finds that such longer period is necessary to encourage exploration and development in areas because of unusually deep water or other

unusually adverse conditions, and as long after such initial period as oil or gas is produced from the area in paying quantities, or drilling or well reworking operations as approved by the Secretary are conducted thereon;

(iii) require the payment of amount or value as determined by one of the bidding systems envisaged by the Act;

(iv) entitle the lessee to explore, develop, and produce the oil ad gas contained within the lease area, conditioned upon due diligence requirements and the approval of the development and production plan required by the Act;

(v) provide for suspension or cancellation of the lease during the initial lease term or thereafter in accordance with the provisions of the Act;

(vi) contain such rental and other provisions as the Secretary may prescribe at the time of offering the area for lease; and

(vii) provide a requirement that the lessee offer twenty per cent of the crude oil, condensate, and natural gas liquids produced on such lease, at the market value and point of delivery applicable to federal royalty oil, to small or independent refiners.

6.1.3.6 Oil and Gas Development and Production

Prior to development and production pursuant to an oil and gas lease in any area of the outer continental shelf other than the Gulf of Mexico the lessee shall submit a development and production plan to the Secretary for approval. A plan shall be accompanied by a statement describing all the facilities and operations other than those on the outer continental shelf, proposed by the lessee and known by him that will be constructed or utilized in the development and production of oil and gas from the lease area. The Secretary shall submit such plan and statement to the Governor of any affected State.

No oil or gas lease may be issued pursuant to this Act in any region of the outer continental shelf, other than the Gulf of Mexico, unless such lease requires that development and production activities be carried out in accordance with a plan which complies with the applicable provisions of the Act. A plan may apply to more than one lease and shall set forth all information on work to be performed and on facilities and operations on the outer continental shelf which are proposed by the lessee or known by him to be directly related to the proposed development and their location; environmental safeguards to be implemented and safety standards to be met; an expected rate of development and production and a time schedule for performance and such other relevant information as the Secretary may by regulation require.

The Secretary may declare the approval of development and production plan in any area of the outer continental shelf to be a major Federal action and if so declared shall determine that procedures under the National Environmental Policy Act of 1969 should commence. If not so declared the Governor of any affected State may submit comments and recommendations.

After reviewing the record of any public hearing held with respect to the approval of a plan pursuant to the National Environmental Policy Act of 1969, or of the comments and recommendations of the Governor concerned, as the case may be, the Secretary shall, within sixty days after the release of the final environmental impact statement, or sixty days after the period allowed for comment and depending on the circumstances of the case, approve, disapprove or require modifications of the plan. Under certain circumstances, depending on the grounds for disapproval, the term of a lease shall be extended after disapproval of the plan and at any time within five years after such disapproval the lessee may reapply for approval of the same or a modified plan and the Secretary shall approve, disapprove, or require modifications of such plan. Upon expiration of the five-year period or in the Secretary's discretion, at an earlier time upon request of a lessee, if the Secretary has not approved a plan, the Secretary shall cancel the lease and the lessee shall be entitled to receive compensation. The Secretary may approve any revision of an approved plan proposed by the lessee if he determines that such revision will lead to greater recovery of oil and natural as, improve the efficiency, safety, and environmental protection of the recovery operation, is the only means available to avoid substantial economic hardship to the lessee, or is otherwise not inconsistent with the provisions of the Act, the extent such revision is consistent with protection of the human, marine, and coastal environments.

Whenever the owner of any lease fails to submit a plan or fails to comply with an approved plan, the lease may be cancelled. Termination of a lease because of failure to comply with an approved plan, including required modifications or revisions, shall not entitle a lessee to any compensation.

If any development and production plan submitted to the Secretary provides for the production and transportation of natural gas, the lessee shall contemporaneously submit to the Federal Energy Regulatory Commission (FERC) that portion of the plan which relates to production of natural gas and the facilities for transportation of natural gas. The Secretary and the FERC shall agree as to which of them shall prepare the required environmental impact statement.

6.1.4 Union State Petroleum Legislation

Union States (States) enact petroleum laws with respect to privately owned land and state owned land on which or beneath which petroleum has accumulated, except in cases where the federal government had been the previous owner of the land and has kept the mineral property for itself before transferring the ownership of the land. Making rules and regulations for state-owned land and its petroleum resources is based on the State being the owner of this resource. Making rules and regulations for privately owned land and its petroleum resources is based on the police power of the State. Within this setting it is any State's priority and duty to ensure that the exploration for, development and production of petroleum seen as an industrial activity are carried out in a safe manner, without causing damage to the environment or endangering the health of the population. But state intervention in petroleum operations has gone further and has extended to the technical methods in which the petroleum resources discovered beneath the land are being exploited by their owner or lessee.

In the early days of the industry exploitation of a petroleum deposit consisted in drilling as many wells as could be accommodated within the space available to the landowner or its lessee. This approach resulted in a poor recovery of petroleum because over-produced wells went prematurely to water and a significant part of the petroleum remained behind in the reservoir. Moreover the driving mechanism (also referred to as reservoir energy) was spoilt and wasted. As a form of self-regulation, the buyers of the oil in those days, i.e., the refining companies also referred to as the oil companies, frustrated by the production fluctuations caused by the wasteful extraction methods adopted by the oil producers, stepped in and tried to acquire as many leases and surrounding producing property as possible. In this way the average size of a lease increased, which allowed to apply more efficient production methods requiring less wells to be drilled. This resulted in a saving of capital and at the same time in the improvement of the ultimate recovery of petroleum. Nevertheless, States were generally not satisfied with the way petroleum resources were exploited by their owners and lessees and in the course of time all of the major producing States enacted conservation laws that allowed intervention in the way private petroleum property was exploited, thus directly affecting the economic interests of the landowner and its lessee.

Note. Such intervention goes far beyond the police power of States. It represents an infringement of the property rights of the owner or its lessee. It appears that the only right left to the owner or its lessee is the right to decide not to exploit the petroleum deposit beneath his land.

The principles of these conservation laws were broadly the same: any waste of oil and gas was prohibited. 'Waste' includes the waste of reservoir energy and the location and spacing and operation of oil and gas wells in a manner in which reduces the ultimate recovery of oil or gas from a particular petroleum deposit. A state authority (a Commissioner of Conservation) is appointed which is authorized to limit and pro-rate the production of oil or natural gas from any field (pool); to regulate the spacing of wells, and to establish drilling units. In the context of state conservation laws a drilling unit is understood to mean the maximum area which may be efficiently and economically drained by one well. The unit is considered to constitute a developed area as long as a well is located thereon which is capable of producing oil and gas in paying quantities. Separate ownerships within such a drilling unit have to be integrated (and are made subject to pooling orders). The owner of each tract is assigned a share of the production of the well draining the unit. If demand for oil or gas (in any particular Union State) exceeds production capacity, the production of any well shall be restricted to the Maximum Efficient Rate (MER) of production, in order to prevent that reservoirs are being damaged by excessive production. The MER is based on the physical characteristics of the producing reservoir. In contrast, if production capacity exceeds demand production limitation will be introduced based on the total production allocated to each Union State. The total applicable to an individual State will be distributed over its individual producers whereby exceptions will be made and exemptions be granted for special cases, such as secondary recovery projects and marginal wells, which may continue to produce at full capacity.

Note 1. Since many years demand for oil in the US far exceeds production capacity. The deficit has to be made up by imports from abroad. Nevertheless there has been a period (first half of 1990) that domestic production, in particular the small producer, was in danger of having to shut down because of the low prices prevailing on the global oil market as a result of over-supply. In 2006 US domestic oil production, including the continental shelf, amounted to 6.9 million b/d, in contrast to a US oil consumption of 20.6 million b/d. Due to the elevated level of the prices on the global oil market, any domestic producer can nowadays produce at full capacity (further discussion in Chapters 2 and 3).

Note 2. The principles of the conservation laws, except for the imposition of a MER, are a thing of the past. Pooling orders and well spacing have been replaced by modern unitization agreements (discussed in Chapter 11).

6.2 THE MIDDLE EAST PETROLEUM AGREEMENTS OF THE 1930s

6.2.1 Background

During the 1930s the governments of seven Middle East Gulf States negotiated and concluded a series of petroleum agreements with major international oil companies. By these agreements (also referred to as 'concessions' as the context may allow), exclusive petroleum rights were granted in respect of large-sized areas, in many instances covering the whole or an important part of the land territory and territorial waters of the Gulf State concerned, and for a very long period (e.g., seventy-five or sixty years). They were further characterized by the absence of mandatory relinquishment of acreage and by the fact that up to their first fiscal revision in 1951 the concessionaires were exempted from paying any taxes and duties, including income taxes, apart from having to pay signature and discovery bonuses, annual or monthly sums and royalties, the latter being tied to minimum annual payments. In no petroleum agreement was provision made for state participation.

Note. Exceptionally, the 1901 D'Arcy concession provided for the payment of sixteen per cent of the annual net profits, whereas the in 1933 revised Anglo-Persian concession required a net profit sharing in addition to the payment of a royalty of four gold shillings per ton. The net profit sharing consisted of a twenty per cent tax on dividends distributed to the shareholders of the concessionaire in excess of 671,250 UK pounds (GBP). Payments on account of royalty and dividend tax were subject to a minimum annual payment of GBP 750,000 (see Chapter 8).

An early concession had on 28 May 1901 been granted by the then Persian government to William Knox D'Arcy. The concession covered the whole territory of Iran with exception of the five northern provinces. In 1908, oil was discovered in the concession area at Masjid-i-Suleiman (after an earlier discovery near Chiah Surkh in 1904). In 1909 the D'Arcy concession was taken over by the newly incorporated

Anglo-Persian Oil Company that went public in order to raise the finance needed for the continuation of the venture. For strategic reasons, the UK government acquired in August 1914, a few days after the outbreak of the First World War, a controlling interest of 51.7 per cent in the concessionaire. In November 1932 the concession was unilaterally cancelled by Reza Shah but after negotiations a revised concession was in April 1933 granted to Anglo-Persian, the former concessionaire. The latter was in 1935 renamed the Anglo-Iranian Oil Company. The newly negotiated concession contained terms which were more in line with those accepted and applied under the petroleum agreements made at around the same the time with the governments of the Arab Gulf States. On this occasion the area covered by the new concession was much reduced and this time comprised an area of one hundred thousand square miles in the South-Western part of Iran. The concession lasted eighteen years. On 20 March 1951 it was nationalized and taken over by the newly established National Iranian Oil Company (NIOC). A new agreement referred to as the Consortium Agreement took in 1954 its place. In its turn this agreement was replaced with effect from (w.e.f.) 21 March 1973 by the 'Sale and Purchase' agreement.

Another early concession was a concession granted by the government of the Ottoman Empire on 28 June 1914 to the Turkish Petroleum Company (TPC). A few years earlier this company had been set up in London by British and German interests through the intermediary of the Armenian oil investor Gulbenkian. On 19 March 1914 agreement was reached on a restructuring of the company allowing for the participation of Anglo-Persian in the coming venture. The shareholding in the reconstituted TPC was fixed as follows: Anglo-Persian Oil Company: 50 per cent; Anglo-Saxon Oil Company (Royal Dutch Shell): 25 per cent; German interests represented by the Deutsche Bank: 25 per cent. The Anglo-Persian Oil Company and the Anglo-Saxon Oil Company each transferred a 2.5 per cent beneficiary shareholding to the Armenian oil investor Gulbenkian (since then known as Mr Five Percent). After the dissolution of the Ottoman Empire the concession that on 28 June 1914 had been granted became invalid, but TPC received in compensation for the loss of its concession a new concession from the government of the newly formed Iraqi State. In 1920 the German interest in TPC was transferred to the Compagnie Francaise des Petroles (CFP). On 15 October 1927 oil was discovered in this concession at Baba Gurgur near Kirkuk. In 1928 the concessionaire was restructured in order add a number of American companies, first seven later reduced to two, as shareholders. The newcomers received a combined share of 23.75 per cent at the expense of Anglo-Persian's shareholding. On this occasion the concessionaire was renamed Iraq Petroleum Company (IPC). On 24 March 1931 IPC received a concession, covering the eastern part of Iraq (i.e., the part situated east of the river Tigris). Some years later two more concessions covering the territory of Iraq west of the Tigris were granted respectively transferred to affiliated companies of IPC: in 1938 the one concession was granted to the Basrah Petroleum Company Ltd and in 1942 the other was transferred to the Mosul Petroleum Company Ltd. The concession held by IPC was nationalized by the government on 1 June 1972. The other concessions were acquired in steps by the government, but were finally in 1975 completely in its possession.

In 1925 a concession was granted by the government of Bahrain to a British individual, Major Holmes. On 12 June 1930 this concession was transferred to the Bahrain Petroleum Company, an affiliate of a US oil company incorporated in Canada. In this concession oil was discovered in 1932.

This discovery awakened the interest of the European and American oil companies in the oil potential of the Arabian Peninsula and led to the application for and subsequent grant of concessions in respectively SaudiArabia (25 May 1933, later substantially extended on 21 July 1939) and Kuwait (23 December 1934). It took some time before oil was discovered in these concessions: Burgan field, Kuwait, 23 February 1938; Damman field, Saudi Arabia, a few weeks later. In 1975 these concessions were completely taken over by the respective governments.

Other States on the Arabian Peninsula followed suit: concessions were granted in the late 1930s in Qatar, Oman and Abu Dhabi. In Qatar the government took possession of both the onshore and offshore concession in respectively September 1976 and February 1977. In Oman and Abu Dhabi each government acquired w.e.f. 1 January 1974 a sixty per cent state participation in its concessions.

At the time the petroleum agreements were negotiated none of the Middle East Gulf States possessed any petroleum legislation. All aspects of petroleum operations and the conduct thereof had to be provided for in the agreements themselves in a complete and self-contained form, since it was not possible to make reference to other legislative sources except for the constitutional law or more basically the general sovereign powers of the Head of State.

Most provisions comprised in the petroleum agreements were established in bilateral negotiations between the government of the ME Gulf State concerned and the foreign entity that at the time applied for a concession. These provisions are hereinafter referred to and described as the 'original provisions'. Later on, new provisions were introduced in the petroleum agreements as a result of a series of revisions dealing with government take (i.e., gross profit sharing, income taxes and royalties) and state participation. The first revision in this context was the (fiscal) revision of the Saudi Arabian concession in December 1950 whereby the 50/50 gross profit sharing principle was introduced. On account of the 'stability-of-terms' provision comprised in the other petroleum agreements, the 50/50 principle was also in the course of 1951 adopted in these agreements, with exception of the Iranian concession of 1933. As mentioned before, this concession was in 1951 nationalized and in 1954 replaced by a new agreement referred to as the Consortium Agreement.

After the founding on 1 September 1960 of the Organization of Petroleum Exporting Countries (OPEC) revisions regarding government take and state participation were the outcomes of collectively conducted negotiations involving the governments of the OPEC Gulf States on the one side and their respective concessionaires on the other. Oil exporting States, that took no part in these negotiations, but whose concessions or leases comprised a 'most favoured nation' clause, could by virtue of that clause implement the collectively agreed revisions in their concessions or leases. Among those States were Oman, Nigeria, Brunei and Malaysia.

For a review of the original provisions three different petroleum agreements have been selected as example, viz.:

(i) the agreement (Kuwait agreement) made on 23 December 1934 between the Shaikh of Kuwait and the Kuwait Oil Company;
(ii) the agreement (Oman agreement) made on 24 June 1937 between the Sultan of Muscat and Oman and an affiliated company of the aforementioned IPC; and
(iii) the agreement (Abu Dhabi agreement) made on 11 January 1939 between the Ruler of Abu Dhabi and another affiliated company of IPC.

The three petroleum agreements have each been revised from time to time, in particular with respect to the government take and state participation. Only the Kuwait agreement was in the end on 4 March 1975 completely taken over by the State.

The 1937 Oman agreement was on 7 March 1967 replaced by a new agreement (referred to as the Main Agreement). This Main Agreement was meant as a modernization and was made co-terminus with the original agreement. It ends therefore on 24 June 2012, but an extension beyond that date has not been excluded. As part of the revision a modern feature was added consisting in a schedule stipulating the maximum number of square miles that the leased area was allowed to contain at certain successive dates. This schedule was repeatedly changed in the ensuing years by virtue of a number of successive Supplemental Agreements. Furthermore all the fiscal conditions (royalty and income tax) that had been agreed between the governments of the Gulf States and their concessionaires up to the time of the revision of the 1937 agreement by the Main Agreement were incorporated in the latter. The presence of the 'most favoured nation' clause in the Main Agreement allowed the Omani government to keep the Main Agreement 'up to date' so to speak by incorporating the beneficial results of negotiations conducted by the other Arab Gulf States within OPEC on the subjects of government take and state participation. On this basis, the government acquired w.e.f. 1 January 1974 a sixty per cent participating interest in the Main Agreement.

The Abu Dhabi agreement as amended and revised from time to time is still in existence, expiring in 2014. Similarly as in Oman, the Abu Dhabi government acquired with effect from 1 January 1974 a sixty per cent participating interest in this agreement, a situation that so far has been maintained.

6.2.2 Original Provisions

6.2.2.1 Obligatory Payments (up to the Fiscal Revision of 1951)

'Kuwait agreement':

(i) Signature bonus: the sum of 475,000 rupees (Rs);
(ii) Annual payments (per anniversary of date of signature): before declaration of commercial discovery Rs 95,000; after declaration of commercial

discovery, either Rs 250,000 or a royalty of Rs 3 per English ton (2,240 lb) of petroleum won and saved by the company, whichever is the greater sum.

'Abu Dhabi agreement':

(i) Signature bonus: the sum of Rs 300,000;
(ii) Upon commercial oil discovery: Rs 200,000;
(iii) Annual payments (per anniversary of date of signature): before commercial oil discovery: Rs 100,000. This annual payment is increased by twenty-five per cent if the company fails to discover a drilling location within four years from the date of signature but does not surrender the agreement. Thereafter at the end of each three-year period this annual payment before discovery is further increased by twenty-five per cent until the commencement of drilling or until the company surrenders the agreement; after commercial discovery: during the first two years after commercial discovery, either Rs 250,000 or a royalty of Rs 3 per English ton (2,240 lb) of petroleum won and saved by the company, whichever is the greater sum.

'Oman agreement':

(i) Signature bonus: the sum of Rs 100,000;
(ii) Monthly payments during the five-year option period of Rs 5,000, making a total sum of Rs 300,000 (any balance is payable by the company if the company exercises its option to reminate the agreement during the option period);
(iii) Upon the company declaring during the option period to take up the concession (acceptance date): Rs 100,000;
(iv) Annual payments (per anniversary of acceptance date), during the first five years: Rs 84,000 or a royalty of Rs 3 per English ton (2,240 lb) of petroleum won and saved by the company, whichever is the greater sum; during the following years: Rs 96,000 or a royalty of Rs 3 per English ton (2,240 lb) of petroleum won and saved by the company, whichever is the greater sum.

Note. At the time the petroleum agreements were concluded one Rupee was equivalent to USD 0.38 and the market value of one barrel of crude oil was USD 1.00. The royalty rate of Rs 3 per ton is therefore equivalent to USD 0.15 per barrel or fifteen per cent of the market value. The highest signature bonus was paid under the Kuwait agreement, namely Rs 475,000, which at the time was equivalent to USD 180,000 or 180,000 barrels of oil.

6.2.2.2 Ownership of Petroleum

'Kuwait/Abu Dhabi agreements': there is no express provision vesting the ownership of petroleum in situ in the State or in the Head of State, but the concessionaire is granted the exclusive ownership of all petroleum produced and won by it within the State.

'Oman agreement': there is no express provision vesting the ownership of petroleum in situ in the Sultan or in the Sultanate, but it may be concluded that having been granted the exclusive right to produce and win petroleum, the concessionaire becomes the owner of the petroleum if and when produced by it.

6.2.2.3 Concession Area

'Kuwait agreement': the area of the Kuwait agreement comprises the State of Kuwait, including all islands and territorial waters belonging to Kuwait. There is no requirement to surrender any part of the agreement area during the validity of the agreement.

'Oman agreement': the area of the agreement (the leased area) comprises the whole territory of the Sultanate of Muscat and Oman, including all islands and territorial waters belonging thereto. By amendment of the agreement in 1951 the Province of Dhofar was excluded from the leased area. In 1969 this Province was again added thereto, restoring the original situation i.e., as existing before 1951. There is no requirement to surrender any part of the leased area during the validity of the agreement.

'Abu Dhabi agreement': the area of the agreement is the whole territory subject to the rule of the Head of State of Abu Dhabi and all its islands and territorial waters. There is no requirement to surrender any part of the agreement area during the validity of the agreement.

6.2.2.4 Duration and Basic Rights

'Kuwait agreement': the agreement is valid for seventy-five years as of the date of signing. During this period the concessionaire has the exclusive right within the agreement area to explore, drill for, produce and win petroleum (described as natural gas, asphalt, ozokerite, crude petroleum and their products and cognate substances); the exclusive ownership of all petroleum produced and won by the concessionaire; the right to refine, transport, sell for use within the State of Kuwait or for export or otherwise deal with or dispose of any and all such petroleum. At any time after the concessionaire has drilled the mandatory four thousand feet or after the expiry of two years from the date of the agreement, whichever is the later date, the concessionaire may terminate the agreement at one year's notice.

'Oman agreement': the agreement is valid for seventy-five years from the date of signing, comprising a so-called 'option period' of five years. During this option period the concessionaire has the exclusive right within the leased area to explore for and produce petroleum (described as natural gas, asphalt, ozokerite, crude petroleum and their products and cognate substances). At any time during the option period the concessionaire has the right to terminate the agreement or to declare its intention to take up the concession over the leased area. If the concessionaire states such intention and makes the required payments, the Sultan shall grant to the concessionaire for the remainder of the period of the agreement the exclusive right to explore, produce, refine, transport, sell, export and otherwise deal with or dispose of the petroleum and to do all things necessary for all or any of the aforesaid purposes within the leased area. At any time after three years from the date of signature of the agreement the concessionaire may terminate the agreement at six month's notice.

'Abu Dhabi agreement': the agreement is valid for seventy-five years from the date of the agreement. There are no sub-divisions of this period. During the validity of the agreement the concessionaire has the sole right within the agreement area to explore for, discover, drill for and produce mineral oils and their derivatives and allied substances, (in the agreement collectively referred to as 'the substances' or 'oil', which indicates that these terms were not meant to include or refer to non-associated natural gas), and the sole right to the ownership of all the oil produced and to the free disposal thereof both inside and outside the territory: provided that the export of oil shall be from the territory of the concession direct without passing across any adjacent territory. At any time after three years from the date of signature of the agreement the concessionaire may terminate the agreement at six month's notice.

6.2.2.5 Regulation of the Exploration Operations

'Kuwait agreement': within nine months from the date of the agreement the concessionaire is obliged to commence geological exploration. The following drilling programme is imposed at places to be selected by the concessionaire and in terms of total aggregate depths:

(i) four thousand feet prior to the end of the third year;
(ii) twelve thousand feet prior to the end of the ninth year; and
(iii) thirty thousand feet prior to the end of the nineteenth year.

If the concessionaire fails to make within twelve years of the date of the agreement a declaration that petroleum has been found in commercial quantities the concessionaire must either pay the Head of State the minimum annual payment required under the agreement (viz. Rs 250,000) or surrender the agreement.

'Oman agreement': there are no provisions specifically dealing with exploration operations. No specific exploration work programme is prescribed.

'Abu Dhabi agreement': the concessionaire is obliged to take the necessary steps as rapidly as possible to discover structures in which the existence of oil in commercial quantities may be probable. Such structures are referred to as drilling locations. For this purpose the company is obliged to commence its geological exploration within six months of the date of signing of the agreement and is further obliged to commence drilling on one of the drilling locations before the end of the second year following the discovery of the location concerned. If the concessionaire does not succeed in discovering a drilling location within four years from the date of signature then it must either surrender all its rights under the agreement or must increase by twenty-five per cent the annual amounts (viz. Rs 100,000) which the company is obliged to pay to the Head of State until the discovery of oil in commercial quantities.

6.2.2.6 Regulation of the Development of Discoveries and Production Schedules

'All agreements': there are no provisions dealing specifically with the development of discoveries or calling for the preparation and design of a plan of development. There are no provisions concerning production schedules and the regulation of production rates.

6.2.2.7 Conduct of Operations

'Kuwait/Oman agreements': the concessionaire is obliged to conduct its operations in a workmanlike manner and by appropriate scientific methods and must take all reasonable measures to prevent the ingress of water to any petroleum bearing strata and must duly close any unproductive holes drilled by it and subsequently abandoned. The concessionaire must keep the Shaikh and his London representative/the Sultan informed generally as to the progress and its result of its drilling operations but such information is treated as confidential.

'Abu Dhabi agreement': the concessionaire is obliged to carry out its operations in an energetic manner and by suitable technical methods, and must at all times keep the Head of State generally informed as to the progress of its drilling operations and the results thereof for each well separately, month by month. Similarly the concessionaire is obliged to submit its geological maps of Abu Dhabi territory to the Head of State or his representatives before they are printed, and the Head of State may, if he so desires, take copies of such maps on condition that all such information remains confidential with the Head of State and his representative.

6.2.2.8 Ancillary Rights

'Kuwait/Oman agreements': for the purpose of its operations the concessionaire has the right without hindrance to construct, maintain and operate power stations, refineries, pipelines and storage tanks, facilities for water supply including drilling for water, telegraph and telephone lines and wireless installations, roads, railways, buildings, ports, harbours, harbour works, aircraft buildings and landing places for aircraft, wharves and jetties and any other facilities or works, which the company may consider necessary, and for such purposes to use free of all payments any stone, sand, gravel, gypsum, lime, clay or similar materials or water which may be available, provided always that the inhabitants of the leased area are not prevented from taking their usual requirements of these materials and that the water supply of the local inhabitants and nomad population who may depend on the same is not endangered. The concessionaire may select at its discretion ('Kuwait agreement':

but in consultation with the Head of State) the position of any such works. The concessionaire may likewise install and operate without hindrance all such means of transportation by land, air and water as may be necessary for the effective conduct of its operations. The concessionaire is granted free of cost the unrestricted use and occupation of and surface rights over all uncultivated lands of the concession area, which the concessionaire may need for the purpose of its operations and in particular the concessionaire has the right to select ('Kuwait agreement': in consultation with the Head of State) an area or areas of land chosen by the concessionaire with exclusive surface rights upon which to erect drilling rigs, pump stations, oil refineries, storage, terminal, shipping and aircraft facilities and other works required for the company's operations. The concessionaire may buy or lease ('Kuwait agreement': with the cognizance of the Head of State) for the concessionaire's operations any lands, houses or buildings with the consent of and on conditions to be arranged with the proprietors thereof, provided that the terms of such purchase or lease are not in excess of those ordinarily current in their respective localities. The concessionaire may acquire only such lands, houses and buildings as are necessary for its operations. The concessionaire must inform the Head of State from time to time of the land, houses and buildings, which it requires to occupy for its operations. Land, houses and buildings previously acquired by the concessionaire from the Head of State, but found no longer necessary for its operations, must be returned to the Head of State free of charge.

'Abu Dhabi agreement': for carrying out the works connected with its enterprise the concessionaire has the right to build and construct all sorts of buildings, installations and engineering works of every description on condition of informing the Head of State thereof prior to the commencement of construction. The company may also use all the means of transport and communication. If it be proposed to erect buildings designed specially for the protection and defence of the concessionaire's property and personnel, then the Head of State has the right to supervise the design of the proposed buildings and to direct the administration thereof. The concessionaire is granted free of charge the unrestricted right to make use of such uncultivated lands within the territory as it may require from time to time for its operations.

6.2.2.9 Disposition of Production

'Kuwait/Oman agreements': the concessionaire is free to dispose of its production as it sees fit. There are no provisions qualifying or restricting this right in any way. There is no obligation to supply the domestic market.

'Abu Dhabi agreement': the concessionaire has the sole right freely to dispose of the oil produced both inside and outside the territory, provided that the export of oil shall be from the territory of the concession direct without passing across any adjacent territory. There is no obligation to supply the domestic market, except that, interestingly, after the discovery of oil in commercial quantities the company is obliged to give the Head of State quantities of petrol sufficient for his own private motor-cars.

6.2.2.10 Termination

'All agreements': unless the concessionaire is excused by reason of *force majeure* the Head of State may terminate the agreement and take without payment all the property of the concessionaire in the agreement area if:

(i) the company fails to make any of the payments prescribed in this agreement within six months of the date on which they fall due; or
(ii) the concessionaire is in default in respect of an arbitration award; or
(iii) ('Kuwait agreement': the concessionaire fails to fulfil its obligations in respect of geological exploration or drilling).

6.2.2.11 Stability of Terms

'All agreements'. Except for a case of concessionaire's default the Head of State is committed not to annul the agreement by general or special legislation or by administrative measure or by any other act whatever. No alteration shall be made in the terms of the agreement by either the Head of State or the concessionaire, except in the event of the parties jointly agreeing that it is desirable in the interest of both parties to make certain alterations, deletions or additions thereto.

Note. After the establishment of OPEC, the Member States and other non-Member oil exporting States, such as Oman, Malaysia and Brunei, introduced in the concessions or leases under their jurisdiction a clause which became known as the 'most favoured nation' clause. The standard text of this clause (taken for the purpose from the agreement signed on 15 January 1961 by the Head of State of Kuwait and Kuwait Shell Petroleum Development Company Lid with regard to Kuwaiti continental shelf) reads as follows:

> If as a result of changes in the terms of concessions in existence on the effective date or as a result of the terms of concessions granted thereafter an increase in benefits to governments in the Middle East should generally to be received by them the concessionaire will consult with the Emir whether in the light of all relevant circumstances including the conditions in which operations are carried out and taking into account all payments made, any alteration to the terms of this agreement would be equitable to the parties.

6.2.2.12 Settlement of Disputes

'All agreements'. Disputes arising between the parties during the validity of the agreement concerning the interpretation or execution of the agreement or anything contained therein or the rights or liabilities of either party thereunder must, failing agreement to settle such disputes in any other way, be referred to arbitration by two arbitrators and a referee, each of them to be appointed by the British Political Resident in the Persian Gulf, if so needed.

'Kuwait agreement': if at any time during the validity of the agreement any dispute arises regarding the accuracy of the accounts of the concessionaire in connection with the amounts of the royalty and of other payments due to the Head of State under the agreement, the Head of State has the right to appoint, in consultation with the British government, a registered firm of auditors to examine the books of the company on behalf of the Head of State in Kuwait and or in London as hc may consider necessary. All expenditure incurred in conncction with such auditing shall be paid by the Head of State. The concessionaire is obliged to provide the registered firm of auditors appointed by the Head of State the necessary facilities to enable them to check the books and registers of the concessionaire and to render every assistance to enable the auditors to thoroughly examine such accounts and must in every way assist them to safeguard the interests of the Head of State.

6.2.2.13 Straddling Structures

'All agreements'. As could be expected in view of the large concession areas involved and their unexplored status, there are no provisions dealing with the possible occurrence of straddling geological structures taken up in any of the agreements.

Note. But there exists a large oil field, the huge Rumaila oil field, that straddles the interstate boundary between Iraq and Kuwait. The field is for the larger part located in Iraq and for a much smaller part in Kuwait. On both sides of the boundary, the field is being produced (in Kuwait at Ratga) albeit the production on the Iraqi side is much larger. Up to now, both States appear not to be interested in or in a position to making an agreement on joint development of the field.

6.2.2.14 Employment of Personnel

'All agreements': the concessionaire is obliged to employ subjects of the Head of State for all work for which they are suited under the supervision of the concessionaire's skilled employees. Should however the local supply of labour in the judgment of the concessionaire be inadequate or unsuitable the concessionaire has the right, with the approval of the Head of State, which shall not be unreasonably withheld, to import labour, preference being given to labourers from neighbouring Arab countries who will obey the local laws. The concessionaire has also the right to import skilled and technical employees. The concessionaire must pay to the workmen it employs a fair wage, and this wage to be decided and stated by the concessionaire at the time the workmen are engaged. The Head of State shall give to the concessionaire and its employees and property all the protection in his power from theft, highway robbery, assault, wilful damage and destruction (*Note*. Oman agreement: within the areas of the leased area which has been declared safe, or may be so declared at any time during the validity of the agreement, for representatives of the concessionaire to enter).

'Kuwait agreement': any employee imported by the concessionaire, who shall by misconduct cause a breach of peace or public disturbance, shall at the request of the Head of State be dismissed and shall, if it is within the power of the concessionaire to do so, be sent out of Kuwait.

'Abu Dhabi agreement': any employee of the concessionaire, proved guilty of an offence that might lead to a breach of peace or interference in the politics of Abu Dhabi, shall, if he is a foreigner, at the request of the Head of State be expelled by the concessionaire from the country.

Note. These provisions clearly show the respective governments' concern about political interference from the side of the major Western oil companies in their countries. About forty years later, this concern which was shared by many other developing countries would be addressed in the 1974 Charter of Economic Rights and Duties of States (see Chapter 10).

'Kuwait agreement': the concessionaire is obliged to provide free of charge medical service for its employees and the Head of State and his family shall be entitled to such medical service and necessary medical supplies free of charge. The Head of State has the right to appoint as from the effective date of the agreement a representative in London (the London representative) to represent the Head of State in all matters relating to the agreement with the concessionaire in its London office. The London representative must be given full access to the production records of the concessionaire, including the agenda of the Board meetings and is entitled to attend the Board's meetings at which the Head of State's interests are discussed.

'Abu Dhabi agreement': the concessionaire shall pay its labour at suitable rates, accord good treatment to the workers of the country, give them the opportunity of practical training and afford free medical treatment to anyone who may require such treatment in the course of his duties.

6.2.3 State Participation

6.2.3.1 Background

The petroleum agreements concluded in the 1930s did not provide for any form of state participation. But this changed quickly when in 1957 Iran issued its new petroleum law, the Petroleum Law of 29 July 1957. Under the terms of this Act the National Iranian Oil Company, NIOC, was authorized to enter into joint venture agreements with foreign entities but only with respect to acreage situated outside the territory of the Consortium. Under the terms of such an agreement a joint company was established having NIOC and the foreign entity as shareholders. On the basis of this joint venture concept the first series of state participation agreements were concluded in 1957 and 1958 and a second series followed in 1964 and 1965. In Egypt the first state participation agreement was also concluded in 1957. The authorization therefore was given under the terms of the Law No. 86 of 1956. In May 1960 the Egyptian government announced that with respect to operations in the Western Desert it would opt for fifty per cent state participation accompanied by the formation of an Egyptian joint stock company. In 1963 and 1964 the Egyptian government signed a series of participation agreements, known as the Western Desert participation agreements, on the basis of these principles. In Kuwait state participation was introduced in 1961, Saudi Arabia followed suit in 1965. In both cases the acreage given in concession (with state participation) was situated on the continental shelf involved. Libya followed the Egyptian example by offering foreign entities the possibility of entering into joint ventures with the Libyan National Oil Company (LNOC), whereby LNOC would be the holder of the exclusive licence underlying the venture concerned.

6.2.3.2 The General Agreement on Participation

After having settled with their concessionaires various fiscal and pricing issues within the framework of the Teheran Agreement of 14 February 1971, the governments of the Arab Gulf States within OPEC turned their attention to the issue of acquiring for themselves a participation in the existing and ongoing concessions under their jurisdiction. At an extraordinary meeting of the Conference, held on 22 September 1971, a resolution was adopted calling upon the Member States to open negotiations with their concessionaires in order to reach agreement on the desired participation.

The Conference resolution of 22 September 1971, calling for state participation with respect to the petroleum agreements (and similar long-term licences or leases in other Member States), was aimed at the acquisition of a state interest in existing and ongoing operations.

Note. The Conference was of course well aware of the fact that the participation sought by the governments was different from the state participation ventures established in the 1950s and 1960s in Iran, Egypt, Saudi Arabia and Libya. In those States the joint ventures entered into with the national oil company had been restricted to areas falling outside the territory of existing concessions and were negotiated before any investment was made or before any petroleum had been found in the area opened-up for the intended agreement.

Ensuing negotiations on this matter between governments and their concessionaires led to the signing in Riyadh on 20 December 1972 of an agreement named the ‘General Agreement on Participation’. Parties to this General Agreement were the same Gulf States and oil companies which had signed the 1971 Teheran Agreement, with the exception of Iran.

Under the terms of the General Agreement each Arab Gulf State within OPEC acquired with effect from 1 January 1973 a twenty-five per cent undivided interest in the petroleum agreement falling under its jurisdiction and the right to increase this initial interest after five years, as from 1 January 1978, in annual increments of five per cent and with a final increment of six per cent, to a maximum percentage of fifty-one per cent. According to this schedule this final percentage should have been reached on 1 January 1982.

The compensation finally agreed amounted to the updated book value of the concessions. The updated book value was calculated by totalling over the past period of the concession year by year the annual differences between on the one hand capital and exploration expenditures and on the other the amount by which the income tax revenues of the government concerned had been diminished by deductions of capital depreciations and exploration expenditure. By paying its percentage share of the updated book value the Gulf State concerned effectively paid back its share of the updated past capital and exploration expenditures made by the concessionaire, while withholding the tax relief that the concessionaire should have obtained with respect to that updated expenditure.

In order to allow the concessionaires to adjust themselves to the loss of production capacity the Arab Gulf States were obliged to sell to their co-concessionaires a certain portion of their initial participation share: this portion, referred to as 'bridging oil' was fixed at seventy-five per cent for the first year, at fifty per cent for the second year and at twenty-five per cent for the third year. Besides any Gulf State had the right (but not the obligation) to sell additional volumes to its co-concessionaire. These extra volumes were referred to as 'adjustment oil' or 'put oil'. The Gulf States had been accorded this right in order to provide them with a guaranteed outlet for their share of production in the event they would experience difficulties with the disposal thereof. The volumes of put oil were fixed on an annual basis and expressed as certain annual percentages of the initial interest as well as of each of the successive incremental interests. The concessionaires had to pay for the bridging oil as well as the put oil, if any, the half-way price, i.e., a price, which was equal to the tax paid cost plus half the difference between the tax paid cost and the posted price. The General Agreement has only been effective during the year 1973 and only with respect to the Arab Gulf States Kuwait, Qatar, Abu Dhabi and Saudi Arabia.

6.2.3.3 The Sixty Per Cent to One Hundred Per Cent State Participation

Meanwhile, state participation at the much higher level of fifty-one per cent was agreed in Libya and Algeria and nationalizations took place in Iraq.

Note. In Algeria state participation at a level of fifty-one per cent was introduced as from the moment the Ordinance No. 71-22 of 12 April 1971 became effective. The state participant was Sonatrach, the Algerian state oil company.

Furthermore, Iraq refused to implement the General Agreement. In the light of these events those OPEC Gulf States, that had implemented the General Agreement, now demanded a reopening of the negotiations that had led to that Agreement. The ensuing negotiations resulted in a series of bilateral agreements between individual Arab Gulf States and their respective concessionaire(s). According to these bilateral agreements state participation in the respective concessions was fixed at sixty per cent with effect from 1 January 1974. As compensation the concessionaires were paid a certain fixed sum of money, not derived from a formula or explained in any other manner but, taken together with the compensation already paid for the initial twenty-five per cent interest, total compensation proved to be no more than the fiscal book value for the whole sixty per cent interest that in the end was transferred. It was certainly much lower than sixty per cent of the updated book value. However this may be the bilateral state participation agreements did long survive.

By December 1974 the government of Qatar had already announced its intention to proceed with a complete takeover of the two concessions (onshore and offshore) under its jurisdiction. After protracted negotiations with the respective concessionaires, in particular with regard to the matter of compensation that should be paid, and the future role of the former concessionaires, it was agreed, in September 1976 and February 1977 respectively, that each concessionaire would transfer its forty per cent interest to the government. Apart from the paying of a compensation the settlement included a five-year oil purchase contract at the official government selling price and a service contract in accordance with which affiliates of the former concessionaires rendered technical and personnel services at a fee.

The government of Kuwait cancelled its bilateral 60/40 agreement in March 1975. With the former concessionaires (British Petroleum and Gulf Oil), a settlement was reached along the lines of the Qatar settlement.

The government of Saudi Arabia also decided in 1975 to proceed with a complete takeover of the concession under its jurisdiction. It took a long time (until 1980) before the contractual arrangements were completed.

The government of Abu Dhabi chooses to maintain the sixty per cent state participation as agreed with respect to the two concessions (onshore and offshore) under its jurisdiction. The State was represented by its national oil company, the Abu Dhabi National Oil Corporation (ADNOC), owner of the sixty per cent interest. The forty per cent interest in each concession was held by a company (ADPC, onshore, and ADMA, offshore) jointly owned by a number of foreign oil companies as shareholders. Initially, ADPC and ADMA were charged with the operation of their respective concessions under the supervision of a management committee in which their individual shareholders and ADNOC were represented. Later the organizational set-up was changed in the sense that ADPC and ADMA lost their role as operator. With respect to each concession this task was taken over by a newly established operating company (ADCO), in which ADNOC took a sixty per cent shareholding. The remaining forty per cent of the shares in ADCO were distributed over the individual shareholders of ADPC (onshore) or ADMA (offshore). ADCO is supervised by a joint management committee, in which ADNOC and the individual shareholders of ADPC or ADMA as the case may be are represented. Interestingly, ADPC itself is no longer represented in the organization of the joint venture.

Although not a party to the negotiations on state participation and even not a member of OPEC the Omani government (represented by the Sultan) nevertheless acquired, by virtue of the 'most favoured nation' clause comprised in the Main Agreement w.e.f. 1 January 1974 a sixty per cent in that agreement. A few years later, and by means of an agreement referred to as the Transfer Agreement dated 15 May 1980, the state participation was implemented by transferring all assets and operating rights and obligations under the Main Agreement to a new operating company jointly owned by the government and the individual shareholders of the concessionaire. The new company was named Petroleum Development Oman LLC (PDO). The concessionaire itself was renamed Private Oil Holdings Oman Limited (POHOL). The new organizational structure resembles to a certain extent the structure chosen for the Abu Dhabi 60/40 state participation but there are significant differences. The interests transferred to PDO comprised all the interests of the government and the concessionaire in the Main Agreement, including the ownership of all assets belonging thereto, except for the ownership of 'petroleum produced at cost' and the disposal thereof, the financial rights and obligations under the Main Agreement, and the right to terminate the Main Agreement on one year's notice as provided thereunder. Therefore, PDO can be described as an asset owning operating cost company which has to produce and deliver oil (for natural gas special arrangements exist) at cost to the government and POHOL. The government and POHOL are obliged to supply to PDO, if and when required, in a 60/40 proportion, all funds required for the authorized operations. The acquisition of new assets to be owned by PDO is financed by the government and POHOL by providing loans in a 60/40 proportion. The loans have to be repaid by PDO out of funds generated from annual allowances for depreciation included in the cost of operations.

6.2.3.4 The Special Situation in Iran

Iran was not involved in the negotiations concerning state participation and the complete takeover of the concessions because its own concession agreement of April 1933 had (already) been nationalized on 20 March 1951 and been taken over by the newly established National Iranian Oil Company (NIOC). It had been the then revolutionary government's intention to follow the Mexican example (*Note*. In Mexico the licences and other assets of the foreign oil industry had been nationalized in March 1938) and entrust the domestic petroleum operations to NIOC without allowing any involvement of foreign oil companies. However the revolution failed and after three and a half years of negotiations a new agreement (referred to as the 'Consortium Agreement') was signed on 25 October 1954 by NIOC and Iranian Oil Participants Ltd, a UK corporation. The latter is usually referred to as the Consortium, because its shareholders formed a consortium of major foreign oil companies, among which the former concessionaire, viz. Anglo-Iranian Oil Company, on that occasion renamed British Petroleum, with a percentage share of forty per cent. Pursuant to this new agreement the petroleum operations were continued by the Consortium, acting on behalf of NIOC.

The Consortium Agreement was in substance not very different from the petroleum agreements of the 1930s but nevertheless formally seen as a risk bearing service contract (see Chapter 5). From this point of view, Iran excluded itself from the negotiations leading to the General Agreement on Participation and did not become a party to that agreement. Nonetheless the Iranian government's interest and concern was now focused on revising the Consortium Agreement along the lines of the General Agreement (as if this Agreement applied to Iran) in order to prevent that the state participation demanded and pursued by the other OPEC Gulf States would bring the latter into a better position than Iran in respect of government take and production shares. Subsequently, the Iranian government negotiated and concluded, with effect from 21 March 1973 with the Consortium a new agreement, referred to as the 'Sale and Purchase Agreement', in replacement of the Consortium Agreement. The parties remained the same, i.e., NIOC and Iranian Oil Participants, Ltd.

Note. To translate the terms of the General Agreement into those of the Consortium Agreement was not an easy matter, in particular not because no loss of concession rights had taken place for which compensation should be paid. In fact, the reimbursement for past expenditures that was agreed on the basis of the principles of the General Agreement represented compensation for the loss of the Consortium's right to acquire total production at tax paid costs. The agreed amount of the reimbursement would be set off against the forty per cent contributions which the Consortium had to make to capital expenditures and the price the Consortium had to pay for the crude oil production it would buy from NIOC. This price was fixed at a level that would secure a financial outcome of the Sale and Purchase Agreement no less favourable than that obtainable under the General Agreement. In fact, Iran was obliged to sell to the member companies of the Consortium over a period of twenty years the available production, by which was meant all crude oil which would be produced from the territory that hitherto had been exploited by the Consortium under the terms of the Consortium Agreement with exception of:

(i) annual quantities, which were set aside for export by NIOC and which were specified to increase from 200,000 b/d in 1973 to 1,500,000 b/d in 1981 and thereafter as would be in proportion to current production levels; and
(ii) the quantities that were needed for the supply of the Iranian domestic market.

These reserved quantities represented in effect the state participation shares acquired by the Gulf States pursuant to the General Agreement. For the first five years of the agreement, i.e., up to 1 January 1978, the date upon which the initial participation under the General Agreement would be increased for the first time (by five per cent to thirty per cent), NIOC and the Consortium would contribute to capital expenditures in a 60/40 proportion. To the extent that the forty per cent contribution demanded from the Consortium was below the level of contributions required to be paid by concessionaires, to whom the terms of the General Agreement were applicable, interest had to be paid on the amount of the difference. Since with effect from 1 January 1974 the General Agreement was replaced by bilateral sixty per cent state participation agreements, no interest was due after that date.

In reaction to the emergence of bilateral 60/40 state participation agreements, the Sale and Purchase Agreement was initially adjusted in order to copy the effects of the former, but some time thereafter negotiations were started to revise the agreement in a more fundamental manner. The intention was to achieve a combination of a fee paying service contract and purchase contract at discounted prices. The negotiations about such novel contract, actually a non-risk bearing service contract, had to be aborted as a result of the outbreak of the revolution in December 1978.

6.2.3.5 The Special Situation in Iraq

Already in 1961 the Iraq Petroleum Company (IPC) had been forced to relinquish 99.5 per cent of the area of its concession (which had been awarded on 24 March 1931). On 1 June 1972 the concession was nationalized by the government, some months before the latter signed the General Agreement, which however it refused to implement. This nationalization measure was formally settled on 28 February 1973. IPC's main asset consisted of the Kirkuk field. IPC had to pay GBP 141 million to Iraq and the latter had to deliver 111 million barrels of Kirkuk crude oil over a period of fourteen months.

At the end of 1973 Iraq nationalized forty-three per cent of the share capital of the Basrah Petroleum Company (BPC), which held a concession in the South-Western part of the country. This nationalization of share capital was politically motivated and was connected to the outbreak of the Yom Kippur war (also called the Ramadan war) in the Middle East on 6 October 1973. In January 1975 Iraq concluded with BPC an agreement under which the sixty per cent participation became applicable with retroactive effect from 1 January 1974. As a result thereof Iraq's total interest (direct and indirect) in the operations of BPC amounted to 77.2 per cent with effect from that date. Later in 1975 Iraq decided to take over BPC's remaining forty per cent share in the concession. Negotiations on a settlement took several years.

6.3 IRAN (OUTSIDE CONSORTIUM TERRITORY)

On 29 July 1957 a new petroleum law was introduced. Under the terms of this new Law the government represented for the purpose by NIOC was authorized to enter into petroleum agreements with foreign entities. The intended agreement envisaged the establishment of a joint company having NIOC and the foreign entity as shareholders, but a non-incorporated joint venture was also possible. The percentage shareholding of NIOC in a joint company should be no less than thirty per cent. Agreements were not to exceed twenty-five years in duration, but could be extended by three periods of five years each.

Note. The first agreement on the basis of the Law was concluded on 8 November 1957 between NIOC and Agip Mineraria, the latter being an affiliate of the Italian state-owned oil company ENI. The area of the agreement comprised about twenty-three thousand square kilometres in three different zones, largely situated on the continental shelf. Agip Mineraria was responsible for the conduct and financing of the exploration work and in this connection assumed the obligation to spend an amount of

USD 22 million within a period of twelve years. Upon making the first commercial discovery a profit making joint company to be named the Iran-Italian Petroleum Company (SIRIP) would be established having NIOC and Agip Mineraria as equal shareholders. The joint company was responsible for carrying out the operations following the first commercial discovery. To each field that was taken in production a development/production period was assigned of twenty-five years from start of commercial production, which period could be extended by three periods of five years each if Agip Mineraria would so desire. The costs of the exploration work that led to the first commercial discovery and that had been paid by Agip Mineraria were repaid by the joint company, however not to Agip Mineraria alone but to both shareholders, each shareholder receiving fifty per cent of these costs. Since NIOC had not contributed to these costs it meant that these cost would not be shared 50/50 between NIOC and Agip, a so-called carry of NIOC. Costs of any unsuccessful exploration work were not repaid. The joint company was obliged to establish posted prices for its exports and, in accordance with the 50/50 profit sharing principle then prevailing in the Middle East, had to pay fifty per cent of its profits to the State. The other fifty per cent was divided equally between the two shareholders. The joint company was financed by its two shareholders through their respective contributions to the share capital and by means of providing loans, but further exploration work was financed by loans solely provided by Agip. These loans for exploration work were for only fifty per cent repaid by the joint company but only insofar as this further exploration work proved successful. Most interestingly, the agreement contained a kind of 'most favoured nation' clause in that it was stipulated in the agreement that on the second and third renewal thereof Agip Mineraria had to accept the terms of other agreements considered on the whole more favourable to Iran than the existing agreement.

In 1958 NIOC concluded two more agreements, this time with respect to acreage exclusively situated on the Iranian continental shelf. Apart from the particular conditions of these agreements (such as size and location of agreement area, exploration work commitments and bonus payments) they were identical to the agreement concluded with Agip Mineraria. Parties were (apart from NIOC) respectively Pan American Petroleum Corporation and Sapphire Petroleum Ltd.

In 1964 and 1965 there followed a new series of participation agreements, also with respect to the Iranian continental shelf. In contrast with the earlier agreements the joint company that was established, was no longer run for profit, but was transformed into a cost company with solely operational responsibilities, this time including all exploration work. The shareholders had to offtake the production from the joint company albeit any portion of NIOC's fifty per cent share of production that NIOC was not able to sell had to be bought by the foreign participant at the half-way price, i.e., a price equal to the tax paid cost plus half the difference between the tax paid cost and the posted price.

The Petroleum Act of 29 July 1957 was succeeded by the Petroleum Act of 1974. This Act entered into force a few years before the outbreak of the revolution in December 1978. The new Act authorized NIOC to conclude risk contracts with foreign oil companies and was meant to regulate petroleum operations outside the Consortium area. The contract that was made available followed the pattern of the risk contract made between NIOC and an affiliate of ERAP in 1967 (see Chapter 5).

6.4 KUWAIT (GULF CONTINENTAL SHELF)

In Kuwait a petroleum agreement made in 1961, and covering the Kuwaiti part of the continental shelf, provided for optional minority state participation. Pursuant to this agreement the State was given the option to participate for twenty per cent in the agreement as from the moment a commercially viable deposit was discovered. If the State would exercise its option, it would have to reimburse twenty per cent of the expenditure made by the concessionaire up to the effective date of state participation. (*Note*. The envisaged participation was never realized, since the concession agreement was terminated before a discovery was made.)

6.5 SAUDI ARABIA (RED SEA CONTINENTAL SHELF)

In 1965 the government of Saudi Arabia entered into a state participation agreement with respect to acreage situated on its part of the Red Sea continental shelf. Parties were Petromin, the Saudi Arabian state oil enterprise, and Auxirap, an affiliate of the French state-owned oil company ERAP. Petromin's participating share amounted to forty per cent; Auxirap's share was sixty per cent. Auxirap was responsible for the conduct and the financing of the exploration work. As soon as a commercial discovery was made a profit-making joint company would be established. The company was liable to pay fifty per cent income tax calculated on the basis of prices required to be posted by the company. The company had to pay royalty at higher rates than were customary at the time (a rate of fifteen per cent increasing to twenty per cent depending on production performance, the latter expressed in daily production rates). The royalty was not credited against income tax but had to be 'expensed', i.e., had to be treated as a cost element deductible from income to arrive at taxable income, in accordance with the then in Saudi Arabia prevailing fiscal treatment of royalties. The joint company would sell the production to its shareholders at a price to be determined by the joint company. At the request of Petromin, Auxirap was obliged to sell Petromin's share. Auxirap had to pay Petromin the price realized in the market but this price was not to be lower than ninety per cent of the company's price.

6.6 EGYPT

On 14 September 1914 a UK company named, Anglo-Egyptian Oilfields Ltd, concluded an agreement with the Egyptian government under which the company acquired over the years a growing and large number of 'old style' thirty-year petroleum mining leases (British colonial model). In 1956 these leases and all other assets of this Company were sequestrated and in 1961 formally expropriated.

After World War II, Egyptian petroleum legislation, as set out in Section II *On Raw Combustibles* (*Note*: raw combustibles include petroleum) of the Law No. 66 of 1953 *On Mines and Quarries* provided for a licensing regime based on the granting of renewable prospecting licences which in turn could be converted into one or more long-term (thirty + fifteen years) mining leases. Under the conditions of the mining lease, the lessee had to pay a royalty at a rate of at least fifteen per cent or twenty-five per cent, depending on which half of the prospecting licence area the mining lease derived from. Upon renewal of the mining lease the rate was fixed at twenty-five percent. The royalty value was defined as the annual average price of similar oil in a recognized market. The government had the preferential right to purchase up to twenty per cent of the petroleum produced from the lease at a price ten per cent less than the recognized market price. (*Note*. Under the preceding Law No. 137 of 12 August 1948 *On Mines and Quarries* the granting of licences for petroleum exploitation was restricted to Egyptian companies only.)

When Law No. 66 was repealed by Law No. 86 of 1956 *On Mines and Quarries*, the former's Section II was upheld. Of special interest is Article 50 of the Law pursuant to which the minister responsible for mining matters could be authorized by a special law to enter into an agreement with a company for prospecting for minerals and exploitation of mines subject to special conditions in derogation of the provisions of the Law as shall be determined in the special law granting the authority.

Note. The principle of Article 50 was not new, it was already contained in Article 69 of the predecessor Law No. 66 but, because this article was placed in Section III of that Law and hence repealed, the principle had to be restated.

Article 51 declared the provisions of Article 50 to be applicable to raw combustibles and therefore to petroleum. As a consequence, the government was authorized to regulate petroleum operations by means of entering into agreements approved by a special law, in derogation of the provisions of the Law, in this case represented by the provisions of the 'saved' Section II of Law No. 66. In order to represent the government in such agreements, a national oil company named the General Petroleum Authority (GPA) was established in 1956. In February 1957 the first state participation agreement was entered into with International Egyptian Oil Company (IEOC), an affiliate of the Italian state-owned oil company ENI.

In May 1960 the government announced that with respect to acreage in the Western Desert it would opt for fifty per cent state participation accompanied by the formation of an Egyptian joint stock company. In 1963 and 1964 the government concluded a series of participation agreements, known as the Western Desert participation agreements. In these agreements the following principles were incorporated:

(i) An exclusive licence for the exploration and production of petroleum is granted to EGPC and the foreign oil enterprise in equal proportions.

(ii) The exclusive licence is granted for a period of thirty years with the possibility of extension by fifteen years.
(iii) The initial licence area comprised an exploration area that was divided into exploration blocks.
(iv) After making a commercial discovery, the exploration block concerned would be converted into a development area.
(v) Over a period of fifteen years, non-converted exploration blocks had to be surrendered.
(vi) The foreign participant was responsible for the execution of the operations and for the funding thereof till the first commercial discovery was made. Thereafter a joint operating company (JOC) would be established, the shares of which would be held by EGPC and the foreign participant in equal proportions.
(vii) The JOC would be solely responsible for the development of the discovery and for the continuation of the exploration. The JOC would be funded by its shareholders in equal proportions. The JOC would not generate income; it did not sell any production but delivered the same to its shareholders in equal proportions.
(viii) Obligatory payments consisted in the payment of a royalty and a fifty per cent State gross profit share, which comprised the royalty, income taxes, duties and make-up payments. (*Note*. The fifty per cent State profit share represented the 50/50 profit sharing principle that as of December 1950 had been adopted in the Middle East petroleum agreements.)
(ix) The government received an option to buy up to twenty per cent of the production.

Later on, the principles of the Western Desert state participation agreements were also applied with respect to acreage in the Gulf of Suez (e.g., the Morgan field, Egypt's most prolific field, which in 1965 had been discovered in that area).

As late as 1969, state participation in concession agreements still belonged to the *Preference Items for Concessions in the U.A.R.* as published by the government. Later, in the early 1970s, when state participation was generally implemented in the Middle East Gulf States and in Libya, and in the majority of cases had been followed by nationalization or a complete takeover of concession agreements, the Egyptian government decided to adopt a new concept consisting in merging the principle of state participation with the more exotic principle of production sharing known from Indonesia (see Chapter 7).

6.7 LIBYA

In 1955 a petroleum law was enacted, the Petroleum Law of 1955, which was the first example of a modern petroleum law appearing in the Northern Africa/Middle East region. It provided from the outset the legal framework for petroleum operations in the country which were just about to begin. The law provided for a straightforward licensing regime based on exclusive concessions rather similar to the licensing regime applied at the time in Egypt (see Section II *On Raw Combustibles* of the Law No. 66 of 1953). Likewise, the Libyan law did away with some peculiarities of the traditional concessions such as the large sized concession areas, covering almost whole countries or anyway a large part of the latter, the extra long duration and the absence of relinquishment obligations.

Note. A special feature of the Petroleum Law was a clause (viz. Clause 16) guaranteeing the stability of terms. This Clause 16 of the standard conditions (after the amendment of the Petroleum Law in 1966) reads as follows:

> The government of Libya will take all steps necessary to ensure that the concessionaire enjoys all the rights conferred by this concession. The contractually rights expressly created by this concession shall not be altered except by mutual consent of the parties. This concession shall throughout the period of its validity be construed in accordance with the Petroleum Law and the Regulations in force on the date of execution of the agreement of amendment by which this paragraph (2) was incorporated into this concession agreement. Any amendment to or repeal of such Regulations shall not effect the contractual rights of the concessionaire without its consent.

Clause 16 was supported and strengthened by the standard Clause 28 which, after the amendment of the Petroleum Law in 1966, reads as follows:

> This concession shall be governed by and interpreted in accordance with the principles of the law in Libya common to the principles of international law and, in the absence of such common principles, then by and in accordance with the general principles of law, including such of those principles as may have been applied by international tribunals.

In 1969 (even before the installation of a new government, the Revolutionary Command Council, under the leadership of Muammar al-Qaddafi) the Libyan government followed again the Egyptian example (namely the Western Desert state participation agreements) and offered foreign oil companies the possibility of entering into joint ventures with the Libyan National Oil Company (LNOC) whereby LNOC would be the holder of the exclusive licence underlying the venture concerned.

Note. Taking an actual agreement made in May 1969 as an illustrative example the terms and rules of the joint venture were as follows. The respective participating interests in the venture of the two participants are not fixed for the duration of the agreement, but made dependable on production performance. Until production exceeds 260,000 b/d, the foreign participant holds a seventy-five per cent interest and LNOC twenty-five per cent, Thereafter, the foreign participant shall transfer a five per cent interest to LNOC at each increase of the production rate by 60,000 b/d, until each participant holds fifty per cent (at 500,000 b/d). All expenditure incurred by the joint venture is borne by the participants in proportion to their interests. However, exploration expenditure made prior to any commercial discovery is borne by the foreign participant and will not be reimbursed by LNOC. All expenditure incurred in respect of the development of a particular field, which is defined as expenditure incurred between the completion of the discovery well and the commencement of commercial production, is financed by the foreign participant and reimbursed with 6.5 per cent interest by LNOC in proportion to its participating interest. The reimbursement takes place by monthly payments in amounts equal to fifty per cent of LNOC's percentage share of total joint production (all producing fields). If the reimbursement is not completed within five years, the balance outstanding shall be paid immediately. In the event that LNOC's participating interest share is increased, it shall reimburse the relevant percentage of all expenditure incurred between the date of completion of the discovery well and the date on which LNOC's percentage was increased, with exception of production operating expenses. From the expenditure to be reimbursed shall be deducted amortization at the rate of ten per cent per year. The reimbursement shall be effected in two equal semi-annual instalments. The foreign participant undertook to arrange for the purchase of LNOC's share of the production to the extent that it does not exceed twenty-five per cent of total production, less any quantities LNOC requires for its own market outlets. The agreed FOB price for such purchase was agreed to be the average market value of Libyan oil less an offtaker's discount of two per cent. For as long as LNOC would not have fully reimbursed the foreign participant, the latter is given the option to nominate a buyer who must be accepted by LNOC for fifty per cent of LNOC's total oil entitlement.

When some years later, production sharing agreements were adopted in Libya as the standard for petroleum authorizations, many elements of the joint venture agreements were taken over by and incorporated in these new agreements. The same phenomenon i.e., borrowing features from preceding state participation agreements, could be observed in Egypt when the government of that country switched to production sharing agreements after a period of adhering to state participation.

With regard to the issue of state participation in its existing concessions, the new Libyan government did not take part in the negotiations leading to the General Agreement on Participation that was signed in Riyadh on 20 December 1972, and did not join that Agreement at a later stage. Instead, following its own policies in this matter, it offered its concessionaires the choice, which had to be made between August 1973 and May 1974, between accepting fifty-one per cent state participation (to be compensated at the fiscal book value) or to be nationalized. The fifty-one per cent share of the production could be bought back at elevated prices. The concession rights and other assets of the concessionaires, which refused to accept fifty-one per cent, were in 1974 nationalized.

Note. Preceding these events, viz. in 1971 and again in 1973, other nationalizations had taken place in Libya. These nationalizations had a political background. In this context British Petroleum fifty per cent interest in Concession No. 65 (the Sarir field) was nationalized on 7 December 1971. This nationalization was settled by an agreement signed on 20 November 1974. Pursuant to this settlement agreement Libya paid the nationalized concessionaire by way of compensation the sum of GBP 17.4 million in cash. On 11 January 1973 Bunker Hunt's fifty per cent interest in the same concession was nationalized. This concession was now completely owned by the Libyan State. A further instance of complete nationalization occurred in 1974 when by virtue of Law No. 11 of 1974

(Decree of Nationalization of 11 February 1974) the concessions held by the US oil company Amoseas, a company jointly owned by California Asiatic Oil Company and Texaco Overseas Petroleum Company, were nationalized. In 1973 the concessionaire had already lost fifty-one per cent of its rights by Law No. 66 of 1973 (Decree of Nationalization of 1 September 1973). The nationalized concessions had been awarded under the terms of the Petroleum Law of 1955. The concession conditions included provision for international arbitration. In accordance with this provision the oil companies submitted the nationalization measures to arbitration, in which however Libya refused to participate on the grounds that the taking of nationalization measures belonged to the sovereign rights of the State and could therefore not be submitted to arbitration. The oil companies obtained an award, in which it was stated that the nationalizations as being in contravention of international law had to be undone [*restitutio in integrum*]. Actually, the award and thereby the restitution was never implemented. The oil companies settled the matter with Libya and under this settlement received as compensation for the loss of their concessions and other assets a volume of crude oil with a value of USD 152 million to be delivered over a period of fifteen months.

After these events, concessions as provided for by the Petroleum Law of 1955 were no longer awarded to foreign oil companies, albeit the Law itself was not repealed. Foreign oil companies were only allowed to undertake petroleum operations on the basis of production sharing contracts to be entered into with the National Oil Corporation (NOC), the designated state oil enterprise which had been established in 1970. NOC is the concessionaire in terms of the Petroleum Law. The first production sharing agreements were signed by NOC on 7 February 1974 (see Chapter 7).

6.8 MALAYSIA

The government of Malaysia had incorporated over the years the results of Middle East negotiations in its own licensing regime (i.e., British Colonial Model, more or less similar to the licensing regime applied at the time in respectively Nigeria, Brunei and, for that matter, in Egypt before the introduction of the state participation agreements in that country). As in Nigeria and Brunei the interests of the Malaysian State were formally protected by the presence a 'most favoured nation' clause in its mining leases and licences. However, after the signing on 14 February 1971 of the Teheran Agreement by the governments of the OPEC Gulf States the Malaysian government developed and pursued its own concession policies. This new approach was introduced by the Petroleum Development Act 1974 (Act of 1974), which came into force on 1 October 1974 and by virtue of which:

(i) petroleum agreements were cancelled with effect from 1 April 1975; and
(ii) ownership in and the exclusive right of exploring for, exploiting, winning and obtaining petroleum, whether onshore or offshore, was vested into a state-owned corporation, which was named Petroliam Nasional Berhad (Petronas).

Note. In vesting the ownership of petroleum into Petronas the new Act constituted a radical departure from the Continental Shelf Act 1966, since the latter had vested all rights with respect to the continental shelf and its natural resources into the Federation to be exercisable by the Government of the Federation.

Since then petroleum operations are regulated by the Act of 1974, as amended by the Petroleum Development (Amendment) Acts of 1975 and 1977, and the Regulations made thereunder. The preceding Petroleum Mining Act 1966 itself was not repealed, merely made non-applicable to Petronas, with exception of the provisions regarding getting permission to enter upon lands for carrying out petroleum operations (see Chapter 7).

6.9 NIGERIA

6.9.1 Background

Originally, petroleum operations in Nigeria were regulated by the Minerals Oils Act of 1958. This Act was an enabling petroleum law providing for a licensing regime of which the terms and conditions as applicable to the licences and leases were set out in the licence or lease document itself. This licensing regime was referred to as the British Colonial Model and was also applicable at the time in Malaysia, Brunei and Egypt. The Mineral Oils Act 1958 was repealed and replaced by the Petroleum Act 1969 which was formerly known as Decree No. 51 of 1969 and issued as such by the then Federal Military Government with effect from 27 November 1969. When the Federal Military Government was succeeded by the Federal Government some consequential changes to the text had to be made: the 'Commissioner' became the 'Minister' and so on.

The new Act contains four Schedules and was accompanied by the Petroleum (Drilling and Production) Regulations.

The Act with its Schedules in combination with the aforesaid Regulations was meant to modernize the previous licensing regime without affecting its concept and principles. (*Note*. It will be recalled that a similar modernization exercise had been undertaken in Oman in 1967, leading to the Main Agreement, see elsewhere in this chapter.) As part of the modernization and in contrast to the preceding Mineral Oils Act they comprise taken together all the relevant (standard) terms and conditions applicable to the licences and leases. Most importantly the Act empowers the Minister to amend regulations or to make and issue new regulations at any time regarding any aspect of petroleum operations carried out under licences or leases granted by him under the Act. In particular the aforesaid Petroleum (Drilling and Production) Regulations are from time to time amended by the Minister to introduce or change important subjects such as rates of royalty and the utilization of associated gas. Another aspect of the modernization exercise is providing for only a rather restricted duration of prospecting licences and oil mining leases.

6.9.2 The Act and Its Schedules

The Act (Principal Provisions):

- Section 1. The entire ownership and control of all petroleum in, under or upon any lands to which this Section applies shall be vested in the State. This Section applies to all land (including land covered by water) which is in Nigeria or is under the territorial waters of Nigeria or forms part of the continental shelf; (*Note*. Petroleum is defined in Section 15 of the Act).
- Section 2. The Minister (of Petroleum Resources) may grant licences to explore for petroleum (exploration licences), licences to prospect for petroleum (prospecting licences) and leases to search for, win, work, carry away and dispose of petroleum (oil mining leases). The provisions of the First Schedule to this Act shall have effect in relation to licences and leases granted under this Section;
- Section 3. No refinery shall be constructed or operated in Nigeria without a licence granted by the Minister.
- Section 4. No person shall import, store, sell or distribute any petroleum products in Nigeria without a licence granted by the Minister.
- Section 7. In the event of a state of national emergency or war (of the existence of which the Minister shall be the sole judge) the Minister shall have the right of pre-emption of all petroleum and petroleum products obtained, marketed or otherwise dealt with under any licence or lease granted under this Act.
- Section 8. The Minister (a) shall exercise general supervision over all operations carried on under licences and leases granted under this Act; (b) shall report annually to the Federal Government... (c) shall have access at all times to the areas covered by licences and leases and to all refineries... for the purpose of inspecting the operations conducted therein and enforcing the provisions of this Act and any Regulation made thereunder... (d) may arrest without warrant any person whom he finds committing, or whom he reasonable suspects of having committed any offence under this Act or any regulations made thereunder, and shall hand over any person so arrested to a police officer with as little delay as possible... (f) may direct in writing that operations under a licence or lease granted under this Act shall be suspended in any area until arrangement have been made which in his opinion are necessary to prevent danger to life and property; (g) may direct in writing the suspension of any operation which in his opinion are not being conducted in accordance with good oil field practice; and (h) may direct in writing the suspension of any operations where in his opinion a contravention of this Act or any Regulations made thereunder has been or may have been or is likely to be committed.
- Section 9. The Minister may make regulations (a) prescribing anything requiring to be prescribed for the purpose of this Act; (b) providing generally for matters relating to licences and leases granted under this Act and

operations carried on thereunder, including (i) safe working, (ii) the conservation of petroleum resources, (iii) the prevention of pollution of water courses and the atmosphere, the making of reports and returns (including the reporting of accidents) . . . (vi) the keeping and inspection of records, books, statistics, accounts and plans; (viii) the measurement of production, and (viii) the measurement of oil delivered to refineries; (c) regulating the construction, maintenance and operation of installations used in pursuance of this Act; (d) regulating refineries and refining operations . . . (e) regulating the importation, handling, storage and distribution of petroleum, petroleum products and other flammable oils and liquids . . . (h) providing for such other matters as in his opinion may be necessary or desirable in order to give proper effect to this Act.

- Section 11. Where by any provision of this Act or any regulation made thereunder a question or dispute is to be settled by arbitration, the question or dispute shall be settled in accordance with the law relating to arbitration in the appropriate State and that provision shall be treated as a submission to arbitration for the purposes of that law. In this Section 'the appropriate State' means the State agreed by all parties to a question or dispute to be appropriate in the circumstances or, if there is no such agreement, the Federal Capital Territory, Abuja.
- Section 14. (3) The transitional and saving provisions in the Fourth Schedule to this Act shall have effect notwithstanding any other provision of this Act.
- Section 15. In this Act: 'explore' means to make a preliminary search by surface geological and geophysical methods, including aerial surveys but excluding drilling below 91.44 meters; 'petroleum' means mineral oil (or any related hydrocarbon) or natural gas as it exists in its natural state in strata, and does not include coal or bituminous shales or other stratified deposits from which oil can be extracted by destructive distillation; 'prospect' means search for by all geological and geophysical methods, including drilling and seismic operations.

The First Schedule to the Act (relating to licences and leases):

- 'Oil exploration licences'. 1. An oil exploration licence shall apply to the area specified therein and shall authorize the licensee to undertake exploration for petroleum in the area of the licence, excluding land in respect of which the grant of an oil prospecting licence or oil mining lease has been approved by the Minister and land in respect of which an oil prospecting licence or oil mining lease is in force. 2. An oil exploration licence shall not confer any exclusive rights over the area of the licence, and the grant of an oil exploration licence in respect of any area shall not preclude the grant of another oil exploration licence or of an oil prospecting licence or oil mining lease over the same area or any part thereof. 3. An oil exploration licence shall terminate on the 31 December next following the date on which it was granted but the licensee shall have the option to renew the licence for one further year if . . . 4. An oil exploration licence shall not confer any right to the grant of an oil prospecting licence or of an oil mining lease.

- 'Oil prospecting licences'. 5. The holder of an oil prospecting licence shall have the exclusive right to explore and prospect for petroleum within the area of his licence; 6. The duration of an oil prospecting licence shall be determined by the Minister, but shall not exceed five years (including any periods of renewal) . . .
- 'Oil mining leases'. 8. An oil mining lease may be granted only to the holder of an oil prospecting licence who has (i) satisfied all the conditions imposed on the licence or otherwise imposed on him by this Act; and (ii) discovered oil in commercial quantities; 9. For the purposes of paragraph 8 of this Schedule, oil shall be deemed to have been discovered in commercial quantities by the holder of an oil prospecting licence if the Minister, upon evidence adduced by the licensee, is satisfied that the licensee is capable of producing at least 10,000 b/d of oil from the licensed area. 10. The term of an oil mining lease shall not exceed twenty, but may be renewed in accordance with this Act; 11. Subject to this Act and any special terms or conditions imposed under paragraph 34 of this Schedule, the lessee of an oil mining lease shall have the exclusive right within the leased area to conduct exploration and prospecting operations and to win, get, work, store, carry away, transport, export or otherwise treat petroleum discovered in or under the leased area; 12. Ten years after the grant of an oil mining lease one-half of the area of the lease shall be relinquished. 13. The lessee of an oil mining lease shall be entitled to apply in writing to the Minister, not less than twelve months before the expiration of the lease, for a renewal of the lease, either in respect of the whole of the leased area or any particular part thereof; and the renewal shall be granted if the lessee has paid all rent and royalties due and has otherwise performed all his obligations under the lease;
- 'Restrictions on assignments'. 14. Without the prior consent of the Minister, the holder of an oil prospecting licence or an oil mining lease shall not assign his licence or lease or any right, power or interest therein or thereunder; 15. The prescribed fee shall be paid on an application for an assignment under paragraph 14 of this Schedule and the Minister's consent for the assignment may be given on payment of such other fee or such premium, or both, and upon such terms as he may decide provided that the Minister may waive payment of that other fee or that premium, or both, if he is satisfied that the assignment is to be made to a company in a group of which the assignor is a member, and is to be made for the purpose of re-organization in order to achieve greater efficiency and to acquire resources for more effective petroleum operations; 16. The Minister shall not give his consent to an assignment unless he is satisfied that (i) the proposed assignee is of good reputation, or is a member of a group of companies of good reputation, or is owned by a company or companies of good reputation; (ii) there is likely to be available to the proposed assignee (from his own resources or through other companies in the group of which he is a member or otherwise)

sufficient technical knowledge and experience and sufficient financial resources to enable him effectually carry out a programme satisfactory to the Minister in respect of operations under the licence or lease which is to be assigned; and (iii) the proposed assignee is in all other respects acceptable to the Federal Government.

- 'Terminations'. 17. (i) The holder of an oil prospecting licence or oil mining lease may, at any time, terminate his licence or lease by giving to the Minister not less than three months notice in writing to that effect; (ii) When notice is given under this paragraph, no rent paid shall be refunded and the termination shall otherwise be without prejudice to any obligation or liability imposed by or incurred under the licence or lease before the effective date of the termination; 18. (i) Without prejudice to paragraph 12 of this Schedule, the holder of an oil prospecting licence or oil mining lease shall be entitled at any time on giving three months notice in writing to the Minister to surrender the licence or lease in respect of any particular part of the licensed or leased area. (ii) Paragraphs 19 to 22 of this Schedule shall apply when a surrender is made under this paragraph; 19. The shape and size of the area to be retained and of the area to be relinquished shall be approved by the Minister;
- 'Power of revocation (no reciprocal treatment)'. 23. (1) The Minister may revoke any oil prospecting licence or oil mining lease if the licensee or lessee becomes controlled directly or indirectly by a citizen of, or subject of, or a company incorporated in, any country which is (a) a country other than the licensee's or lessee's country of origin; and (b) a country the laws of which do not permit citizens of Nigeria or Nigerian companies to acquire, hold and operate petroleum concessions on conditions which in the opinion of the Minister are reasonably comparable with the conditions upon which such concessions are granted to subjects of that country;
- 'Power of revocation (standard causes)'. 24. (1) The Minister may revoke any oil prospecting licence or oil mining lease if in his opinion the licensee or lessee (a) is not conducting operations continuously and in a vigorous and businesslike manner and in accordance with good oil field practice; or (b) has failed to comply with any provision of this Act or any regulation or direction given thereunder or is not fulfilling his obligations under the special conditions of his licence or lease; or (c) fails to pay his due rent or royalties, whether or not they have been demanded by the Minister, within the period specified by or in pursuance of this Act; or (d) has failed to furnish such reports on his operations as the Minister may lawfully require;
- 'Obligatory payments: Fees, rents and royalties'. 30. There shall be paid in respect of licences and leases to which this Schedule applies such application fees as may be prescribed; 31. There shall be paid in respect of licences and leases to which this Schedule applies such rents as may be prescribed; 32. Royalties shall be paid at the prescribed rates or, where rates are specified in special terms and conditions attached to the relevant licence or lease, at the

rates so specified, provided that the Minister may, on giving notice in that behalf, call for and receive any royalty (or part thereof) in kind; (*Note*. The fees, rents and rates and values of royalties are prescribed in the Petroleum [Drilling and Production] Regulations, as amended from time to time.)

- 'State participation'. 34. If he considers it to be in the public interest the Minister may impose on a licence or lease to which this Schedule applies special terms and conditions not inconsistent with this Act including terms and conditions as to (a) participation by the Federal Government in the venture to which the licence or lease relates, on terms to be negotiated between the Minister and the applicant for the licence or lease;
- 'Natural gas'. 35. If he considers it to be in the public interest the Minister may impose on a licence or lease to which this Schedule applies special terms and conditions not inconsistent with this Act including terms and conditions as to (b) special provisions applying to any natural gas discovered; which provisions shall include (i) the right of the Federal Government to take natural gas produced with crude oil by the licensee or lessee free of cost at the flare or at an agreed cost and without payment of royalty; (ii) the obligation of the licensee or lessee to obtain the approval of the Federal Government as to the price at which natural gas produced by the licensee or lessee (and not taken by the Federal Government) is sold; and (iii) a requirement for the payment by the licensee or lessee of royalty on natural gas produced and sold; (*Note*. The utilization of associated natural gas is regulated in the Petroleum [Drilling and Production] Regulations as amended from time to time.)
- 'Payment of compensation for damage caused'. 36. The holder of an oil exploration licence, oil prospecting licence or oil mining lease shall, in addition to any liability for compensation to which he may be subject under any other provision of this Act, be liable to pay fair and adequate compensation for the disturbance of surface or other rights to any person who owns or is lawful occupation of the licensed or leased lands;
- 'Employment of personnel'. 37. The holder of an oil mining lease shall ensure that (a) within ten years from the grant of his lease (i) the number of citizens of Nigeria employed by him in connection with the lease in managerial, professional and supervisory grades (or any corresponding grades designated by him in a manner approved by the Minister) shall reach at least 75 per cent of the total number of persons employed by him in those grades, and (ii) the number of citizens of Nigeria in any one such grade shall be not less than 60 per cent of the total, and (b) all skilled, semi-skilled and unskilled workers are citizens of Nigeria;
- 'Indemnity against third party claims'. 38. The holder of any licence or lease to which this Schedule applies shall at all times indemnify and keep harmless the Federal Government, the Minister and every officer in the civil service of the Federation or the States (and their agents) against all actions, costs, charges, claims and demands whosoever which may be made or brought by any third party in relation to any matter or thing done or purported to be done in pursuance of this Act.

- '*Force majeure*'. 40. (1) Failure on the part of the holder of a licence or lease to which this Schedule applies to fulfil any of the terms or conditions of the licence or lease shall not (except as may be otherwise provided for in or in relation to the licence or lease) give the Minister any claim against the licensee or lessee, or be deemed a breach of the licence or lease, if the failure arises from causes beyond the control of the licensee or lessee; (2) If from any such cause the fulfilment by any such licensee or lessee of any term or condition of his licence or lease or of any provision of this Act is delayed, the period of delay shall be added to the period fixed for the fulfilment of the term or condition;
- 'Arbitration'. 41. In any question or dispute arises in connection with any licence or lease to which this Schedule applies between the Minister and the licensee or lessee (including a question or dispute as to the payment of any fee, rent or royalty), the question or dispute shall be settled by arbitration unless it relates to a matter expressly excluded from arbitration or expressed to be at the discretion of the Minister. (*Note*. No further details are provided.)

The Second Schedule to the Act (on the Rights of Pre-emption)

Note. The provisions of this Second Schedule comprise the details of and shall have effect in relation to the right of pre-emption given to the Minister under Section 7 (1) of the Act.

The Third Schedule to the Act (on Repeals)

Note. The Third Schedule contains a list of preceding Acts that are repealed by this Act.

The Fourth Schedule (on Transitional and Savings Provisions)

1. Any licence or lease granted under an enactment repealed by this Act shall continue in force notwithstanding the repeal, but shall be subject to this Act and to any regulations made thereunder except as regards the duration of the licence or lease, the rent and royalties payable in respect thereof and any term an condition as to which the Minister certifies that the justice of the case requires that the term or condition in question shall continue to be effective notwithstanding this Act.

By letter dated 15 September 1972 the holders of pre-1969 oil mining leases were informed that the powers of the Commissioner to save any clause in these leases would not be exercised.

Note. On the effective date of the Act quite a number of oil mining leases for land and delta acreage were in existence. The main leaseholder at the time was the Shell-BP Petroleum Development Company of Nigeria Ltd. (hereinafter referred to as Shell-BP). Its land oil mining leases had originally been granted under the Mineral Oils Act of 1958 (now repealed) for a period of thirty years w.e.f. 1 January 1960. Similarly, its delta oil mining leases had been granted for thirty years w.e.f. 1 January 1962. The land and delta leases could, according to their terms (which were incorporated in the lease document concerned), be renewed for a period of thirty years, but when the time for renewal arrived the question arose of whether in the light of the abovementioned letter of 15 September

1972 from the Commissioner this condition was still valid. This would be the case if either the thirty years' renewal condition of the pre-1969 leases could be considered to have been upheld by the Fourth Schedule as having to do with or being part of 'duration', or the thirty years' renewal condition would not contradicted by any condition of the Act or its Regulations (see Paragraph 10 of the First Schedule to the Act from which it may be concluded that the period of renewal has been left open). In the end both types of leases were renewed for the period applied for, i.e., for thirty years, w.e.f. 1 July 1989, all now expiring on 30 June 2019, and against payment of a renewal fee of USD 5 million per lease, the same fee that had to be paid when the offshore oil mining leases were renewed the year before. At the effective date of their renewal (1 July 1989) the land and delta mining leases had still been valid: the land leases would have expired on 31 December 1989 and the delta leases on 31 December 1991. This premature renewal in July 1989 was considered necessary in order to prepare the ground for the later assignment of interests by the Nigerian National Petroleum Corporation (NNPC) under the terms of the Assignment Agreement of 30 June 1989 (discussed hereinafter). Oil mining leases for the offshore were originally granted to Shell-BP on 23 March 1979 subject to this Act for a period of twenty years but retroactively w.e.f. 1 December 1968. These leases were renewed for twenty years w.e.f. 1 December 1988, expiring now on 30 November 2008. A renewal fee of USD 5 million per lease was required to be paid.

6.9.3 The Petroleum (Drilling and Production) Regulations

The Petroleum (Drilling and Production) Regulations provide for all the necessary details to complete the Nigerian petroleum legislation. It contains the application procedures, ancillary rights and restrictions, recruitment and training of Nigerians, methods and practices of working and conservation, rules on field development (i.e., all fields, structures, reservoirs and other oil traps shall be developed and produced in strict accordance with a field development programme, which shall be submitted for the prior approval of the Director of Petroleum Resources and shall give details of the estimated size of the pool, the known physical parameters of the pool, reservoirs or structures at the time of drawing up the programme, the intended drilling pattern if any, the production or drainage pattern, and the anticipated drive mechanism), rules on keeping records, reports, accounts and measurement of production, etc. The Regulations also provide for the method of calculation and the rates of fees, rents and royalties. The Regulations have been and will be amended from time to time and can therefore immediately be adapted to changed circumstances Early amendments were introduced by the Petroleum (Drilling and Production) (Amendment) Regulations 1973 (as to natural gas and posted prices), by those of 1979 (as to royalty rates) and by the Associated Gas Re-Injection Decree 1979 (as to the utilization of associated gas). The latter Decree was itself amended w.e.f. 1 January 1985 by the Associated Gas Re-Injection (Amendment) Decree 1985 with subsidiary legislation contained in the Associated Gas Re-injection (Continued Flaring of Gas) Regulations 1984.

6.9.4 State Participation

Although Nigeria had become a member of OPEC in July 1971, it was not directly involved in the negotiations that had started end 1971 between the governments of the Arab Gulf States within OPEC and their concessionaires on the subject of state participation in their existing and ongoing concessions. Nigeria did also not join later in the General Agreement on Participation that in the end was signed in Riyadh on 20 December 1972. There was however no need for the Nigerian government to take part in the Middle East negotiations on participation. The government's interests were adequately protected by a 'most favoured nation' clause included in its financial agreement of 25 April 1967 made with Shell-BP.

Note. The financial agreement referred to above introduced the expensing of royalties on crude oil exported and the calculation of taxable income on the basis of posted prices less certain to be agreed allowances (see Chapter 8). The 'most favoured nation' included in the financial agreement had been preceded by a sideletter agreement of 1 June 1959 giving the same undertaking on the part of Shell-BP but this time referring more specifically to any percentage change in the 50/50 profit sharing arrangement that was established by a Deed of Covenant of the same date. Apart from being more specific the 1959 sideletter made only reference to the continent of Africa and did not mention the oil exporting States of the Middle East as the 'most favoured nation' clause in financial agreement of 1967 did.

Nevertheless, as a Member State of OPEC, the Nigerian government closely followed the negotiations on state participation and in the course of time implemented the results obtained under the first stage multi-lateral and second stage bilateral negotiations by concluding with Shell-BP a series of agreements to this purpose. Said series opened with the Heads of Agreement made on 11 June 1973. These Heads were based on the General Agreement on Participation albeit the Heads provided for an initial participating interest of thirty-five per cent instead of twenty-five per cent. By a Letter Agreement dated 16 April 1974, the terms of the heads were revised to the effect of increasing the State's participating interest and fixing it, w.e.f. 1 April 1974 at fifty-five per cent (instead of sixty per cent w.e.f. 1 January 1974 as had been agreed with the Arab Gulf States within OPEC. The State's participating interest was vested in NNPC by the Nigerian Petroleum Corporation Act of 1977 and by the same Act NNPC was authorized to acquire an additional five per cent participating interest w.e.f. 1 July 1979. (*Note*. Nigeria followed in this matter the example of Oman and Abu Dhabi.)

Pursuant to the Acquisition of Assets (British Petroleum Company Ltd.) Act of 1979 whereby BP's shares in Shell-BP were extinguished and this restructured company, now a Shell company) was left in possession of a participating interest of twenty per cent, NNPC acquired a further direct twenty per cent interest in the licences and leases w.e.f. 1 August 1979.

Note. This partial nationalization measure was politically motivated and is very reminiscent of the equally politically motivated nationalization measure taken at the end of 1973 by the government of Iraq with respect to the Basrah Petroleum Company (BPC). In the latter case the measure nationalized forty-three per cent of the share capital of

BPC and resulted in the introduction of a new shareholder, not in a restructuring of the concessionaire.

The 80/20 division of interests between NNPC and the restructured leaseholder and the issues that in the meantime had arisen between the parties involving the proper compensation for the initial thirty-five per cent and the additional five per cent interest and for quantities of buy-back oil with respect to 1973 and 1974 were finally settled in a Participation Agreement dated 22 August 1984. Subsequently, by the Assignment Agreement of 30 June 1989, NNPC transferred out of its eighty per cent interest a ten per cent interest to Shell and a five per cent interest to each of two new participants, Elf and Agip, thereby creating a new joint venture comprising four participants. At a later stage, NNPC sold another five per cent interest to Elf, retaining fifty-five per cent for itself.

6.10 VENEZUELA

Although one of the founding Member States of OPEC, Venezuela did not take part in the negotiations on the subject of state participation in existing and ongoing concession that were organized in 1971/72 by the Arab Gulf States within OPEC. Without going through the intermediate stages of partial state participation as envisaged by the multi-lateral General Agreement on Participation and the subsequent bilateral 60/40 state participation agreements, the government terminated all concessions w.e.f. 1 January 1976. Following the examples set by Kuwait, Qatar and Saudi Arabia, agreements were made with the former concessionaires which comprised service agreements and oil purchase contracts with respect to the latter's former concession areas.

6.11 UNITED KINGDOM

6.11.1 Scope and Structure

Petroleum legislation in the United Kingdom consists of Acts (formal laws) and subsidiary legislation in the form of implementing and supplementary Regulations made under such Acts. The ownership of petroleum in situ in the sub-soil of the seaward areas, comprising the territorial waters and the continental shelf, is vested in the Crown (Petroleum [Production] Act 1934 and the Continental Shelf Act 1964 with respect to the territorial waters and the continental shelf respectively). Both Acts have to be categorized as enabling petroleum laws, authorizing the making of comprehensive Regulations and the granting of licences. The Regulations provide the model clauses for incorporation in licences to be granted. Once a licence has been granted, its terms and conditions can only be altered through enactment of an Act, which provides for a change of the model clauses incorporated in the licence. Such changes with retroactive effect were enacted by the Petroleum and Submarine Pipelines Act 1975 and by the Oil and Gas (Enterprise) Act 1982.

Note. The Petroleum and Submarine Pipelines Act 1975 introduced a new system of operating in accordance with approved development plans and production limitations and rules concerning the use and construction of submarine pipelines. The Act also created the British National Oil Corporation (BNOC), which, in compliance with the directions of the Secretary, was authorized to explore for and obtain petroleum in any part of the world. The Act had not made it obligatory for a licensee to accept BNOC as a participant in a licence nor was BNOC with retrospective effect given a participating interest in the licences preceding the effective date of the Act, but during the period of its existence which ended in 1985, the requirements concerning BNOC's participation in the licence were stipulated as particular conditions in the Gazette notice of the licensing round concerned (see hereinafter).

New regulations can amend or entirely replace the foregoing regulations, but cannot change the terms and conditions of licences granted under these regulations. Regulations are made by the Secretary of State for Trade and Industry hereinafter 'the Secretary of State' (in the model clauses for licences referred to as the Minister) and made subject to the control of Parliament. However under the Acts of 1934 and 1964 the Secretary of State has been given complete freedom to introduce, before granting a licence, any condition he thinks fit in the particular circumstances. The Secretary's discretionary power makes it possible to design terms and conditions that are suitable for particular cases such as for instance exploring for petroleum in deep water or other hostile areas. As a rule, the petroleum industry will be consulted before any such particular conditions are prescribed.

Licences can be applied for landward areas or for seaward areas (the offshore). These two areas are separated by the low water line. The seaward areas comprise the territorial sea and such part of the continental shelf as falls to the United Kingdom under international law. The UK North Sea continental shelf has been demarcated in accordance with boundary treaties made with the opposite North Sea coastal States. For the purposes of licensing the seaward areas are covered by a grid consisting of units measuring twelve minutes longitude by ten minutes latitude (about 200 to 250 square km), each unit comprising 120 sections. Licences are granted with respect to tranches of blocks as well as or in addition to individual blocks. Each block is bounded by minute lines of latitude and longitude and therefore comprising a whole number of sections.

As example for a description of applicable legislation including model clauses for offshore production licences (Schedule 4) is taken the Petroleum (Production) (Seaward Areas) Regulations 1988 (the Principal Regulations). In 1995 these Regulations were amended by the Petroleum (Production) (Seaward Areas) (Amendment) Regulations 1995 in order to implement the relevant provisions of EC Directive 94/22/EC of 30 May 1994 (hereinafter referred to as the EC Directive). Details of the blocks to be licensed, the latest date upon which applications for production licences are to be made and the period within which licences are to be granted are to be published in the Official Journal of the European Communities.

The 1995 Amendment Regulations also introduced (in accordance with paragraph 4 of Article 3 of the EC Directive) a procedure whereby licensees may apply for a licence for an area contiguous to the area or areas covered by its existing licence where the Secretary of State decides that geological or production considerations justify the grant of such a licence and accordingly so notifies the licensees involved. The previously existing possibility of non-invited applications has been omitted. Schedule 3 (form of application for a licence) and Schedule 4 (model clauses for production licences in seaward areas) of the Principal Regulations were also revised.

The Principal Regulations were further amended with the coming into force on 16th December 1996 of the Petroleum (Production) (Seaward Areas) (Amendment) Regulations 1996. The amendments completed the implementation of the EC Directive. Apart therefrom the amendments enable the Secretary of State to offer for licence tranches of blocks (a tranche is defined as a group of contiguous blocks) as well as or in addition to individual blocks and reduce the fee payable upon application for a production licence and extend the particulars required to support an application for a licence. On this occasion Schedule 3 and Schedule 4 were further revised. The amendment of Schedule 4 involved the duration of a licence, the possibility of extension beyond the initial term of three years and the size of the licensed area to be retained by the licensee into the third term.

Summarizing: every application for a production licence shall relate to a block or a tranche described in a notice published in the Official Journal of the European Communities or to a number of blocks or tranches so described. The notice shall provide for applications relating to any of the blocks or tranches so described to be made and determined in competition with others, specifying a date before which any such applications are to be made, being a date at least ninety days after the date on which the notice is published, and a date on which, or a period within which licences will be granted to successful applicants. In case of any block or tranche in respect of which provision for competing applications was made on a previous occasion in a notice in said Official Journal and that provision did not result in the grant of a licence the notice may provide for applications to be made and determined at any time. Where the Secretary of State decides that geological or production considerations justify the granting of a production licence in respect of any area to the holder of a licence in respect of a contiguous area and notifies him and any other holders of licences in respect of areas contiguous to the area in question accordingly, any of them may apply for a licence in respect of the area in question within whatever period the Secretary of State considers sufficient for this purpose and specifies in the notification.

As set out in Schedule 3 to the Principal Regulations (as amended) an application for a production licence shall contain inter alia:

(a) reference numbers of the block(s). the tranche(s) or the blocks and tranches in respect of which the application is made and, if the application is made by tender, the consideration by way of initial payment which the applicant is prepared to offer for each such block or tranche;

(b) an analysis of the geology of the applied for area identifying in particular petroleum prospects;
(c) the technical data on which the analysis of the geology is based;
(d) the work programme for evaluating the potential petroleum production from the applied for area which the applicant would be prepared to undertake in the initial term of the applied for licence;
(e) evidence of the applicant's technical and financial capacity to undertake that work programme, including the number of staff the applicant intends to assign for that purpose and any relevant technical qualifications held by those staff;
(f) a statement of the general environmental policy of the proposed operator in respect of similar offshore licensed activities together with a summary of his management systems for the implementation of that policy and a summary of how those management systems will be applied to the aforementioned work programme.

Note. The above provisions stem from the implementation of the EC Directive.

Offshore licensing is organized in successive rounds of licensing. Should the Secretary of State wish to introduce new model clauses for incorporation in the licences to be granted under a new round of licensing, then the prevailing regulations must have to be amended or completely be replaced by new regulations before announcing the round. In fact, each of the rounds held so far has been preceded by the issue of amendment regulations or by new regulations. If the Secretary of State decides to introduce particular conditions not contained in the regulations applicable to the licensing round, these conditions will now have to be included in the notice published in the Official Journal of the European Communities.

Note regarding cash bonuses. As to paying a cash bonus for licences when put up for sale, it may be mentioned that on the occasion of the 4th round (1971/1972), the 8th round (1982/1983) and the 9th round (1984/1985) each time fifteen blocks were tendered. Highest bid in the 4th round auction amounted to GBP 21 million for block 211/21 (i.e., GBP 105,567 per square km). A total sum of almost GBP 33 million was received for the seven blocks auctioned under the 8th round. Under the 7th round (1980/1981) forty-two blocks were awarded under the condition of an initial payment of GBP 5 million per block).

The duration of an offshore production licence is determined in the Petroleum (Production) (Seaward Areas) Regulations, applicable to the round under which the licence is granted. In the past, i.e., before the implementation of the EC Directive, a particular duration was stipulated in the Gazette Notice, if modifications were considered desirable in a special case.

Note. Up to and including the 10th round (1986/1987) the term of an offshore licence was six years (called the initial term). If all conditions, but in particular the execution of the mandatory work programme, have been fulfilled, the term of the licence is prolonged by thirty years. Under the 9th round deep water frontier area blocks were treated as a particular case. The initial term was eight years and the extension was forty years. Under the 11th round (1988/1989) and 12th round (1990/1991), which were governed by the Principal

Regulations as amended in 1990, the offshore licence was given a 'standard' initial term of six years, which initial term could be prolonged by thirty years in two stages: a first extension of twelve years and a second extension of eighteen years. The last prolongation is granted if before the expiry of the first extension a development programme has been approved or has been imposed or if the Secretary of State in his discretion has so decided. Under the 11th round, an exception was made in respect of licences comprising deep water blocks. The periods of such a licence were extended as follows: after an initial term of eight years, a first prolongation of sixteen years and a second prolongation of twenty-four years. For frontier area blocks, treated as a particular case under the 12th round, the periods granted were 9, 15 and 24 years respectively. Under the rules of Schedule 4 of the Principal Regulations (as amended) the first nine years were split into an initial term of three years and a second term of six years.

State participation, although still respected under the 1994 EC Directive, will not be required as a particular condition because the subject is no longer pursued in the UK.

Note regarding BNOC. After the Petroleum and Submarine Pipe-lines Act 1975 had created the British National Oil Corporation (BNOC), the requirements regarding participation by the corporation in the licences were stipulated as a particular condition in the *London Gazette* notice of the offshore licensing round concerned. Under the 5th round (1976/77) and in the 6th round (1978/79) a fifty-one per cent participation of BNOC in the licence, if so desired by this corporation, was made obligatory. Under the 7th (1980/1981), 8th (1982/83) and 9th (1984/85) rounds the licensee was required to give BNOC the option to buy at market price up to fifty-one per cent of any petroleum produced under the licence. The same purchase option was acquired by BNOC with respect to the offshore licences which had been granted under the first four offshore licensing rounds which preceded the corporation's establishment under the Act of 1975. On 13 March 1985 the UK government announced its intention to terminate the existence of BNOC. The main reason for the government's decision was of a political nature: the then Conservative government did not believe in state participation. But there was also a pragmatic reason for the decision. The corporation encountered increasing difficulties in selling its oil production. Its share of UK North Sea oil production had increased to about sixty per cent, the sum of 12.5 per cent royalty oil (licensees were obliged to pay royalty in kind if so demanded by the State), 51 per cent participation oil and 51 per cent purchase-option oil. In 1984 UK production amounted to 2.3 million b/d. A sizable part of the corporation's share had to be sold back to the licensees who had a buy back option, but at that time still about 800,000 b/d had to be sold by the corporation on the open market. BNOC's total sales of about 1.3 million b/d, including 250,000 b/d royalty oil, put the corporation in a strong position to influence the development and level of the price of oil on the global market. The official price established by BNOC for the North Sea marker crude, Brent-Blend (end 1983/beginning 1984 fixed at USD 30/b) had a direct impact on the price of the crude oil, Bonny-Light, exported from Nigeria and through the Bonny-Light price the BNOC price (to the discomfort of the corporation) became linked to the OPEC price system. In the second half of 1984 OPEC requested the UK government to persuade BNOC to maintain the price of Brent-Blend at its prevailing high level in order to give Nigeria no cause to breach the OPEC price arrangements. However, BNOC experienced increasing problems in selling its oil against the price at which it has bought the oil from its licensees (the latter received from BNOC the official price). When then, in October 1984, Statoil, the Norwegian state oil company, started to sell its oil at a discount, BNOC in its turn was forced to lower the official price of Brent-Blend from USD 30/b to USD 28.65/b (meanwhile the open market price had dropped to USD 27/b). Nigeria felt obliged to follow suit and, in contravention to the OPEC agreement on pricing, lowered the price of its Bonny-Light by USD 2/b to USD 28/b. A price crisis was the result (see Chapter 3). BNOC did not succeed in solving its difficulties, whereupon the government decided to withdraw the corporation as intermediary trader and allow the licensees to sell their oil themselves. BNOC had suffered grate losses on the reselling of the oil, but about eighty per cent were compensated, at least as far as the UK Treasury was concerned, by the fact that the licensees had to pay income taxes calculated on the official price they were receiving from BNOC.

A few years before taking the decision to terminate the existence of BNOC, the government had already taken away some important privileges that the corporation had received when it was established. The Act of 1975 had designated BNOC as the official government adviser in petroleum affairs. BNOC was given a preferential position in acquiring any licence interest that was intended to be transferred and no decision on approving such assignment was given before BNOC had made it known whether it would be interested in acquiring for itself the interest involved. Moreover, the Act exempted BNOC from paying the Petroleum Revenue Tax. In 1979 all these privileges were abolished.

The termination of BNOC had been anticipated and the ground therefore was prepared by the Oil and Gas (Enterprise) Act of 1982. Pursuant to this Act, BNOC was authorized to establish 'equity oil subsidiaries', to transfer its petroleum exploration and production rights to such subsidiaries and to dispose of its shares in them. Under this procedure a subsidiary under the name 'Britoil' was established, which later was privatized. In 1988 Britoil was bought by British Petroleum.

After the existence of BNOC was terminated in 1985 (Oil and Pipelines Act 1985), the existing purchase options were inherited by the Oil and Pipeline Agency (OPA) and the arrangement in the form of a particular condition was continued under the 10th round, but only in respect of petroleum other than natural gas, i.e., the licensees under that round were required to grant OPA an option to buy up to fifty-one per cent of the petroleum other than natural gas produced under the licence. This condition was not repeated in later rounds, since the government in June 1988 put an end to all forms of all state participation.

Note regarding British Petroleum. In August 1914, just after the outbreak of the First World War, the UK government was authorized to purchase 51.7 per cent of the shares in the Anglo-Persian Oil Company. The latter was a private sector company that had gone public on 19 April 1909. The company was the sole owner of a concession in former Persia that on 18 May 1901 had been granted to William Knox D'Arcy. The acquisition of the shares, for which the government paid GBP 2 million, afforded the government a controlling interest in the company, but at the time it was stressed by the government that the acquisition was made for strategic reasons only, namely to secure the supply of fuel oil to the Royal Navy, and that it had no intention of interfering in commercial matters. In 1933 this concession was unilaterally cancelled by the government and a new concession was negotiated and finally granted in April 1933 to the former concessionaire which in 1935 was renamed the Anglo-Iranian Oil Company. In the wake of a revolution under the leadership of Mossadegh, the 1933 concession was nationalized on 20 March 1951 and taken over by the newly established National Iranian Oil Company (NIOC). It had been the then revolutionary government's intention to entrust the domestic petroleum operations to NIOC without allowing any involvement of foreign oil companies. However the revolution failed and after 3.5 years of negotiations a new agreement was signed on 25 October 1954 by NIOC and Iranian Oil Participants Ltd, a UK corporation, also known as the Consortium. The Anglo-Iranian Oil Company obtained a share of forty per cent in the consortium and was on that occasion renamed British Petroleum. Much later, the UK government started gradually to reduce its shareholding in British Petroleum. In 1977 an interest of 35.2 per cent was sold. In 1987 the government tried to sell the remaining shareholding, but did not completely succeed. In December 1995 the remaining 1.8 per cent was sold for GBP 518 million.

UK petroleum operations are not only regulated by the primary petroleum laws and the government regulations based thereon but also by the Petroleum Act 1987 (concerning the abandonment of offshore installations, the establishment of safety zones around installations and pipelines on land). Furthermore, there have been enacted a number of Acts, which are relevant to the conduct of offshore petroleum operations, such as the Mineral Workings (Offshore Installations) Act 1971, as amended, and its regulations, which regulate activities on offshore installations, the Act, which provides for the prevention of pollution caused by oil, and the Gas Act 1986 (concerning the supply of gas in the UK through pipes). The latter Act has to be brought into line with the EU Directive concerning the completion of the internal market for natural gas (the EC Directive 98/30/EC of 22 June 1998). Finally, the United Kingdom is a Member State of the EU and Party to a number of Conventions concerning the protection of the marine environment (see further Chapter 9). As such the United Kingdom has introduced and from time to time will have to introduce new petroleum and environmental legislation in order to implement corresponding EC Directives (e.g., the 1994 EC Directive discussed elsewhere in this chapter) and a few EC Directives concerning environmental matters) and decisions taken under the respective Conventions. In this connection are OSPARCON and the decisions taken by the Commission established under its rules of particular importance.

6.11.2 The 1988/1996 Model Clauses (Schedule 4)

6.11.2.1 Exclusive Rights

An offshore production licence is granted by the Secretary of State for Trade and Industry (referred to as the Minister) in exercise of the powers conferred upon him by the Act of 1934 and the Act of 1964 and grants to the licensee the exclusive right for the duration of the licence and subject to the provisions thereof to search and bore for and get petroleum in the seabed and subsoil under the seaward area comprising an area of a certain number of square kilometres more particularly described in Schedule 1 to the licence in terms of a number of square km and particular blocks.

6.11.2.2 Duration, Termination and Surrender of Acreage

The initial term is three years. If the work programme offered and accepted for the initial term has been duly performed and if the licensee has submitted a work programme for further exploration, the licence may be continued for a further term of six years (called the second term). If this programme is duly performed and if the licensee has drilled at least one well in the initial licensed area, the licence may be continued, if the licensee so desires, for a term of fifteen years (the third term) in respect to the continuing part of the licensed area. The size of the said continuing part depends on licensee's drilling performance. The size shall be one third of the initial licensed area or, if greater, 120 sections, if the licensee has since the grant of the licence drilled one well in the initial licensed area, whereas the size shall be two-thirds of the initial licensed area, or, if greater, 120 sections, if two or more wells have been drilled.

Note. In the Gazette Notice concerning the 12th round, a special arrangement was made with respect to licences comprising frontier area blocks. The licensee concerned may retain at the end of the initial term (which in this case has a duration of nine years) one-quarter of the licensed area by drilling one exploration well, one-half of the area by drilling two wells and may keep the whole licensed area by drilling three or more exploration wells.

After the third term, the licence may be continued for a period of twenty-four years if the licensee so desires, and if before the expiry of the third term a development programme has been approved or has been imposed or if the Minister in his discretion has so decided. Where a licence has continued for a total period of twenty-four years after the expiry of the third term, the Minister, on application, may in his discretion agree with the licensee that the licence shall continue for a period to be further agreed.

The licensee may on any anniversary of the date of the commencement of the initial term, terminate the licence or surrender any part of the licensed area. Any surrendered acreage (whether surrendered voluntarily or surrendered obligatorily on the occasion of the licence entering onto the third term), shall be bounded by minute lines of latitude extending not less than two minutes of longitude and minute lines of longitude extending not less than two minutes of latitude, consist of not less than thirty sections and have boundaries which either coincide with the corresponding boundaries of the block or are not less than two sections distant from those boundaries.

Note 1. The licence duration provisions as described above (i.e., four consecutive terms of 3, 6, 15 and 24 years with an obligatory surrender of acreage at the end of the second term) applied to the 17th round of licensing. For the 18th and 19th rounds the conditions were changed as follows: the first term of the licence became six years, followed by a second term of twelve years and, provided a field development programme has been approved for at least one field by the end of the second term, a final term of eighteen years would be granted. At the end of the first term fifty per cent of the licence area has to be surrendered or leaving at least thirty sections. For the next rounds, from round 20 up to and including round 23, the first term was four years, the second term again four years and the third and final term eighteen years. At the end of the first term, fifty per cent of the licence area has to be surrendered or leaving at least thirty sections.

Note 2. On the occasion of the 21st round of licensing (February 2003), a new type of licence was introduced: viz. a 'Promote Licence'. The promote licence allows the holder thereof to assess and study the prospectivity of the licence area without having to accept burdensome work commitments. The duration of the licence is two years but may be converted into a regular licence if the licensee offers a substantive work programme.

Note 3. On the occasion of the 22nd round of licensing (February 2004) again a special type of licence was introduced, viz. the 'Frontier Licence'. This licence was meant for exploration activities in the Atlantic Margin west of the Shetlands. A frontier licence is granted for relatively large areas for the first term of two years. If the licensee wants to continue after the first term, seventy-five per cent of the area has to be surrendered and a full work programme has to be offered. For entering a third term of six years, a further part of the area has to be surrendered.

6.11.2.3 Obligatory Payments

The licensee is obliged to make the following payments to the Minister:

(i) in respect to each half year (the chargeable period) payments of royalty of an amount equal to the percentage specified in a Schedule 3 to the licence of the value of the petroleum (this value for the chargeable period being determined by reference to schedule 2 of the Oil Taxation Act 1975); and

(ii) payments in accordance with Schedule 2 to the licence (i.e., initial payment and annual payments measured by the number of square km comprised in the licensed area).

Note with reference to (i). With effect from 1 January 2003 royalty payments have been abolished. Further details can be found in Chapter 8.

6.11.2.4 Working Obligations

The licensee shall before the expiry of the initial term of the licence carry out such scheme of prospecting including any geological surveys and test drilling as may be set out in Schedule 4 to this licence.

At any time the Minister may serve a notice on the licensee requiring him to submit to him an 'appropriate' exploration work programme with respect to a specified period that should fall within the term of the licence.

Note. For the purpose of this provision an appropriate programme is described as one which any person who, if had the competence and the resources needed to exploit the rights granted by his licence to the best commercial advantage and were seeking to exploit those rights in this manner, could reasonably be expected to carry out during the period specified in the notice served upon him by the Minister.

If the Minister notifies the licensee that he is of the opinion that the programme submitted by the licensee does not satisfy the relevant requirements, the latter must either submit the matter to arbitration or prepare a further programme that satisfies the relevant requirements. The licensee must carry out any programme that is approved by the Minister or has been determined in arbitration to satisfy the relevant requirements.

6.11.2.5 Development and Production Programmes

(1) The licensee shall not:

(a) erect or carry out any relevant works for the purpose of producing petroleum from the licensed area or for the purpose of transporting to a place on land petroleum produced from that area, or
(b) produce petroleum from that area (otherwise than in the course of searching for petroleum or drilling wells), except with the consent in writing of the Minister or in accordance with a programme which the Minister has approved or served on the licensee in pursuance of the following provisions.

(2) The licensee shall prepare and submit to the Minister in respect of such period during the term of this licence as the Minister may direct a programme specifying:

(a) the relevant works which the licensee proposes to erect or carry out for either of the purposes mentioned above under (a);
(b) the proposed locations of the works and other particulars; and
(c) the maximum and minimum quantities of petroleum in the form of gas and the maximum and minimum quantities of petroleum in other forms which the licensee proposes to produce as mentioned above under (1)(b) in each calendar year during the specified period.

(3) If the Minister directs the licensee (a) to prepare different programmes in respect of petroleum from such different parts of the licensed area as are specified in the direction; or (b) where a programme approved or served relates to a particular period during the term of this licence, to prepare a programme or programmes in respect of a further period or further periods during that term, the licensee shall comply with the direction.
(4) After the licensee has submitted its proposed programme, the Minister shall give notice in writing stating:

(a) that the Minister approves the programme;
(b) that the Minister approves the programme subject to the condition that certain specified works shall not be used before the expiration of a specified period or shall not be used without his consent; or

(c) that the Minister rejects the programme on one or both of the following grounds namely (a) that the carrying out of any of the proposals included in the proposed programme would be contrary to good oil field practice; (b) that the proposals included in the programme in respect of the maximum and minimum quantities of petroleum are, in the opinion of the Minister, not in the national interest.

(5) If the Minister rejects the programme prepared by the licensee as being contrary to good oilfield practice he shall include in the notice a statement clarifying the matter. If the Minister rejects the programme prepared by the licensee as being not in the national interest, he shall include in the notice the rates at which he considers that, in the national interest, petroleum should be produced from the area. If the programme proposed by the licensee is rejected by the Minister, the licensee shall prepare modifications of the programme, which ensure that the carrying out of the programme with those modifications would not be contrary to good oil field practice or, as the case may be, which ensure the production of petroleum from the area at the rates specified by the Minister and which are not such that the carrying out of the programme with those modifications would be contrary to good oilfield practice.

(6) If the licensee fails to submit a modified programme, and save for arbitration in a dispute in the matter of good oil field practice, the Minister may instead of revoking the licence in consequence of the failure serve on the licensee such a programme as the Minister considers that the licensee should have submitted to him. The licensee must carry out any programme approved by the Secretary or imposed upon it by the latter.

(7) Where the Minister gives notice approving an original programme, with or without constraints, or approving a modified programme (prepared after the rejection of the original programme), or imposing a different programme, or where it is determined by arbitration in the matter of good oil field practice that the licensee is not required to modify its (initially rejected) proposed programme in this respect, the Minister may give to the licensee a so-called 'limitation notice'. A limitation notice authorizes the Minister by a further notice given to the licensee from time to time after the expiration of the period specified in that behalf in the limitation notice, to provide that the programme to which the limitation notice relates shall have effect while the further notice is in force with the substitution for any quantity of petroleum or any period specified in the programme relating to the maximum and minimum quantities of gas and the maximum and minimum quantities of petroleum in other forms of a different quantity of petroleum or a different period specified in the further notice. A quantity or period specified in such a further notice as that to be substituted for a quantity or period which is specified in the programme in question shall be within the limits specified in the limitation notice as those applicable to that quantity or period specified

in the programme and those limits shall be such as to secure that the expenditure to be incurred by the licensee in complying with the further notice, in case where an effect of the notice is to increase the quantity of petroleum which the licensee is required to produce from the licensed area in any period, is less than the cost of drilling a new well in the licensed area at the time when the further notice is given.

(8) The Minister shall not give such a further notice of which an effect is to increase the quantity of petroleum which the licensee is required to produce from the licensed area during any period, unless the Minister is satisfied that the notice is required by reason of a national emergency and shall not give any other such further notice unless he is satisfied that the notice is required in the national interest. Any question arising under the above provisions as to what is, is not or what is required in the national interest or as to what is, or is required by reason of a national emergency shall be determined by the Minister.

Note. The limitation notice procedures regulating production levels are in conformity with Article 6.2 of the 1994 EC Directive concerning the freedom of Member States to include conditions in authorizations. In this case the said procedure can be considered to fall under 'the planned management of hydrocarbon resources (for example the rate at which hydrocarbons are depleted or the optimization of their recovery').

6.11.2.6 Commencement and Abandonment and Plugging of Wells

The licensee shall not commence or after abandoning, shall not recommence the drilling of any well and shall not abandon any well without the consent in writing of the Minister. The licensee shall ensure compliance with any conditions subject to which any such consent is given.

6.11.2.7 Distance of Wells from Boundaries of Licensed Area

No well shall, except with the consent in writing of the Minister, be drilled or made so that any part thereof is less than 125 metres from any of the boundaries of the licensed area.

6.11.2.8 Control of Development Wells

The licensee shall not suspend work on the drilling of a development well, or having suspended it, shall not begin it again, except with the consent in writing of the Minister and in accordance with the conditions, if any, subject to which the consent is given.

6.11.2.9 Provision of Storage Tanks, Pipes, Pipelines and Other Receptacles

The licensee shall use methods and practice customarily used in good oilfield practice for confining the petroleum obtained from the licensed area in tanks, gas holders, pipes, pipelines or other receptacles constructed for that purpose.

6.11.2.10 Avoidance of Harmful Methods of Working

The licensee shall execute all operations in or in connection with the licensed area in a proper and workmanlike manner in accordance with methods and practice customarily used in good oilfield practice and shall take all steps practicable in order:

(i) to control the flow and to prevent the escape or waste of petroleum discovered in or obtained from the licensed area;
(ii) to conserve the licensed area for productive operations;
(iii) to prevent damage to adjoining petroleum bearing strata;
(iv) to prevent the entrance of water to petroleum bearing strata except for the purposes of secondary recovery; and
(v) to prevent the escape of petroleum into any waters in or in the vicinity of the licensed area.

The licensee shall not flare any gas from the licensed area; or use gas for the purpose of creating or increasing the pressure by means of which petroleum is produced from that area, except with the consent in writing of the Minister and in accordance with the conditions, if any, of the consent.

6.11.2.11 Appointment of Operators

The licensee shall ensure that another person (including, in the case where the licensee is two or more persons, any of those persons) does not exercise any function of organizing or supervising all or any of the authorized operations unless that other person is a person approved in writing by the Minister.

6.11.2.12 Fishing and Navigation

The licensee shall not carry out any of the authorized operations in or about the licensed area in such a manner as to interfere unjustifiably with navigation or fishing in the waters of the licensed area or with the conservation of the living resources of the sea.

6.11.2.13 Unit Development

If at any time in which the licence is in force the Minister shall be satisfied that the strata in the licensed area or any part thereof form part of a single geological petroleum structure or petroleum field (hereinafter referred to as 'an oil field') other parts whereof are formed by strata in areas in respect of which other (similar) licences are then in force and the Minister shall consider that it is in the national interest in order to secure the maximum ultimate recovery of petroleum and in order to avoid unnecessary competitive drilling that the oil field should be worked and developed as a unit in cooperation by all persons including the licensee whose licences extend to or include any part thereof the following provisions shall apply:

(i) upon being so required by notice in writing by the Minister the licensee shall cooperate with such other licensees in the preparation of a scheme (hereinafter referred to as 'a development scheme') for the working and development of the oil field as a unit by the licensee and the other licensees in cooperation, and shall, jointly with the other licensees submit such schemes for the approval of the Minister;

(ii) if no development scheme shall be submitted to the Minister within the period stated in the notice, or if the development scheme so submitted shall not be approved by the Minister, the latter may himself prepare a development scheme which shall be fair and equitable to the licensee and all other licensees, and the licensee shall perform and observe all the terms and conditions thereof; and

(iii) if the licensee objects to any such development scheme prepared by the Minister he may refer the matter to arbitration.

6.11.2.14 Directions as to Oil Fields across International Boundaries

Where the Minister is satisfied that any strata in the licensed area or any part thereof form part of an oil field, other parts whereof are in an area to which the Minister's powers to grant licences do not apply and the Minister is satisfied that it is expedient that the oil field should be worked and developed as a unit in cooperation by the licensee and all other persons having an interest in any part of the oil field, the Minister may from time to time by notice in writing give to the licensee such directions as the Minister may think fit as to the manner in which the rights conferred by the licence shall be exercised.

6.11.2.15 Disposal of Production

The licensee shall ensure that all petroleum won and saved from the licensed area other than petroleum used therein for the purpose of carrying on drilling and production operations or pumping to field storage and refineries shall be delivered on shore in the United Kingdom, unless the Minister gives notice of his consent in writing to delivery elsewhere and in such case the licensee shall ensure compliance with any conditions subject to which that consent is given. Any conditions imposed by the Minister on such a consent may, without prejudice to the generality of the Minister's right to impose conditions of any nature, include provision:

(i) as to the place of delivery;
(ii) as to the price to be obtained for the petroleum to which such consent relates;
(iii) as to the time within which and the manner in which payment of the price is to be made; and
(iv) requiring payment to be made to a person resident in the United Kingdom.

6.11.2.16 Reports to Be Treated as Confidential

All records, etc., and information (referred to as 'the specified data') which the licensee is or may be from time to time required to furnish under the provisions of the licence shall be supplied at the expense of the licensee and shall not be disclosed (except with the consent in writing of the licensee which shall not be unreasonably withheld) be disclosed to any person not in the service or employment of the crown, provided that the Minister shall be entitled at any time:

(a) to make use of any of the specified data for the purpose of preparing and publishing such returns and reports as may be required of the Minister by law;
(b) to furnish any of the specified data to some specified Institutions involved in geological activities.

The Minister and those Institutions shall be entitled at any time to prepare and publish reports and surveys of a general nature using information derived from any of the specified data and to publish any of the specified data of a geological, scientific or technical kind either after the expiration of the period of five years beginning with the date when the data was due to be supplied to the Minister in accordance with specific provisions of the licence, or, if earlier, the date when the Minister received the data or after the expiration of such longer periods as the Minister may determine after considering any representations made to him by the licensee about the intended publication.

6.11.2.17 Indemnity against Third Party Claims

The licensee shall at all times keep the Minister effectually indemnified against all actions, proceedings, costs, charges, claims and demands whatsoever which may be made or brought against the Minister by any third party in relation to or in connection with this licence or any matter or thing done or purported to be done in pursuance thereof.

6.11.2.18 Restrictions on Assignment, etc.

The licensee shall not, except with the consent in writing of the Minister and in accordance with the conditions, if any, of the consent, do anything whatsoever whereby under the law any right granted by this licence or derived from a right so granted becomes exercisable by or for the benefit of or in accordance with the directions of another person.

6.11.2.19 Power of Revocation

If any of the events specified in the licence shall occur then and in any such case the Minister may revoke the licence and thereupon the same and all the rights thereby granted shall cease and determine but subject nevertheless and without prejudice to any obligation or liability incurred by the licensee or imposed upon him by or under the terms and conditions of the licence. The specified events include among others: any payment of consideration for the licence being in arrear or unpaid for two months; any breach or non-observance by the licensee of any of the terms and conditions of this licence; the bankruptcy of the licensee; any breach or non-observance by the licensee of the terms and conditions of a development scheme; if the licensee is a company, the licensee's ceasing to direct and control either its operations under the licence or any commercial activities in connection with those operations from a fixed place within the United Kingdom; if the licensee is a company and there is a change in the control of the licensee; and the Minister serves notice in writing on the licensee stating that the Minister proposes to revoke the licence unless such a further change in the control of the licensee as specified in the notice takes place within the period of three months beginning with the date of service of the notice and that further change does not take place within that period.

6.11.2.20 Arbitration

If at any time any dispute, difference or question shall arise between the Minister and the licensee as to any matter arising under or by virtue of this licence or as to their respective rights and liabilities in respect thereof then the same shall, except where it is expressly provided by this licence that the matter or thing to which the same relates is to be determined, approved, or consented to by the Minister, be referred to arbitration in accordance with the Arbitration Act 1950 by a single arbitrator who, in default of agreement between the Minister and the licensee and, in case of arbitration in relation to a development scheme, other licensees affected by that scheme, as to his appointment, shall be appointed by the Lord Chief Justice of England for the time being.

6.12 NORWAY

6.12.1 Scope and Structure

Petroleum legislation in Norway consists of Acts (formal laws) and subsidiary legislation in the form of implementing and supplementary Regulations.

The Act currently in force is the Act of 29 November 1996, No. 72 as amended. With effect from 1 July 1997, this Act replaced its predecessor, the Act of 22 March 1985, No. 11. Like its predecessor, the Act is a comprehensive petroleum law exclusively focused on the regulation of offshore petroleum operations.

Note. Before the present Act came into force offshore petroleum operations were regulated by the Act of 22 March 1985 No. 11. In turn the latter Act was preceded by the Act of 21 June 1963 No. 12 and the Regulations issued under the respective Acts. The 1963 Act was of the nature of an enabling law, leaving the main provisions of the petroleum regime to be set out in regulations. The regulations concerned were the Royal Decrees of 9 April 1965 and 8 December 1972 respectively, both 'relating to exploration for and exploitation of petroleum on the Norwegian continental shelf'. After the 1985 Act entered into force, the scope of the Act of 1963 was restricted to 'scientific research and exploration for and exploitation of sub-sea natural resources other than petroleum resources' and the said Royal Decrees were both terminated. In contrast with the 1963 Act, the 1985 Act was a comprehensive petroleum law, containing the main provisions of the petroleum regime and leaving only details to be settled by regulation to be issued by the King. The numerous licences, among which there are very important ones as far as production volumes are concerned, granted under either the 1965 Royal Decree or the 1972 Royal Decree remained in existence but became 'automatically' governed by the provisions of the 1985 Act, except as was provided in the said Act itself with respect to their duration, in respect of which the terms of the preceding Decrees were maintained. The same procedure was followed when the present Act was introduced. Licences are granted in licensing rounds. Up to and including the 3rd round licences were granted under the 1965 Royal Decree. As from the 4th round up to and including the 9th round licences were granted under the 1972 Decree. As from the 10th round licences were granted in accordance with the 1985 Act and as from 1 July 1997, the effective date of the present Act licences are granted in accordance with said Act.

The main reason to replace the Act of 1985 was for Norway to implement the relevant provisions of the EC Directive 94/22/EC of 30 May 1994 (hereinafter referred to as the EC Directive). The occasion was used to pay more attention to procedures relating to plans for development and production and plans for decommissioning and disposal of facilities, to clarification and codification of existing practices. Nonetheless, the 1996 Act itself has on a number of occasions and on a number of points been amended. The following Acts were involved: the Act of 14 December 2001 No. 98; the Act of 28 June 2002 No. 61; the Act of 20 December 2002 No. 88; the Act of 27 June 2003 No. 68; and finally the Act of 1 July 2006.

Note. Although not a Member State of the EC or EU, Norway was committed to said implementation by reason of its membership of the Agreement on the European Economic Area (EEA) of 2 May 1992. The latter was negotiated between the then European Economic Community, the European Coal and Steel Community, and their Member States, on the one hand, and the individual Member States of the European Free Trade Association (EFTA), on the other and was signed in Oporto on 2 May 1992. The Agreement provides for an EEA Joint Committee which is empowered to decide to extend Community legislation, such as the aforementioned Directive of 30 May 1994, to the EEA.

In as much the Norwegian petroleum legislation operates under a comprehensive petroleum law the standard rights and obligations of a licensee are mainly found in the Act and to a lesser extent in the regulations made under the Act and in the special legislation concerning safety, working conditions and pollution control.

The few conditions particular to the individual licence are to be found in the licence itself (e.g., the actual size and geographic location of the licensed area, the duration, the actual obligatory work programme and the adjustment of the undivided interests of the individual licensees in case of state participation) and further in the associated agreements. The standard conditions of a licence will change and new terms will be imposed, if and when the Act and the regulations based thereon will be changed provided always that conditions may be saved under the transitional provisions of any new Act. Changing or replacing the prevailing Act is within the power of the Norwegian Parliament (the *Storting*). Changing or replacing regulations supplementing or implementing the Act is delegated by the Act to the King but any new regulations made by the King may also contain saving provisions. Still it may be expected that all persons, who undertake petroleum activities on the Norwegian continental shelf, shall at any time more or less operate under the same terms and conditions, the only differences between them consisting in the aforesaid particular conditions of their licences, among which most prominently the percentage state participation.

Apart from national legislation, there exists relevant international legislation. In the first place should be mentioned the 1982 Convention on the Law of the Sea the rules of which determine Norway's rights and obligations as a coastal State. Next in importance comes the 1992 Convention for the Protection of the Marine Environment of the North-East Atlantic (OSPARCON), to which Norway is a party. The requirements and rules of this Convention and the decisions taken by the Commission established under it are of direct relevance to the conduct of the Norwegian offshore petroleum operations which take place within the Area of the Convention in as much these requirements, rules and decisions will have to be implemented and made applicable to the licences and their holders through regulations made under the Act.

6.12.2 The 1996 Act (as Amended)

Chapter 1 Introductory Provisions. Section 1-1 The right to subsea petroleum deposits and resource management. The Norwegian State has the proprietary right to subsea petroleum deposits and the exclusive right to resource management.

Section 1-2 Resource management. Resource management is executed by the King in accordance with the provisions of this Act and decisions made by the Storting (Parliament). Resource management of petroleum resources shall be carried out in a long-term perspective for the benefit of the Norwegian society as a whole. In this regard the resource management shall provide revenues to Norway and shall contribute to ensuring welfare, employment and an improved environment, as well as to the strengthening of Norwegian trade and industry and industrial development, and at the same time take due regard to regional and local policy considerations and other activities.

Section 1-3 Requirements relating to licence, etc. None other than the State may conduct petroleum activities without the licences, approvals and consents required pursuant to this Act. Provisions otherwise in the Act and regulations issued pursuant to the Act shall apply to such activities insofar as they are appropriate.

Section 1-4 Scope of application. The Act applies to petroleum operations in connection with sub-sea petroleum deposits under Norwegian jurisdiction. The Act also applies to petroleum activities inside and outside the Kingdom and the Norwegian continental shelf to the extent that the application follows from international law or from agreement with a foreign State. The Act does not apply to Svalbard, including its internal waters and territorial sea.

The Act shall apply to utilization of produced petroleum that takes place on Norwegian land territory or seabed subject to private property rights only when such utilization is necessary to or constitutes an integrated part of production or transportation of petroleum.

When a pipeline located in Norwegian internal waters, in Norwegian territorial sea or on the continental shelf starts outside Norwegian jurisdiction, the King may, to the extent this follows from international law, decide which provisions of this Act shall apply to such pipeline with associated equipment.

As regards liability for pollution damage according to Chapter 7 and damage arisen as a result of pollution and waste according to Chapter 8, the special provisions laid down in those chapters shall apply.

This Act does not apply to Svalbard, including its internal waters and territorial sea.

The King may issue further regulations to supplement or delimit the provisions of this Section, i.e., with regard to what utilization is considered necessary to or constitutes an integrated part of production or transportation of petroleum as mentioned in the second paragraph. In case of doubt, the king may decide whether a facility or activity shall be comprised by the Act or not.

Note 1. The Norwegian North Sea continental shelf is demarcated up to 64 degrees north latitude pursuant to a boundary treaty made with the United Kingdom (the Agreement of 10 March 1965 relating to the delimitation of the continental shelf between the two States). Up in the North, in the Barents Sea, the territorial waters have been divided between Norway and the former Soviet-Union, but the adjacent continental shelf has not yet been demarcated.

Note 2. The territory of Svalbard to which the 5th paragraph refers includes all islands and their internal waters and territorial sea making up the Archipelago of Svalbard as described in Article 1 of the Treaty of Svalbard signed on 9 February 1920. Said Treaty recognized and confirmed Norway's full and absolute sovereignty over the Archipelago and the waters belonging thereto but made this sovereignty subject to an equal rights and treatment regime in favour of ships and nationals of the Contracting States in respect of hunting and fishing and in respect of maritime, industrial, commercial, mining activities.

From the moment the concept of the continental shelf had been formulated and described under the terms of the 1958 Convention on the Continental Shelf the Norwegian government took the position that the Treaty did not apply to the continental shelf of Svalbard as defined in the said Convention. When later the continental shelf was described in terms of the 1982 Convention of the Law of the Sea the Norwegian government shifted its position and claimed that the continental shelf of Svalbard formed part of the Norwegian continental shelf and for this reason was beyond the territorial reach of the Svalbard Treaty.

Section 1-5 Other Norwegian law. Norwegian law other than this Act, including provisions relating to licences, consents or approvals required according to the legislation, shall also be applicable to petroleum activities. This applies unless otherwise warranted by an Act, a decision by the King, international law or agreement with a foreign State. Notwithstanding the above, other Norwegian law shall not apply to mobile facilities under foreign flag other than those that are permanently placed, unless otherwise stipulated by an Act or by a decision made by the King in Council.

Section 1-6 Definitions. In this Act the following definitions shall apply:

a) Petroleum, all liquid and gaseous hydrocarbons existing in their natural state in the subsoil, as well as other substances produced in association with such hydrocarbons.
b) Petroleum deposit, an accumulation of petroleum in a geological unit, limited by rock characteristics by structural or stratigraphic boundaries, contact surface between petroleum and water in the formation, or a combination of these, so that all the petroleum comprised everywhere is in pressure communication through liquid or gas. In cases of doubt, the Ministry will determine what shall be regarded as a petroleum deposit.
c) Petroleum activity, all activities associated with subsea petroleum deposits, including exploration, exploration drilling, production, transportation, utilization and decommissioning, including planning of such activities, but not including, however, transport of petroleum in bulk by ship.
d) Facility, installation, plant and other equipment for petroleum activities, however not supply and support vessels or ships that transport petroleum in bulk. Facility also comprises pipeline and cable unless otherwise provided.
e) Exploration, geological, pertophysical, geophysical, geochemical and geotechnical activities, including shallow drilling, as well as operation and use of a facility to the extent it is used for the purpose of exploration.
f) Exploration drilling, drilling of wildcat and appraisal wells, as well as operation and use of a facility to the extent it is used for the purpose of exploration drilling.
g) Production, production of petroleum, including drilling of production wells, injection, improved recovery, treatment and storage of petroleum for transport, and shipment of petroleum for transport by ship, as well as the construction, placing, operation and use of a facility for the purpose of production.

h) Transportation, shipment of petroleum by pipeline as well as the construction, placing, operation and use of a facility for the purpose of transportation.
i) Utilization, cooling in order to liquefy gas, refining and petrochemical activity, production and transmission of electric power and other use of produced petroleum, storage of petroleum as well as the construction, placing, operation and use of a facility for the purpose of utilization.
j) Licensee, physical person or body corporate, or several such persons or bodies corporate, holding a licence according to this Act or previous legislation to carry out exploration, production, transportation or utilization activities. If a licence had been granted to several such persons jointly, the term licensee may comprise the licensees collectively as well as the individual licensee.
k) Operator, anyone executing on behalf of the licensee the day to day management of the petroleum activities.
l) Continental shelf, the seabed and subsoil of the submarine areas that extend beyond the Norwegian territorial sea, throughout the natural prolongation of the Norwegian land territory to the outer edge of the continental margin, but no less than 200 nautical miles from the base lines from which the breadth of the territorial sea is measured, however not beyond the median line in relation to another State. (*Note*. This definition conforms to the definition included in the 1982 Convention on the Law of the Sea, see Chapter 9).
m) Upstream pipeline network, any pipeline or network of pipelines operated or constructed as part of an oil or gas production project, or used to convey natural gas from one or more production facilities of this type to a processing plant, a terminal or a final landing terminal. Those parts of such networks and facilities that are used for local production activities of a deposit where natural gas is produced are not regarded as upstream pipeline networks.
n) Natural gas undertaking, any natural or legal person carrying out at least one of the following functions: production, transmission, distribution, supply, purchase or storage of natural gas, including liquefied natural gas (LNG), and which is responsible for the technical tasks or maintenance related to these functions, but shall not include final customers.
o) Eligible customers, gas-fired power generators, irrespective of their annual consumption level, as well as other final customers consuming more than 25 million cubic metres of gas per year and per consumption site and having access to the system in the EEA State in which the party in question is domiciled.

Chapter 2 Exploration Licence. Section 2-1 Granting of exploration licence, etc. The Ministry may grant to a body corporate a licence to explore for petroleum within limited areas of the seabed or its subsoil. Exploration licence may also be granted to a physical person domiciled in an EEA State.

The exploration licence gives the right to explore for petroleum. It does not give exclusive right to exploration in those areas that are mentioned in the licence or any preferential right when production licences are granted. Production licences may be granted to others, or licence may be granted according to Section 4-3 in areas covered by exploration licences, without giving rise to any liability or any obligation to refund fees that have been paid. Exploration licence is granted for three years, unless another period of time is stipulated.

The Ministry may authorize a licensee holding an exploration licence to undertake other exploration activities. The King may issue regulations relating to the contents of an application for exploration licence, the scope of such licence, the further conditions of the licence and the fee to be paid.

Section 2-2 The area covered by the exploration licence. The exploration licence shall state the area covered by the licence. The exploration licence does not give any right to exploration in areas covered by production licences, unless otherwise decided by the Ministry in accordance with Section 3-11.

Chapter 3 Production Licence, etc. Section 3-1 Opening up of new areas. Prior to the opening of new areas with a view to granting production licences, an evaluation shall be undertaken of the various interests involved in the relevant area. In this evaluation, an assessment shall be made of the impact of the petroleum activities on trade, industry and the environment, and of possible risks of pollution, as well as the economic and social effects that may be a result of the petroleum activities.

The opening of new areas is a matter which shall be put before local public authorities, central trade and industry associations and other organizations which may be presumed to have a particular interest in the matter.

Furthermore, it shall be made known through public announcement which areas are planned to be opened up for petroleum activities, and the nature and extent of the activities in question. Interested parties should be given at least three months to present their views.

Section 3-2 Division of the continental shelf. Offshore areas inside the outer boundary of the continental shelf are divided into blocks of fifteen latitude minutes by twenty longitude minutes in size, unless adjacent land areas, common boundaries with the continental shelf of other States, or other circumstances warrant otherwise.

Section 3-3 Production licence. The King in Council may, on conditions to be further stipulated, grant production licences. A production licence may cover one or several blocks or parts of blocks.

Production licence may be granted to a body corporate established in conformity with Norwegian legislation and registered in the Norwegian register of Business Enterprises, insofar as other requirements are not applicable pursuant to international agreements. Production licence may also be granted to a physical person domiciled in a State of the European Economic Area (EEA).

A production licence entails the exclusive right to exploration, exploration drilling and production of petroleum deposits in areas covered by the licence. The licensee becomes the owner of the petroleum which is produced.

The King may stipulate as a condition for granting a production licence that the licensees shall enter into agreements with specified contents with one another.

Note. Production licences are accompanied by model agreements, which have to be executed by the licensee within a certain period after the licence has been granted. The so-called associated agreements are the joint operating agreements, formerly known as the state participation agreements, the personnel training agreements and the frame agreements for research and development.

Section 3-4 Agreements with a view to applying for a production licence. Cooperation agreements entered into with a view to applying for a production licence shall be submitted to the Ministry. The Ministry may require alterations to be made in such agreements.

Section 3-5 Announcement and granting of a licence. Prior to the granting of a production licence, the Ministry shall, as a rule, announce the area for which applications for production licences may be submitted.

The announcement shall be published through notification in *The Norwegian Gazette* (*Norsk Lysingsblad*) and the *Official Journal of the European Communities*. The notification shall stipulate a time limit for the filing of applications of not less than ninety days, and it shall contain such information as decided by the Ministry.

Note. In 2003/2004 licensing rounds were held in which companies were invited to bid for pre-defined acreage (PDAs) in the more mature areas of the Norwegian continental shelf. The 19th licensing round opened in June 2005 and included sixty-four blocks of which thirty in the Barents Sea. Reference: *PEPS*, IHS Inc.- Energy, Country: Norway, Laws, Date: 9 January 2006.

The granting of a production licence shall be done on the basis of factual and objective criteria, and the requirements and conditions stated in the notification. The King is not obliged to grant any production licence on the basis of the applications received.

The King may grant production licences without announcement. Prior to such granting of a production licence, the licensees of production licences in all adjacent areas shall be given the opportunity to apply for a production licence for the area in question. Notification shall be published in *The Norwegian Gazette* (*Norsk Lysingsblad*) and the *Official Journal of the European Communities* indicating the blocks which are affected.

Further regulation about the content of an application for production licence, and about the payment of application fees, are issued by the King.

Note. The above Section 3-5 implements the provisions of Article 3 of the EC Directive 94/22/EC of 30 May 1994.

Section 3-6 State participation. The King may decide that the Norwegian State shall participate in petroleum activities according to the Act (see Section 11-1 to and including Section 11-10).

Note. State participation in licences has a long and complicated history. It was introduced as early as 1969. With respect to licences granted in that year and in 1971 by virtue of the

Royal Decree of 1965, the State was given the choice between either to acquire an undivided interest (varying from five per cent to even forty per cent) in the licence or to receive a percentage share (varying from ten per cent to 17.5 per cent) in the net income of the licensee. After the Royal Decree of 1972 entered into force and after the establishment of *Den Norske Stats Oljeselskap* (Statoil) on 14 July of that year licensees were obliged to accept Statoil as a co-licensee and to enter into an agreement of cooperation with this company, under predetermined conditions. Under most licences granted so far Statoil's (initial) fifty per cent interest share in the licence may be increased at the expense of the undivided interest shares of the other licence holders. A notable exception in this respect forms production licence PL 037, covering the Norwegian part of the Statfjord field (blocks 33/9 and 33/12). This important licence was granted on 10 August 1973, under the last phase of the 2nd round. In this licence Statoil retained its initial and final fifty per cent interest. Where in a licence an increase of Statoil's initial share is provided for, such an increase is within pre-determined limits at the option of Statoil: in this respect distinction should be made between on the one hand pre-9th round 1972 licences where the limits are determined on a sliding scale and on the other 9th round 1972 licences and 1985 licences, i.e., licences issued under the terms of the Act of 1985, where the limit is set by the Ministry on the basis of a firm number stipulated in the licence concerned. Under the rules of the pre-9th round 1972 licences the Statoil interest is increased from fifty to fifty-one per cent upon Statoil joining or concurring with a statement of commerciality. Such statement in turn triggers the declaration of commerciality to be issued by the operator, which marks the start of Statoil's financial contributions. The sliding scale used in these licences relates a range of estimated peak values of the daily production from the licensed area to the value of Statoil's percentage interest (higher than fifty per cent) at any such peak values. Such a sliding scale and the modalities of its implementation are set out in the associated joint operating agreement. Generally, the sliding scale starts at 60, 62, or 65 per cent from a zero daily production and results in a Statoil interest in the licence of respectively seventy-five or eighty per cent at a daily peak production of 200,000 barrels oil equivalent or more. Such increase becomes effective upon the approval of the Ministry of the production profile of the commercial field in which Statoil is going to participate. Under the post-8th round licences the initial undivided interests of the licence holders are adjusted on the basis of a pre-determined number at the option of the Ministry at the time of approving the required development plan as prepared by the licensees including Statoil. Upon the date of approval Statoil's proportional financial contributions will commence. All costs incurred by Statoil's co-licensees prior to the start of Statoil's financial contributions i.e., the cost of exploration, appraisal and evaluation of the first commercial discovery under the aforesaid licences, are for the account of said co-licensees. Meanwhile, a restructuring of state participation had taken place. With effect from 1 January 1985, an arrangement was made between the State and Statoil in accordance with which the State acquired a direct undivided interest in almost all licences, in which Statoil was holding such an interest. From Statoil's initial fifty per cent in such licences, thirty per cent was transferred to the State and twenty per cent retained by Statoil. Any future increase of Statoil's interest in the licence must be completely transferred to the State. The State's share, also referred to as the State Direct Financial Interest (SDFI), was represented and managed by Statoil, which exercised on behalf of the State all rights and obligations attached to the SDFIs, albeit the State remained directly responsible for the funding of its share. This was the situation till Statoil was partially privatized. The partial privatization started in 2001 when 18.2 per cent of Statoil shares were offered to the public. In 2004 a further 4.6 per cent of Statoil shares were sold, and in 2005 about 0.8 per cent, leaving in the end the State with a 76.33 per cent shareholding interest in Statoil, now having become Statoil ASA. Prior to the initial public offering of Statoil shares fifteen per cent of the SDFIs were sold to that company and 6.5 per cent share to Norsk Hydro and other Norwegian offshore operators. The balance of 78.5 per cent was transferred to the custody of a new state company, Petoro A.S., which was established on 9 May 2001 for the purpose of serving as the new manager of the SDFI portfolio, since Statoil, as a consequence of its partial privatization, could no longer act as such. This rearrangement of the management of state participation is provided for by a new Chapter 11 which was introduced into the Act by the Act of 18 May 2001 No. 23. In 2007 Statoil merged with Norsk Hydro and formed a new 67.3/32.7 company by the name of StatoilHydro. In this new company the State through its majority shareholding in both Statoil and Norsk Hydro holds an overall majority interest of 62.5 per cent.

In recent rounds of licensing the modalities of state participation are much simpler and SDFI is a minority interest and sometimes there is no SDFI at all. Still, if Statoil or in the future StatoilHydro participates in a licence the overall state participation is higher than the SDFI, if any, in the licence concerned.

Section 3-7 Operator. When granting a production licence, the Ministry shall appoint or approve an operator.

Change of operator must be approved by the Ministry. When warranted for particular reasons, the Ministry may undertake the change of operator.

If the Ministry appoints or approves an operator who is not a licensee according to the production licence, the provisions concerning the obligations of the licensee established in or pursuant to this Act shall apply correspondingly to the operator unless otherwise specifically provided.

Section 3-8 Work obligation. The King may impose on the licensee a specific work obligation for the area covered by the production licence.

Note. In practice, applicants for a production licence will propose a work programme and the licence will be granted, everything else being equal, to the applicant that made the best proposal in this regard.

Section 3-9 The duration of a production licence, etc. The production licence shall be granted for up to ten years. If the production licence is granted for a shorter period of time, the Ministry may subsequently extend the licence period within the ten-year limit.

A licensee who has fulfilled the work commitment according to Section 3-8 and the conditions otherwise applicable to the individual production licence may demand that the licence shall be extended after the expiry of the period stipulated pursuant to the first paragraph. The extension period shall be stipulated in the individual production licence, and shall as a general rule be up to thirty years, but may in specific cases be up to fifty years.

When granting a licence, the King shall stipulate what part of the area covered by the production licence the licensee may demand an extension for pursuant to the second paragraph. The size of the area stipulated according to the preceding sentence shall as a rule constitute fifty per cent of the area covered by the production licence, notwithstanding that the licensee shall be entitled to keep at least one hundred square kilometres. The Ministry may on application consent to the licensee keeping more than the area stipulated when the licence is granted according to this provision.

The King may issue regulations relating to the delimitation of the areas to be relinquished according to the third paragraph.

The Ministry may, on application from the licensee and when particular reasons so warrant, extend the production licence in excess of the extension according to the second paragraph. Application for extension must have been submitted no later than five years prior to the expiry of the production licence, unless otherwise approved or decided by the Ministry. The Ministry stipulates the conditions for such particular extension.

Section 3-10 Dividing the area of a production licence. The Ministry may on application from a licensee approve that part of the area covered by the production licence is partitioned off and grant a separate production licence for the area partitioned off.

The King may issue regulations relating to delimitation of the area partitioned off.

Section 3-11 Right for others to exploration. The Ministry may, in specific cases, grant to someone other than the licensee the right to carry out exploration in an area covered by a production licence. The Ministry shall determine the kind of exploration that may be carried out and the duration of the activities.

Section 3-12 Right for others to place facilities, etc. The licensee cannot oppose the laying of pipelines, cables or wires of various kinds, or the placing of other facilities on, in or above the area covered by the production licence. Such facilities must not cause unreasonable inconvenience to the licensee.

The provision of the first paragraph applies correspondingly to necessary route and soil surveys prior to such placement.

Section 3-13 Natural resources other than petroleum resources, etc. A production licence does not preclude the granting to others than the licensee of rights to undertake exploration for and production of natural resources other than petroleum resources, provided it does not cause unreasonable inconvenience to the petroleum activities conducted by a licensee pursuant to the production licence. The same applies to scientific research.

Section 3-14 Relinquishment of areas. The licensee may during the period mentioned in Section 3-9, first paragraph, with three months notice, relinquish parts of the area covered by the production licence. Thereafter, relinquishment of parts of the area may take place at the end of each calendar year, provided notice of such relinquishment has been given at least three months in advance. The Ministry may require the obligations stipulated according the production licence and the conditions on which it has been granted to be fulfilled prior to relinquishment.

The King may issue regulations relating to the delimitation of relinquished areas.

Section 3-15 Surrender of a production licence. The licensee may during the period mentioned in Section 3-9, first paragraph, with three months notice, surrender a production licence in its entirety. Thereafter, surrender may take place at the end of each calendar year provided notice of such surrender has been given at least three months in advance. The Ministry may require the obligations stipulated according to the production licence and the conditions on which it has been granted to be fulfilled prior to surrender.

Chapter 4 Production, etc., of Petroleum, Section 4-1 Prudent production. Production of petroleum shall take place in such a manner that as much as possible of the petroleum in place in each individual petroleum deposit, or in several deposits in combination, will be produced. The production shall take place in accordance with prudent technical and sound economic principles and in such a manner that waste of petroleum or reservoir energy is avoided. The licensee shall carry out continuous evaluation of production strategy and technical solutions and shall take the necessary measures in order to achieve this.

Section 4-2 Plan for development and operations of petroleum deposits. If a licensee decides to develop a petroleum deposit, the licensee shall submit to the Ministry for approval a plan for development and operation of the petroleum deposit. The plan shall contain an account of economic aspects, resource aspects, technical, safety related, commercial and environmental aspects, as well as information as to how a facility may be decommissioned and disposed of when the petroleum activities have ceased. The plan shall also comprise information on facilities for transportation or utilization comprised by Section 4-3. In the event that a facility is to be placed on the territory, the plan shall in addition provide information about what applications for licences etc. have been submitted according to other applicable legislation.

The Ministry may, when particular reasons so warrant, require the licensee to produce a detailed account of the impact on the environment, possible risks of pollution and the impact on other affected activities, in respect of a larger defined area.

If the development is planned in two or more phases, the plan shall, to the extent possible, comprise the total development. The Ministry may limit its approval to apply to individual stages.

Section 4-3 Specific licence to install and to operate facilities for transport and utilization of petroleum. The Ministry may on specified conditions grant a specific licence to install and to operate facilities when right to install and to operate facilities does not follow from an approved plan for development and operation pursuant to Section 4-3. Application shall be submitted containing plan for the construction, placing, operation and use of such facilities as mentioned in the first paragraph, including shipment facilities, pipelines, liquefaction facilities, facilities for generation and transmission of electric power and other facilities for transportation or utilization of petroleum. Licence may be granted for a fixed period of time, which may on application from the licensee, be extended by the Ministry.

The provisions of Section 4-2, with the exception of the first and fifth paragraphs, shall apply correspondingly unless otherwise decided by the Ministry.

Section 4-4 Stipulation of production schedule, etc. The Ministry shall, prior to or concurrently with approval pursuant to Section 4-2 or a licence being granted pursuant to Section 4-3, approve the production schedule. A production schedule other than that which follows from Section 4-1 may be stipulated if warranted by resource management considerations or other significant social considerations.

Burning of petroleum in excess of the quantities needed for normal operational safety shall not be allowed unless approved by the Ministry. Upon application from the licensee, the Ministry shall stipulate, for fixed periods of time, the quantity which may be produced, injected or cold vented at all times. An application shall be submitted at such times and shall have such contents as decided by the Ministry. The Ministry shall base this stipulation on the production schedule on which the development plan is based, unless new information on the deposit or other circumstances warrant otherwise.

When necessary for weighty social reasons, the King in Council may, for an individual petroleum deposit or several petroleum deposits, stipulate other production schedules than those stipulated or approved pursuant to the first and third paragraph, and may in this connection order improved recovery. If the decision according to this paragraph is to the effect that production shall be reduced in relation to the production schedule stipulated or approved, the Ministry shall endeavour to apportion to a reasonable extent the reduction proportionally between the relevant petroleum deposits. In the event of such apportionment, special considerations shall be given to long-term agreements for the supply of gas and to petroleum deposits which in part are situated on the continental shelf of another State.

Upon application from the licensee, the Ministry may approve test production of a petroleum deposit. Duration, quantity and other conditions for such test production shall be decided by the Ministry.

The Ministry may require the licensee to produce a report on field related matters, including alternative schemes for production and, if applicable, injection, and the total recovery factor corresponding to various production schedules.

Section 4-5 Postponement of exploration drilling and development. The Ministry may decide that exploration drilling or development of a deposit shall be postponed. The provisions relating to extension of the licence, extension of the time limit set for implementing the work obligation and payment of area fee during the extension period in Section 3-13 third, fourth and fifth paragraph shall apply accordingly.

Section 4-6 Preparation, commencement and continuation of production. The Ministry may make a decision to require preparation, commencement or continuation of production, and hereunder, that on-going production shall be continued or increased, when this is economically beneficial to society, when necessary to develop an efficient transportation system or to ensure efficient utilization of the facilities comprised by Sections 4-2 and 4-3. Such a decision may also be made if it is efficient for reservoir engineering reasons, or when it is desirable that two or more petroleum deposits are produced in conjunction with each other, or when warranted by other significant social reasons.

If the decision according to the first paragraph is to the effect that production shall be prepared or commenced, the licensee shall be given a period of two years to present a plan according to Section 4-2 and a progress schedule. If the decision is to the effect that production shall be continued, the licensee shall be given a period of no less than 6 months to present a plan for implementation.

If the plan according to the second paragraph is not submitted prior to the expiry of the time limit, or if the licensee informs the Ministry that he will not prepare, commence or continue production of the deposit, or if the licensee without plausible reason and in spite of being ordered to such effect fails to take the necessary steps to implement the plan, the Ministry may initiate measures to commence or continue production, and in this connection revoke the production licence or parts of the area of the production licence to the extent this is considered necessary to achieve efficient production. The same shall apply if the licensee presents a plan which the Ministry does not consider to be satisfactory, provided that the licensee shall nevertheless be given a period of at least six months to present an amended plan.

If the production licence is revoked pursuant to the third paragraph, the licensee's costs for exploration and exploration drilling in connection with the deposit in question shall be reimbursed by the State.

Section 4-7 Joint petroleum activities. If a petroleum deposit extends over more than one block with different licensees, or onto the continental shelf of another State, efforts shall be made to reach agreement on the most efficient coordination of petroleum activities in connection with the petroleum deposit as well as on the apportionment of the petroleum deposit. This shall apply similarly when, in the case of several petroleum deposits, joint petroleum activities would obviously be more efficient.

Note. Petroleum operations involving petroleum reservoirs extending into the continental shelf of another State will be regulated by the terms of a specific agreement to be concluded with that other State. With respect to the United Kingdom the framework for such an interstate petroleum development agreement was set out in the Agreement of 10 March 1965. The most important interstate agreement following from this Boundary Treaty was the interstate agreement of 16 October 1979 providing for the joint exploitation of the Statfjord Field Reservoirs (Norwegian production licence No. 037, awarded on 10 August 1973 under the last phase of the 2nd round).

Agreements on joint exploration drilling shall be submitted to the Ministry. Agreements on joint production, transportation, utilization and cessation of petroleum activities shall be submitted to the Ministry for approval. If consensus on such agreements is not reached within reasonable time, the Ministry may determine how such joint petroleum activities shall be conducted, including the apportionment of the deposit.

Section 4-8 Use of facilities by others. The Ministry may decide that facilities comprised by Sections 4-2 and 4-3 and which are owned by the licensee, may be used by others, if so warranted by considerations for efficient operation or for the benefit of society, and the Ministry deems that such use would not constitute any unreasonable detriment of the licensee's own requirements or those of someone who has already been assured the right of use. Nevertheless, natural gas undertakings and eligible customers domiciled in an EEA State shall have a right of access to upstream pipeline networks, including facilities supplying technical services incidental to such access. The Ministry stipulates further rules in the form of regulations and may impose conditions and issue orders relating to such access in the individual case. Any agreement on the use of facilities comprised by Sections 4-2 and 4-3 shall be submitted to the Ministry for approval unless otherwise decided by the Ministry. The Ministry may on approving an agreement according to the first sentence, or in the event that no such agreement is reached within a reasonable period of time, as well as in the case of an order according to the first paragraph, stipulate tariffs and other conditions or subsequently alter the conditions that have been approved or stipulated in order to ensure that implementation of projects is carried out with due regard to considerations relating to resource management and to providing the owner of the facility with a reasonable profit taking into account, among other things, investments and risks.

Section 4-9 Extended operator responsibility upstream pipeline network. The Ministry may appoint someone to assume extended operator responsibility for the overall operation of upstream pipeline network and associated facilities, including undertake change of operator when warranted for particular reasons. The overall operation of upstream pipeline network and associated facilities shall be carried out in accordance with prudent technical and sound economic principles. Whoever has the extended operator responsibility as mentioned in the first paragraph shall act in a neutral and non-discriminatory manner.

The King may issue further rules relating to the responsibility as mentioned in the first and second paragraphs including deciding that whoever has been assigned to assume this responsibility shall also make decisions in respect of access to upstream pipeline network, and may order owners and users of upstream pipeline network and associated facilities and licensees of production licences where petroleum is produced, to adapt their activities. Such order might be given to ensure prudent resource management and efficient operation of the upstream pipeline network in question.

Section 4-10 Area fee, production fee, etc. The licensee shall pay a fee for a production licence, after expiry of the period stipulated pursuant to Section 3-9 first paragraph first sentence, calculated per square kilometre (area fee).

The licensee shall furthermore pay a fee calculated on the basis of the quantity and value of petroleum produced at the shipment point of the production area (production fee). With regard to petroleum which is injected, exchanged or stored prior to being delivered to be taken ashore or used for consumption, the production fee shall be calculated on the basis of the quantity and value of the petroleum at the shipment point for the original production area at the time when the petroleum according to contract is delivered to be taken ashore or used for consumption. Nevertheless production fee shall not be paid for petroleum produced from deposits where the development plan is approved or where the requirements to submit a plan for development and operation are waived after 1 January 1986.

When granting a production licence, a non-recurring fee (cash bonus) may be levied and there may be stipulated a fee which shall be calculated on the basis of production volume (production bonus).

The Ministry may with six months' notice decide that the production fee shall be paid wholly or partly in the form of produced petroleum. The Ministry may in such cases require the licensee to make sure that such petroleum is transported, processed, stored and made available at prices, priority and other conditions which are no less favourable than the terms applicable to the licensee's own petroleum from the relevant production area. The Ministry may, with six months' notice decide that the fee shall again be paid in cash.

Petroleum which the State is entitled to receive as production fee, and the State's right to transportation, processing and storage of such petroleum may be transferred to others. Such transfer shall relieve the State of future obligations.

The King may issue regulations relating to the size of the fees and bonuses mentioned in the first, second and third paragraphs and on the method of calculation, including provisions on stipulation of the value which shall form the basis for the calculation, on metering of the petroleum e.g., for fiscal purpose, and on information which the licensee shall provide about the production.

Section 4-11 Landing of petroleum. The King decides where and in which way landing of petroleum shall take place.

Section 4-12 Supplies to cover national requirements. The King may decide that the licensee shall make deliveries from his production to cover national requirements and provide transportation to Norway. The King may further decide to whom such petroleum shall be delivered.

A price shall be paid for the petroleum delivered, which shall be determined in the same way as the price which forms the basis for calculation of the production fee, with the addition of transportation costs.

Section 4-13 Supplies in the event of war, threat of war, etc. In the event of war, threat of war or other extraordinary crisis, the King may decide that a licensee shall place petroleum at the disposal of the Norwegian authorities.

The provisions of Section 4-12, second paragraph, shall apply correspondingly unless the particular situation warrants otherwise. In such situation the King shall with binding effect determine the price.

Chapter 5 Cessation of Petroleum Activities. Section 5-1 Decommissioning plan. The licensee shall submit a decommissioning plan to the Ministry before a licence according to Section 3-3 or Section 4-3 expires or is surrendered, or the use of a facility is terminated permanently. The plan shall contain proposals for continued production or shutdown of production and disposal of facilities. Such disposal may inter alia constitute further use in the petroleum activities, other uses, complete or part removal or abandonment. The plan shall contain the information and evaluations deemed necessary to make a decision according to Section 5-3. The Ministry may require further information and evaluations, or alternatively require a new or amended plan.

Unless the Ministry consents to or decides otherwise, the decommissioning plan shall be submitted at the earliest five years, but at the latest two years prior to the time when the use of a facility is expected to be terminated permanently. The Ministry may waive the requirement to submit a decommissioning plan.

Section 5-3 Decision relating to disposal. The Ministry shall make a decision relating to disposal and shall stipulate a time limit for implementation of the decision. In the evaluation on which the decision is based, emphasis shall, inter alia, be attached to technical, safety, environmental and economic aspects as well as to consideration for other users of the sea. The Ministry may stipulate specific conditions in connection with the decision. The licensee and the owner are under obligation to make sure that a decision relating to disposal is carried out, unless otherwise decided by the Ministry. The obligation to carry out the decision relating to disposal is applicable if this decision is made or is to be implemented after the expiry of the licence. If the decision is to the effect that the facility shall continue to be used in the petroleum activities or for other purposes, the licensee, owner and user are jointly obliged to make sure that future decisions on disposal are carried out, unless otherwise decided by the Ministry.

Use of a facility for purposes other than petroleum activities, complete or part removal or abandonment cannot be decided under this Act for a facility onshore or on seabed subject to private property rights.

Section 5-4 Liability. Whoever is under obligation to implement a decision relating to disposal according to Section 5-3 is liable for damage or inconvenience caused wilfully or negligently in connection with disposal of the facility or other implementation of the decision.

If the decision is for abandonment the licensee or owner shall be liable for damage or inconvenience caused wilfully or inadvertently in connection with the abandoned facility unless otherwise decided by the Ministry. In the event of decisions for abandonment, it may be agreed between the licensees and the owners on one side and the State on the other that future maintenance, responsibility and liability shall be taken over by the State based on an agreed financial compensation.

Section 5-6 Takeover by the State. The State has a right to take over the licensee's fixed facility when the licence expires, is surrendered or revoked, or when the use of such facility has been terminated permanently. The King decides with binding effect if and to what extent compensation shall be paid for the takeover. In the event of a takeover of a facility onshore or on the seabed subject to private property rights, compensation shall be paid to the extent this follows from otherwise applicable rules.

Chapter 7 Liability for Pollution Damage. Section 7-1 Definitions. Pollution damage means damage or loss caused by pollution as a consequence of effluence or discharge of petroleum from a facility, including a well, and the costs of reasonable measures to avert or limit such damage or such loss, as well as damage or loss as a consequence of such measures. Damage or loss incurred by fishermen as a consequence of reduced possibilities for fishing is also included in pollution damage.

Ships used for stationary drilling are regarded as a facility. Ships used for storage of petroleum in conjunction with production facilities are regarded as part of the facility. The same applies to ships for the transport of petroleum during the time when loading from the facility takes place.

Section 7-2 Scope and applicable law. The provisions of this chapter are applicable to liability for pollution damage from a facility when such damage occurs in Norway or inside the outer limits of the Norwegian continental shelf or affects a Norwegian vessel, Norwegian hunting or catching equipment or a Norwegian facility in adjacent sea areas. With regard to measures to avert or limit pollution damage it is sufficient that damage may occur in such an area.

The provisions of this chapter are also applicable to pollution damage from facilities used in petroleum activities according to this Act, when the damage occurs in onshore or offshore territory belonging to a State, which has acceded to the Nordic Convention on Environment Protection of 19 February 1974.

The King may, irrespective of the provisions contained in this Act, by agreement with a foreign State issue rules relating to liability for pollution damage caused by petroleum activities pursuant to the Act. Such rules shall, however, not restrict the right to compensation according to this Act in respect of any injured party under Norwegian jurisdiction.

Section 7-3 The liable party and the extent of liability. The licensee is liable for pollution damage without regard to fault. The provisions relating to the liability of licensees apply correspondingly to an operator who is not a licensee when the Ministry has so decided in connection with the approval of the operator status.

If there are several licensees under the licence and one of them is the operator, or if the Ministry has made a decision according to the first paragraph, claims for shall initially be directed to the operator. If any part of the compensation is left unpaid on the due date by the operator, this part shall be covered by the licensees in accordance with their participating interest in the licence. If someone fails to cover his share, this shall be allocated proportionately between the others.

If it is demonstrated that an inevitable event of nature, act of war, exercise of public authority or a similar *force majeure* event has contributed to a considerable degree to the damage or its extent under circumstances which are beyond the control of the liable party, the liability may be reduced to the extent it is reasonable, with particular consideration to the scope of the activity, the situation of the party that has sustained damage and the opportunity for taking out insurance on both sides.

In the event of pollution damage from a facility located in an area outside the Norwegian continental shelf, the party who has approval from the competent authority to conduct the activities to which the facility is connected is regarded to be a licensee.

Section 7-4 Channelling of liability. The liability of a licensee for pollution damage may only be claimed pursuant to the rules of the Act.

Liability for pollution damage cannot be claimed against:

(i) anyone who by agreement with a licensee or his contractors has performed tasks or work in connection with petroleum activities;
(ii) anyone who has manufactured or delivered equipment to be used in the petroleum activities;
(iii) anyone who undertakes measures to avert or limit pollution damage, or to save life or rescue values which have been endangered in connection with the petroleum activities, unless the measures are performed in conflict with prohibitions imposed by public authorities or are performed by someone other than public authorities in spite of express prohibition by the operator or the owner of the values threatened; or
(iv) anyone employed by the licensee or by someone mentioned under items (i), (ii) or (iii).

Section 7-5 Recourse. The licensee cannot claim recourse for pollution damage against someone exempted from liability pursuant to the rules of Section 7-4, unless the person in question or someone in his service has acted wilfully or by gross negligence. Recourse liability may be mitigated to the extent that this considered reasonable in view of manifested conduct, economic ability and the circumstances in general.

Section 7-6 Petroleum activities without a licence. If pollution damage occurs in a petroleum activity and the activity has been conducted without a licence, the party that has conducted the petroleum activity shall be liable for the damage regardless of fault. The same liability rests on others who have taken part in the petroleum activity and who knew, or should have known, that the activity was conducted without a licence.

Section 7-8 Legal venue. Legal action for compensation for pollution damage shall be brought before the courts in the court district where the effluence or discharge of petroleum has taken place or where damage has been caused.

The Ministry decides where the action shall be brought if:

(a) the effluence or discharge has taken place or the damage has been caused outside the area of any court district;
(b) it cannot be demonstrated within which court district the effluence or discharge has taken place or damage has been caused;
(c) the effluence or discharge has taken place in one court district and the damage is caused in another court district; or
(d) damage has been caused in more than one court district.

Chapter 8 Special Rules for Compensation to Norwegian Fishermen. Section 8-1 Scope of application and definitions. This chapter applies to compensation for financial losses incurred by Norwegian fishermen as a result of the petroleum activities occupying fishing fields or resulting in pollution and waste, or as a result of damage caused by a facility or actions in connection with the placing of a facility. This chapter does not apply to pollution damage as mentioned in Section 7-1.

Section 8-2 Occupation of a fishing area. In the event of petroleum activities within an area entirely or partly occupying a fishing field the State is obliged, to the extent that fishing becomes impossible or is substantially impeded, to award compensation in respect of any resulting financial loss. The State may claim recovery from the licensee if the licensee ought to have averted the loss.

Section 8-3 Pollution and waste. The licensee is liable, regardless of fault, in respect of financial losses incurred as a result of pollution and waste from the petroleum activities, and the cost of reasonable measures to avert or limit such damage or such loss, including damage or loss as a result of such measures. The liability of the licensee pursuant to the first paragraph also includes damage and inconvenience due to pollution and waste as a result of supply vessel and support vessel traffic, as well as during relocation of the facility to or from the field concerned. The licensee has the right of recourse against the perpetrator actually causing the loss or the shipowner, providing the other prevailing conditions of liability have been fulfilled.

Section 8-4 Joint and several liability. If damage has been caused as described in Section 8-3 and it is not possible to identify who caused the damage, the licensees shall be jointly and severally liable insofar as the damage may be believed to have been caused by petroleum activities in connection with the licence in question.

Section 8-5 Facility, etc., causing damage. If a facility or an action in connection with the placing of such facility causes damage and the injured party does not have a right to compensation pursuant to the provisions of Section 8-2, the licensee shall, regardless of fault, be liable for damages in respect of the financial losses suffered by fishermen as a result of the damage.

Chapter 9 Special Requirements to Safety. Section 9-1 Safety. The petroleum activities shall be conducted in such manner as to enable a high level of safety to be maintained and further developed in accordance with the technological development.

Section 9-2 Emergency preparedness. The licensee and other participants in the petroleum activities shall at all times maintain efficient emergency preparedness with a view to dealing with accidents and emergencies which may lead to loss of lives or personal injuries, pollution or major damage to property.

Section 9-3 Emergency preparedness against deliberate attacks. The licensee shall initiate and maintain security measures to contribute to avoiding deliberate attacks against facilities and shall at all times have contingency plans to deal with such attacks. The licensee shall place facilities at the disposal of public authorities for drills and shall participate in such drills to the extent this is necessary.

Section 9-4 Safety zones, etc. Around and above facilities there shall be a safety zone unless otherwise decided by the Ministry. In the event of accidents and emergencies the Ministry may establish or extend safety zones. The extent of zones referred to in the first and second sentences shall be determined by the King. This provision is not applicable to pipelines and cables.

Note. Article 60 of the 1982 Convention on the Law of the Sea contains rules with regard to safety zones. As discussed in Chapter 9, Article 60 appears exclusively to address installations and structures that – just like artificial islands which are specifically mentioned – extend above the surface of the sea and to exclude any submarine construction such as submarine pipelines and submarine production installations from qualifying for the establishment of a safety zone around them. The provisions of this Section 9-4 appear to be in line with Article 60 in as much as the definition of 'facility' in Section 1-6 does not mention subsea facilities other than pipelines and cables for which however an exception may be made as in fact has been done in this Section.

The King may decide that a safety zone shall extend across the border line onto the continental shelf of another State. Furthermore the King may decide that there shall be a safety zone on the Norwegian continental shelf even if the facility in question is located outside the Norwegian continental shelf. The Ministry may decide that a zone corresponding to the safety zone shall be established in reasonable time prior to the placing of facilities as mentioned in the first paragraph. The Ministry may decide that there shall be a safety zone around and above abandoned or dumped facilities or parts of such facilities. Unauthorized vessels, hovercrafts, aircrafts, fishing gear or other objects must not be present in zones as mentioned in the first, second, third and fourth paragraphs. If fishing can take place in the zone or in parts of the zone without threatening safety or interfering with the exercise of the petroleum activities, the Ministry may nevertheless decide that such fishing can take place.

Section 9-5 Suspension of the petroleum activities, etc. In the event of accidents and emergencies as mentioned in Section 9-2 the licensee or anyone else responsible for the operation and use of the facility shall, to the extent necessary, suspend the petroleum activities for as long as the requirement to prudent operations warrants such suspension.

Chapter 10 General Provisions. Section 10-1 Requirements to prudent petroleum activities. Petroleum activities according to this Act shall be conducted in a prudent manner and in accordance with applicable legislation for such petroleum activities. The petroleum activities shall take due account of the safety of personnel, the environment and of the financial values which the facilities and vessels represent, including also operational availability.

The petroleum activities must not unnecessarily or to an unreasonable extent impede or obstruct shipping, fishing, aviation or other activities, or cause damage or threat of damage to pipelines, cables or other sub-sea facilities. All reasonable precautions shall be taken to prevent damage to animal life and vegetation in the sea, relics of the past on the seabed and to prevent pollution and littering of the seabed and to prevent pollution and littering of the seabed, its subsoil, the sea, the atmosphere or onshore.

Section 10-2 Management of the petroleum activities, bases, etc. The licensee shall unless otherwise decided by the Ministry, have an organization which is capable of managing independently the petroleum activities from Norway. To achieve this, the Ministry may stipulate specific requirements in respect of the organization and the capital of the company.

The licensee shall see to it that the circumstances permit trade union activities to take place among its own employees and the personnel of contractors and subcontractors in accordance with Norwegian practice. The petroleum activities shall be conducted from a base in Norway. The licensee may be ordered to use bases designated by the Ministry.

Section 10-3 Regulatory supervision of the petroleum activities. The Ministry carries out regulatory supervision to see that the provisions laid down in or pursuant to this Act are complied with by all who carry out petroleum activities comprised by the Act. The Ministry may issue such orders as are necessary for the implementation of the provisions laid down in or pursuant to this Act. Expenses related to the regulatory supervision may be required to be covered by the licensee or by the party which the supervision in each case is directed at or where it takes place.

Section 10-4 Material and information concerning the petroleum activities. Material and information which the licensee, operator, contractor etc. possesses or prepares in connection with planning and implementation of petroleum activities pursuant to this Act shall be available in Norway and may be required to be submitted free of charge to the Ministry or to anyone designated by the Ministry. Information which has been provided to the authorities may, in accordance with further provisions issued by the Ministry, be used for the preparation of overview maps and for statistical purposes, among others by the Central Bureau of Statistics of Norway.

Section 10-5 Agreement between affiliated companies. The Ministry may, when particular reasons so warrant, consent to the licensee entering into an agreement which authorizes a parent company or a company with which the licensee is affiliated in a similar manner, to undertake the activities on behalf of the licensee. It shall be set as a condition for the abovementioned consent that the arrangement will not result in less tax revenues to Norway.

Section 10-6 Obligation to comply with the Act and to see to it that provisions are complied with. The licensee and other persons engaged in petroleum activities comprised by this Act are obliged to comply with the Act, regulations and individual administrative decisions issued by virtue of the Act through the implementation of necessary systematic measures. In addition the licensee shall see to it that anyone performing work for him, either personally, through employees or through contractors or subcontractors, shall comply with the provisions laid down in or pursuant the Act.

Section 10-7 Security. Upon granting a licence and subsequently, the Ministry may decide that the licensee shall provide such security as approved by the Ministry for the fulfilment of the obligations which the licensee has undertaken, as well as for possible liability in connection with the petroleum activities.

Section 10-8 Responsibility for commitments. Licensees who jointly hold a licence are jointly and severally responsible to the State for the financial obligations arising out of petroleum activities pursuant to the licence.

Section 10-9 Liability for damage caused. If liability in respect of a third party is incurred by anyone undertaking tasks for a licensee, the licensee shall be liable for damages to the same extent as, and jointly and severally with, the perpetrator and, if applicable, his employer. Liability for pollution damage is governed by the rules of Chapter 7.

Section 10-11 Training. The King may issue rules relating to the licensees' obligation to undertake training of civil servants.

Section 10-12 Transfer. Transfer of a licence or participating interest in a licence for petroleum activities may not take place without the approval of the Ministry. The same applies to other direct or indirect transfer of interest or participation in the licence, including, inter alia, assignment of shareholdings or other ownership shares, which may provide decisive control of a licensee possessing a participating interest in a licence. Transfer of a group of licensees' right of ownership to fixed facilities may not take place without the approval of the Ministry.

Section 10-13 Revocation. In the event of serious or repeated violations of this Act, regulations issued pursuant hereto, stipulated conditions or orders issued, the King may revoke a licence granted pursuant to this Act. If an application for a licence contains incorrect information or if information of significance has been withheld, and it must be assumed that the licence would not have been granted had correct or complete information been available, the licence may be revoked in relation to the licensee concerned.

A licence may be revoked if the security which the licensee is obliged to provide pursuant to Section 10-7 has become significantly weakened, or if the company or other association holding the licence is dissolved or enters into debt settlement proceedings or bankruptcy proceedings.

Section 10-14 Consequences of revocation, surrender of rights or lapse for other reasons. Revocation of a licence, surrender of rights or lapse of rights for other reasons do not entail release from the financial obligations which follow from this Act, regulations issued pursuant hereto or specific conditions. If a work obligation or other obligation has not been fulfilled, the Ministry may demand payment, in full or in part, of the amount which the fulfilment of the obligations would have cost. The amount shall be determined by the Ministry with binding effect.

Section 10-16 Enforcement measures. In the event of serious or repeated violations of acts and regulations, stipulated conditions or orders issued, the Ministry may impose a temporary suspension of the activities. The Ministry may initiate necessary measures for the account and risk of the licensee if orders are not complied with. The costs of such measures are ground for enforcement of distraint.

Section 10-18 Authority to issue regulations and stipulate conditions. The King may issue regulations to supplement and implement this Act, including inter alia provisions relating to working conditions, confidentiality and relating to the licensee's obligation to make information on the activities pursuant to the Act available to the public. The King may also issue regulations relating to the duty to provide information for the fulfilment of Norway's obligations pursuant to the EEA agreement. In connection with individual administrative decisions, other conditions than those mentioned in this Act may be stipulated, when they are naturally linked with the measures or the activities to which the individual administrative decision relates.

Chapter 11 Management of the State Direct Financial Interest.

Note. This chapter was added to this Act by the Act of 18 May 2001 No. 23 (entry into force on 17 June 2001).

Section 11-1 The State's participation in petroleum activities. The State participates in petroleum activities under this Act in that the State reserves a specified share of a licence granted pursuant to this Act and in the joint venture established by a joint operating agreement in accordance with the licence.

The King may decide that the State shall participate in activities under this Act otherwise than mentioned in the first paragraph, and that the State shall participate in other activities connected with activities under this Act. In such case the provisions contained in this chapter shall apply accordingly to the extent they are suitable, unless otherwise specifically decided by the King.

Section 11-2 Management company. The commercial aspects in relation to the participating interests which the State owns or reserves for itself shall be managed by a limited company owned by the State as sole owner.

The company shall be a licensee in respect of the participating interests it manages on behalf of the State. In individual joint venture the company shall have rights and duties as a licensee according to rules stipulated in or pursuant to this Act with appurtenant regulations.

The revenues resulting from the management of the participating interests shall belong to the State. The operating expenses, investments and other expenditure incurred to or relating to the management of the participating interests shall be covered by appropriation from the State. Funds for the operation of the company shall be provided by the State.

Section 11-3 The State's responsibility for the company. The State is directly liable for any obligations incurred by the company by contract or otherwise. Claims against the State shall be made to the company.

Note. The company in question is Petoro AS which was founded on 9 May 2001.

Chapter 12 Entry into Force and Amendment of Laws. Section 12-1 Entry into forcem etc. This Act enters into force as from the time decided by the King. (*Note*. The Act entered into force on 7 July 1997.)

Section 3-9 first, second and third paragraphs do not apply to production licences granted pursuant to the Royal Decree of 9 April 1965. Such production licences are valid for up to forty-six years from the time the licence was granted.

Section 4-5 does not apply to production licences granted prior to 1 July 1985 (*Note*. This refers to production licences granted under the Royal Decree of 1965 and the Royal Decree of 1972 respectively).

Regulations issued pursuant to previous Act of 21 June 1963 No. 12 relating to scientific research etc. or the Act of 22 March 1985 No. 11 relating to petroleum activities, or by virtue of regulations issued pursuant to the said Acts, shall apply insofar as no other provision has been or will be made applicable.

Section 11-2 Repeal and amendments of acts. The Act of 22 March 1985 No. 11 relating to petroleum activities shall be repealed as from the time of entry into force of this Act.

6.12.3 Associated Agreements

The form of each of the three associated agreements, viz. the training agreement, the frame agreement on technological research and development and the joint operating agreement (formerly known as the state participation agreement), is predetermined and the entering into such an agreement is made a condition of the licence. To illustrate the scope and purpose of these agreements, the production licence PL 139 is taken as an example. This production licence was granted on 29 May 1987, in the context of the 11th round of licensing to Statoil (fifty per cent initial interest and operator) and five foreign companies, the latter each with a ten per cent initial interest.

6.12.3.1 The Training Agreement

This agreement gives a national company the right to demand that at the company's expense a reasonable number of its employees shall participate free of charge in internal training programmes of the foreign licensees insofar as such programmes are relevant for petroleum activities on the Norwegian continental shelf. The national company may also demand that a reasonable number of its employees shall be given the opportunity to work in the foreign licensee's organization in

Norway or in project groups which the foreign licensee or its affiliates have established in connection with the work required to be done under the licence. As far as PL 139 is concerned, Statoil is the operator and sole national participant, but in other cases, where a foreign licensee is the operator, the associated training agreement concerned allows the national company to demand to take over from the foreign operator the performance of specific work tasks, studies or other limited functions.

6.12.3.2 The Frame Agreement

This agreement concerning the petroleum operations is an agreement between the Ministry and the operator, acting on behalf of the licensees. Its purpose is to serve as the contractual framework within which the operator shall make use of and develop competence and know-how among Norwegian contractors in order to develop new technology to the mutual benefit of the two Parties. The frame agreement obliges the operator to ensure that at least fifty per cent of the total expenditure, which is spent on research and development work related to the operations under the licence, shall be spent in Norway and in cooperation with Norwegian contractors. In this connection Norwegian contractors must be offered the opportunity to use offshore installations for testing of concepts, processes and equipment, to conduct projects and studies for the operator and to participate in Research and Development (R&D) projects. Contracts made between the operator and a Norwegian contractor within the scope and context of the frame agreement must contain appropriate provisions regarding confidentiality, industrial property rights and remuneration, but it is already stipulated in the frame agreement that inventions and other technology developed under a contract shall be the property of the party who provided to the largest extent the technical concepts and/or information therefore. The need for R&D must be reported to the Ministry whenever a discovery is made and evaluated.

6.12.3.3 The Joint Operating Agreement

This agreement (formerly known as the state participation agreement) is concerned with the protection of minority interests, in particular in connection with the preparation of the required development plan, taking into account the dominant position of Statoil and the likelihood of a divergence of interests emerging regarding such development plan, and further with the modalities of the disposal of an individual licensee's share of the production.

Note. According to Chapter 11, Section 11-2 of the Act, Petoro AS as the new manager of the State's direct financial interests in the licences shall in the individual joint venture have rights and duties as a participant in accordance with the joint operating agreement. This means that any reference to Statoil in the agreements should be taken to be a reference to Petoro AS.

At the time that the Ministry approves the development plan and effective from that moment, the Ministry may amend the initial undivided interests by increasing the combined State/Statoil interest to sixty per cent; and by reducing each of the five other companies' interest to eight per cent. The five other companies (i.e., the foreign licensees) have to contribute two hundred per cent of their participating interest share of all exploration and appraisal expenditure up to the time that it is decided by the management committee to prepare a development plan or until such a plan is approved.

The joint venture is managed by a management committee, in which each licensee is represented and may vote in accordance with its percentage participating interest. Binding decisions are taken if supported by at least three members, representing at least forty per cent interest, but the approval of the annual programmes of exploration and appraisal drilling, the approval of the development plan, including decisions in principle regarding development and transport and the main dimensions of production platforms, transport and storage facilities and decisions to enter into contracts concerning such platforms and facilities require the support of at least two members representing more than fifty per cent interest.

No licensee can be forced to participate in the development of a petroleum reservoir. A reservoir is only developed by those licensees who shall have notified in writing the Ministry and the other licensees that they wish to participate in the development plan adopted by the management committee. Statoil's participation in the plan must be approved by the Ministry. If not all licensees have acceded to the development plan, the licensees who did accede to the plan may propose that the reservoir shall be developed on a sole risk basis. If the management committee did not succeed in adopting a development plan, each licensee may propose that the reservoir shall be developed on a sole risk basis. Each of the other licensees may join in the proposal. Development operations, whether on a joint or sole risk basis, shall take place on behalf of the participating licensees and must be completed without unreasonable delay after the Ministry's approval of the development plan.

Each licensee has the right and the obligation to take in kind and separately dispose of its share of the crude oil produced. Produced natural gas shall be disposed of in accordance with resolutions of the management committee. A negotiating committee, consisting of Statoil and any other national licensee (in the considered case not present), shall, in close contact with the foreign licensees, plan and prepare the exploitation of natural gas and complete negotiations with possible buyers. Any licensee can claim its share of the produced natural gas in kind if it can achieve better conditions by entering into separate contracts or if the gas is to be used in the licensee's or in its affiliated companies' own facilities.

6.13 THE EC DIRECTIVE 94/22/EC OF 30 MAY 1994

6.13.1 Introduction

In the context of pursuing the completion of the internal market for energy products the European Parliament and the EC Council adopted on 30 May 1994 the

Directive 94/22/EC (OJ No. L 164/3) on the conditions for granting and using authorizations for the prospecting, exploration and production of hydrocarbons in the EC Member States. The Directive was aimed at ensuring the non-discriminatory access to and pursuit of activities relating to the prosecution, exploration and production of hydrocarbons under conditions which encourage greater competition in this sector. Such non-discriminatory access and pursuit was expected to favour the best prosecution, exploration and production of resources in Member States and to reinforce the integration of the internal energy market. The Directive was meant to set up common rules for ensuring that the procedures for granting authorizations for the prospecting, exploration and production of hydrocarbons are open to all entities possessing the necessary capabilities; that authorizations are granted on the basis of objective, published criteria; and that the conditions under which authorizations are granted are likewise known in advance by all entities taking part in the procedures. In the Directive it is recognized that Member States must retain the options to limit the access to and the exercise of these activities for reasons justified by public interest and to subject these activities to the payment of a financial contribution or a contribution in hydrocarbons, but it is emphasized that these options must be used in a non-discriminatory way and that steps must be taken to avoid imposing on entities conditions and obligations which are not justified by the need to perform this activity properly.

6.13.2 Principal Provisions

Article 2

1. Member States retain the right to determine the areas within their territory to be made available for the exercise of the activities of prospecting, exploring for and producing hydrocarbons.
2. Whenever an area is made available for the exercise of the activities set out in paragraph 1 Member States shall ensure that there is no discrimination between entities as regards access to and exercise of these activities. However, Member States may refuse on grounds of national security, to allow access to and exercise of these activities to any entity which is effectively controlled by third countries or third countries nationals.

Article 3

1. Member States shall take the necessary measures to ensure that authorizations are granted following a procedure in which all interested entities may submit applications in accordance either with paragraph 2 or 3.
2. This procedure shall be initiated: (a) either at the initiative of the competent authorities by means of a notice inviting applications to be published in the Official Journal of the European Communities, at least 90 days before the closing date for applications; (b) or by means of a notice inviting applications to be published in the Official Journal of the European Communities following submission of an application by an entity without prejudice to Article 2 (1).

Other interested entities shall have a period of at least 90 days after the date of publication in which to submit an application. Notices shall specify the type of authorization, the geographical area or areas in part or all of which an application has been or may be made and the proposed date or time limit for granting authorization.

3. Member States may grant authorizations without initiating a procedure under paragraph 2 where the area for which authorization is requested: (a) is available on a permanent basis; or (b) has been the subject of a previous procedure according to paragraph 2 which has not resulted in the grant of an authorization; or (c) has been relinquished by an entity and does not fall automatically under (a).
4. A Member State may decide not to apply the provisions of paragraph 1 if and to the extent that geological or production considerations justify the granting of the authorization for an area to the holder of an authorization for a contiguous area.
5. The following shall not be considered as the grant of an authorization within the meaning of paragraph 1: (a) the grant of an authorization solely by reason of a change of name or ownership of an entity holding an existing authorization, a change in the composition of such an entity or the transfer of an authorization; (b) the grant of an authorization to an entity having another form of authorization where the possession of the latter authorization implies the right to the grant of the former authorization; (c) the decision of the competent authorities taken within the framework of an authorization and relating to the commencement, interruption, prolongation or cessation of the activities or to the prolongation of the authorization itself.

Article 4

Member States shall take the necessary measures to ensure that: (a) if the geographical areas are not delimited on the basis of a prior geometric division of the territory, the extent of each area is determined in such a way that it does not exceed the area justified by the best possible exercise of the activities from the technical and economic points of view. In case of authorizations granted following the procedures laid down in Article 3(2), objective criteria shall be established to this end and shall be made available to the entities prior to the submission of applications; (b) the duration of an authorization does not exceed the period necessary to carry out the activities for which the authorization is granted. However, the competent authorities may prolong the authorization where the stipulated duration is insufficient to complete the activity in question and where the activity has been performed in accordance with the authorization; (c) entities do not retain exclusive rights in the geographical area for which they have received an authorization for longer than is necessary for the proper performance of the authorized activities.

Article 5

Member States shall take the necessary measures to ensure that:

1. authorizations are granted on the basis of criteria concerning in all cases: (a) the technical and financial capability of the entities; (b) the way in which they

propose to prospect, to explore and/or to bring into production the geographical area in question; and, where applicable; (c) if the authorization is put up for sale, the price which the entity is prepared to pay in order to obtain the authorization; (d) if, following evaluation under the criteria (a), (b) and, where applicable, (c), two or more applications have equal merit, other relevant objective and non-discriminatory criteria in order to make a final choice among these applications. The competent authorities may also take into account, when appraising applications, any lack of efficiency and responsibility displayed by the applicants in operations under previous authorizations. Where the competent authorities determine the composition of an entity to which they may grant an authorization, they shall make that determination on the basis of objective and non-discriminatory criteria. Where the competent authorities determine the operator of an entity to which they may grant an authorization, they shall make that determination on the basis of objective and non-discriminatory criteria;

2. the conditions and requirements concerning exercise or termination of the activity which apply to each type of authorizations by virtue of the laws, regulations and administrative provisions in force at the time of submission of the applications, whether contained in the authorization or being one of the conditions to be accepted prior to the grant of such authorization, are established and made available to interested entities at all times. In the case provided for in Article 3(2)(a) they may be made available only from the date starting from which applications for authorization may be submitted;
3. any changes made to the conditions and requirements in the course of the procedure are notified to all interested entities;
4. the criteria, conditions and requirements are applied in a non-discriminatory manner; [and]
5. any entity, whose application for an authorization is unsuccessful is, if the entity so wishes, informed the reasons for the decision.

Article 6

1. Member States shall ensure that the conditions and requirements referred to in Article 5(2) and the detailed obligations for use of a specific authorization are justified exclusively by the need to ensure the proper performance of the activities in the area for which an authorization is requested, by the application of paragraph 2 or by the payment of a financial contribution or a contribution in hydrocarbons.
2. Member States may, to the extent justified by national security, public safety, public health, security of transport, protection of the environment, protection of biological resources and of national treasures possessing artistic, historic or archaeological value, safety of installations and of workers, planned management of hydrocarbon resources (for example the rate at which hydrocarbons are depleted or the optimization of their recovery) or the need to secure tax revenues, impose conditions and requirements on the exercise of the activities set out in Article 2(1).

3. The rules for payment of contributions referred to in paragraph 1, including any requirement for State participation, shall be fixed by Member States in such a way as to ensure that the independence of management of entities is maintained. However, where the grant of authorizations is subject to the State's participation in the activities and where a legal person has been entrusted with the management of this participation or where the State itself manages the participation, neither the legal person nor the State shall be prevented from assuming the rights and obligations associated with such participation, equivalent to the importance of the participation provided that the legal person or the State shall not be party to information nor exercise any voting rights on decisions regarding sources of procurement for entities, that the legal person or the State in combination with any public entity or entities shall not exercise a majority voting right on other decisions and that any vote by the State or the legal person shall be based exclusively on transparent, objective and non-discriminatory principles, and shall not prevent the management decisions of the entity from being based on normal commercial principles. However, the provisions of the preceding sub-paragraph shall not prevent the legal person or the State from opposing a decision by the holders of an authorization which would not respect the conditions and requirements specified in the authorization, regarding depletion policy and protection of the financial interests of the State. The option to oppose a decision shall be exercised in a non-discriminatory manner, particularly regarding investment decisions and sources of supply of entities. Where the State's participation in the activities is managed by a legal person which also holds the authorizations, the Member State shall put in place arrangements requiring that legal person to keep separate accounts for its commercial role and its role as manager of the State's participation and guaranteeing that there is no flow of information from the part of the legal person responsible for the management of the State's participation to the part of the legal person which holds authorizations in its own right. However, where the part of the legal person responsible for the management of the State's participation engages the part of the legal person which holds authorization as a consultant, the former may make available any information, which is necessary for the consultancy work to be carried out. The holders of all authorizations to which the information relates shall be informed in advance of what information will be given in this way and shall be given sufficient time to raise objections.
4. Member States shall ensure that the monitoring of the entities under an authorization is limited to that necessary to ensure compliance with the conditions, requirements and obligations referred to in paragraph 1. In particular, they shall take the measures necessary to ensure that no entity is required, by any law, regulation or administrative requirement, or by any agreement or undertaking, to provide information on its intended or actual sources of procurement, except at the request of the competent authorities and exclusively with a view to the objectives set out in Article 36 of the Treaty. (*Note*: this Article 36 allows Member States to prohibit or restrict the import or export of goods for reasons of public security, protection of health, etc.).

Note 1. State participation is addressed in the Directive since the modalities thereof could have a discriminating effect and distort competitiveness in the sector. As stated in Article 6 (3) state participation may be imposed as a condition of the authorization but there should be no interference with the management of the other participants. As formulated in the Directive, the State participant should not prevent the management decisions of the other participants from being based on normal commercial principles. Since the French and Italian state controlled oil enterprises have been privatized, there are only two EC Member States viz. Denmark and The Netherlands, and one EEA Member State, viz. Norway, left where state participation is in force, and for which the provisions of the Directive might be relevant. However the Danish regime provides for minority state participation only. In the Netherlands, under the terms of the Mining Law of 31 October 2002, it is either fifty per cent (old regime licences) or forty per cent (new regime licences). In Norway state participation was rearranged under the terms of Chapter 11 of the Petroleum Act of 29 November 1996 No. 72. This chapter was added to this Act by the Act of 18 May 2001 No. 23 (entry into force on 17 June 2001). A new wholly-owned state company named Petoro AS was founded on 9 May 2001 to serve as the manager of the Norwegian State direct financial interest (SDFI) in the licences in replacement of the former Statoil (since 1 October 2007 StatoilHydro).

Note 2. Norway is a Member State of the European Economic Area (EEA), and as such it had been committed to incorporate the provisions of the Directive in its petroleum legislation. The EEA is an association of States established by agreement between the EEG (now EC), the ECSC (ceased to exist on 23 July 2002) and their Member States on the one hand and the Member States of the European Free Trade Association (EFTA) on the other. The association agreement was signed in Oporto on 2 May 1992. Norway used the occasion of having to implement the Directive for making and issuing a complete new Petroleum Act, viz. the Act of 29 November 1996 No. 72, effective from 1 July 1997. The Act has been amended on a number of occasions, for instance, as mentioned above, on 17 June 2001 for the purpose of restructuring state participation.

Article 7

Without prejudice to the provisions concerning or contained in individual authorizations and to the provisions of Article 3(5)(b) legal, regulatory and administrative provisions which reserve to a single entity the right to obtain authorizations in a specific geographical area within the territory of a Member State shall be abolished by the Member States concerned before 1 January 1997.

Note: The preferential position in the matter of obtaining exploration acreage as in the past enjoyed by Agip, the exploration and production subsidiary of ENI, the former Italian state owned enterprise, has also been specifically addressed by the Directive. This preferential position had to be abolished by Italy before 1 January 1997.

Chapter 7

Past and Current Production Sharing Contracts

7.1 INDONESIA

7.1.1 Background

Production sharing contracts were adopted by the Indonesian government in replacement of the exclusive licences that had been terminated by virtue of Government Decree No. 44 of 26 October 1960. This Decree determined that oil and natural gas are part of the national riches under the control of the State. Exploration and exploitation were declared to be the responsibility of the State which for this purpose could delegate this responsibility to national state enterprises. These enterprises had to be established by law. Such laws established their authority to undertake the exploration and exploitation of petroleum. The Decree authorized the competent minister to designate enterprises which should assist the national state enterprise in fulfilling its task, if so needed. Then existing exclusive licences had in fact been nationalized and the former licensees (referred to as concessionaires) were only allowed to continue exercising their rights for a period as short as possible and to be determined by government regulation. The affected concessionaires were accorded a preferential right in obtaining a contract of work to be entered into with the designated state enterprise in respect to the area of their former (and suspended) petroleum agreement.

The present designated state oil enterprise *Perusahaan Pertambangan Minyak dan Gas BumiNegara (Pertamina)* was established by Law No. 8 of 1971, in replacement of the state enterprise P.N. Pertamina. The latter dated from 1968 and was on this occasion liquidated. The stated objective of Pertamina (as laid down in Article 5 of Law No. 8 of 1971) was to develop and carry out the exploitation of oil and natural gas in the widest sense of the word for the maximum prosperity of the people and the State as well as for creating national strength. Pertamina's tasks are described (Article 13) as consisting in carrying out oil and natural gas exploitation for the acquisition of maximum prosperity for the people and the State and in supplying and serving the domestic demand for oil and natural gas. For the implementation of its tasks Pertamina was permitted (Article 12) to cooperate with another party in the form of a 'Production Sharing Contract'. The terms and conditions of such cooperation would be regulated by a government regulation; and any resulting production sharing contract would become effective as of the moment of approval by the President.

In the 'Elucidation' annexed to the Law No. 8 it is stated that in this cooperation the most favourable terms for the State have to be sought. Furthermore it is stated that every production sharing contract, which has been approved by the President, will be notified to Parliament.

Note. The concept of production sharing was not new to Indonesia. Its origin can be traced back to the Netherlands-Indies Mining Law of 1899, as amended in 1919, which law had in the colonial past been in force in Indonesia. In accordance with Article 5a of this colonial mining law, the competent minister could be authorized by special law to enter into contracts with an oil enterprise granting the latter (the contractor) the exclusive right to search for and produce petroleum within a certain territory and for a certain number of years. A contractor was obliged to pay a royalty and a proportional part of the gross profit (revenues less costs and losses carried forwards). The proportion was related to the annual capital expenditures. The government had the right to demand that royalty and profit share be paid in oil or oil products, provided the oil or oil products so received would be used by the government for its own needs. Clearly, it is only a small step from this type of contract to the production sharing contract that was now adopted in Indonesia.

Initially, the production sharing contracts were seen as income tax replacing agreements. The contractor was liable to pay income tax but contractor's income taxes were considered to be part of Pertamina's share of the profit oil and the latter was responsible for its payment. In order to satisfy all formalities, a contractor received tax receipts from the tax authorities enabling the contractor to demonstrate (mainly vis-à-vis the tax authorities of its home country) that it not only had been subject to income tax but that the tax had also been paid.

Note. Indonesia offers an example of a State where nationalization of foreign petroleum rights was not intended as a means of establishing a state controlled and managed domestic petroleum industry (in contrast to the objectives of the nationalization measures that had been taken in Mexico [1938] and in Iran [1951]). For the Indonesian government the expropriation of foreign held rights meant the first step (together with the establishment of national state enterprises) in a restructuring of the petroleum regime whereby the granting of exclusive licences would be replaced by the awarding of contracts of work, in this case production sharing contracts, to foreign oil enterprises.

7.1.2 The Original Indonesian Model

The early Indonesian production sharing contracts stood at the basis of the concept's further development and evolution that took place not only in Indonesia itself but also in all other non-Western States that followed Indonesia's example and made this concept the centre piece of their petroleum regime.

The original model is summarized in Pertamina's *Summary of Indonesian Oil Contracts*, 2nd edition, January 1972, in which the data of all contracts concluded between 1969 and 1971 are represented.

The main provisions are the following:

(i) Duration of contract thirty years including a period of ten years for exploration;
(ii) Contract is automatically terminated if within the ten year exploration period the contractor does not succeed in discovering petroleum;
(iii) During the ten year exploration period, at regular intervals, parts of the contract area have to be relinquished;
(iv) Pertamina is responsible for the management of the operations. The contractor is responsible to Pertamina for the execution of the operations and provides the necessary funds therefore;
(v) If petroleum is discovered which in the judgment of Pertamina and its contractor (alternatively: in the judgement of the contractor after consultation with Pertamina) can be produced commercially then the contractor will commence development;
(vi) The contractor is responsible for the preparation of annual work programmes and corresponding budgets. Any such programme and budget are submitted for approval to Pertamina. Pertamina's agreement to a proposed work programme shall not be unreasonably withheld; (*Note*. In an actual contract it is stated that when considering work programmes and budgets proposed by contractor, Pertamina will give due consideration to the fact that its contractor carries the risk of and provides the necessary funds for the petroleum operations.)
(vii) If Pertamina wishes to propose a revision as to certain specific features of the submitted work programme and budget Pertamina and its contractor shall meet and endeavour to agree on the revisions proposed by Pertamina. In any event, any part of the work programme as to which Pertamina has not proposed a revision shall in so far as possible be carried out as prescribed therein;
(viii) It is recognized by both parties that the details of a work programme may require changes in the light of changing circumstances. The contractor is authorized to make such changes provided they do not change the general objective of the work programme;
(ix) Pertamina shall periodically consult with contractor with a view to the fact that contractor is responsible for carrying out the work programme adopted pursuant to the contract;

(x) The contractor is obliged to execute a work programme in a workmanlike manner and by any appropriate scientific methods;

(xi) Up to a maximum of forty per cent of available crude oil production may be taken by the contractor in repayment of costs incurred. In this context and for this purpose 'costs' mean expenditures as paid, without application of depreciation or amortization rules, even when the purchase or acquisition of capital assets is concerned;

(xii) The balance of crude oil remaining after the contractor has taken his entitlement to cost oil (at least sixty per cent of total production) is split sixty per cent to Pertamina and forty per cent to the contractor (main producing areas) or sixty-five per cent to Pertamina and thirty-five per cent to contractor.

(xiii) Up to twenty-five per cent of total crude oil production had to be supplied to the domestic market. The actual amount that had to be delivered within the said twenty-five per cent limit was a certain fraction of the Indonesian domestic consumption. This fraction was arrived at by dividing total contract production by total Indonesian production (all contract areas taken together). The domestic supply obligation was shared between Pertamina and its contractor in the same proportions as profit oil was divided between them. If the system of dividing daily production in tranches was applicable, up to twenty-five per cent of each tranche had to be delivered to the domestic market. The contractor received USD 0.20 per barrel for crude oil delivered to the domestic market;

(xiv) Pertamina is fully entitled to take its share of profit oil in kind. If Pertamina elects to take any part thereof, it shall so advise contractor not less than ninety days prior to the commencement of each semester of each year provided that such election shall not interfere with the proper performance of any crude oil sales agreement which the contractor has executed prior to the notice of such election. If Pertamina does not take its share of the profit oil in kind the contractor is obliged to market Pertamina's share to the best of its abilities. If however Pertamina can realize a better price for contractor's cost oil, it can exercise an option to market this portion of the production itself;

(xv) The contractor is entitled to retain abroad the proceeds of the sale of all crude oil from the venture, except for the proceeds of Pertamina's share of the profit oil sold by contractor;

(xvi) The valuation of the crude oil production for the purpose of the contract is based on realized prices f.o.b. point of export in Indonesia as paid by third parties, or, if there are no such third party sales and prices, a price shall be derived from prices paid by third parties for crude oil of similar type and quality;

(xvii) The contractor is subject to Indonesian income taxes and all other taxes, including all dividend withholding taxes or taxes imposed on the distribution or remittance of income or profit by contractor. However, all such present and future taxes are to be paid and assumed by Pertamina out of its share of the profit oil. The contractor will receive official receipts from the tax authorities evidencing that all taxes due by him had been paid; (*Note*. In the 1980s the Indonesian government was persuaded to change the income tax paying procedure and to accept direct payment of income taxes by the contractor, due to the difficulties which US oil companies experienced in their dealings with the US Internal Revenue Service (IRS) in getting the tax payments made by Pertamina on behalf of contractor accepted and recognized for the purpose of the US foreign tax credit.)

(xviii) If associated natural gas cannot be commercially exploited it has to be flared or put at the disposal of Pertamina for the latter's account. If natural gas can commercially be exploited, Pertamina and its contractor may decide to participate in a gas project. Costs and revenues will then be treated on the same basis as in the case of crude oil; (*Note*. As was customary in the early contracts there are no specific provisions made for natural gas; in those days markets for natural gas had still to be developed.)

(xix) Pertamina is entitled to request its contractor to offer a five per cent undivided interest in the contractor's side of the contract to an Indonesian enterprise (Indonesian participant) to be designated by Pertamina. Pertamina should make its request within three months after the contractor's first declaration of commercial discovery. The Indonesian participant must repay five per cent of the costs incurred by contractor up to that moment. Repayment should be made in cash or kind. If in kind, fifty per cent of the Indonesian participant's share of production would be reserved for that purpose and the amount to be repaid would be increased by fifty per cent.

7.1.3 Amendments to the Original Model

The first amendment to the original Model took place in reaction to the dramatic increase of the international oil price at the end of 1973, beginning 1974. This amendment concerned the introduction of a system of successive daily production tranches was introduced. Each daily production tranche was divided in cost oil (tied to a maximum of forty per cent of the tranche) and profit oil. For each tranche higher in the series the split of the profit oil portion became more favourable to Pertamina. In fact, the original profit oil split of 65/35 was under the new system only applicable up to 75,000 b/d (the bottom tranche); the profit oil in the next tranche, which ranged from 75,000 b/d to 200,000 b/d, was divided 67.5/32.5; and profit oil in the upper tranche, i.e., that part of the production in excess of 200,000 b/d, was divided 70/30. In addition, an excess profit charge was introduced payable in cash with respect to each barrel of net profit oil. In this context, net profit oil meant the profit oil remaining after satisfying the domestic supply obligation. The excess profit charge amounted to a certain percentage of the difference between the selling price and a base price which was fixed at USD 5 per barrel. The intention was to replace the respective 65/35; 67.5/32.5 and 70/30 proportions, in which net profit oil was divided, by a new 85/15 proportion but only in respect of the aforementioned price difference. As a result the amount of the excess profit charge (surcharge) payable by contractor per barrel net profit oil in respect of the bottom tranche can be calculated as: (.35–.15) × (USD 12–USD 5) = USD 1.40 (assuming a selling price of USD 12 per barrel). For the next tranche it would be (.325–.15) × (USD 12–USD 5) = USD 1.225 and then for the upper tranche (.30–.15) × (USD 12–USD 5) = USD 1.05.

During the 1980s the Indonesian production sharing contract was simplified. According to this new pattern production is not divided in tranches and no excess profit sharing is applied. Total production is made available for cost recovery. Recoverable costs are described in fiscal terms, whereby a distinction is made between costs that have to be written off over a number of years and costs that have to be written off in the year in which they have been incurred. The contractor may claim an investment credit equal in value to twenty per cent of the capital expenditure, and this credit has to be recovered from production by priority. The recovery of the investment credit is restricted. The recovery of the credit is only allowed if the sum of Pertamina's share (sixty-six per cent) of profit oil and fifty-six per cent (the combined tax rate) of the contractor's share (thirty-four per cent) therein (hence in the aggregate eighty-five per cent of profit oil) is at least equal to the value of forty-nine per cent of total production. If a contract concerns development in deep water entailing higher capital expenditures, the contractor receives an incentive in the form of an increased investment credit. The percentage figure of the credit is fifty per cent in respect of any water depth between 200 metres and 350 metres and is one hundred per cent in respect of a water depth of 350 metres or more. The domestic supply obligation (DSO) on the basis of USD 0.20 per barrel is maintained albeit during the first five years of production oil may be sold to the domestic market at market value. The DSO is limited to twenty-five per cent of total production and is shared between Pertamina and its contractor in the same proportion as in which the profit oil is shared between the parties, namely 66/34. The contractor must satisfy its DSO out of its share of the profit oil. If there is not enough profit oil (due to the claim on cost recovery production) then the contractor's DSO is correspondingly reduced.

In 1989 a new model production sharing contract was developed and made available to interested foreign oil companies. Distinction is made between First Tranche Petroleum, a portion amounting to twenty per cent of total production, and the remaining portion of eighty per cent. The eighty per cent portion is meant for recovery of recoverable costs, a basic investment credit of seventeen per cent in respect of expenditure made for offshore production installations and submarine pipelines and an additional investment credit of 110 per cent in respect of offshore installation located in waters deeper than two hundred metres. The previous restriction on the recovery of the investment credit (i.e., a minimum limit on the combined Pertamina and government share of revenue) is abolished. Any production remaining after costs and credits have been recovered from the eighty per cent portion is shared between Pertamina and its contractor in various proportions depending on the particular category (marginal fields, enhanced recovery, etc.), to which the oil production belongs, and, within the various categories, on production volumes. The first tranche petroleum is shared between the parties in the same manner as the profit oil.

7.2 EGYPT

On the basis of Articles 50 and 51 of the Law No. 86 of 1956 *On Mines and Quarries*, the minister responsible for petroleum matters could be authorized by special law to enter into an agreement with a foreign oil company for prospecting for and exploiting petroleum, subject to special conditions in derogation of the provisions of the Law which otherwise would be applicable to petroleum operations (see Chapter 6). Soon after Law No. 86 came into force the government started to make use of its powers to regulate petroleum operations on the basis of petroleum agreements instead of concessions. Initially, the type of agreement chosen was the state participation agreement, meant to regulate petroleum operations in the Western Desert and later also in the Gulf of Suez (see Chapter 6).

The concept of production sharing was introduced in Egypt when in May 1970 a typical Indonesian style production sharing contract was concluded between EGPC and the North Sumatra Oil Development Corporation (NOSODECO), a Japanese company. Thereafter, as of July 1973, EGPC concluded a great number of agreements on this Indonesian pattern, albeit some features of the preceding licensing regime and state participation were retained such as the grant to EGPC of an exclusive concession with respect to the whole country and the establishment of a Egyptian stock company jointly owned by EGPC and contractor upon making the first commercial discovery. At the same time the more important of the earlier state participation agreements (e.g., the participation agreement under the provisions of which the Morgan field in the Gulf of Suez was produced) were converted into production sharing contracts.

Note. The principal features of an Egyptian style production sharing contract are as follows. As the licence holder, EGPC is liable to pay a royalty in respect to each individual production sharing contract, and is made the sole owner of any petroleum produced under such contract. After commercial discovery development leases are created within the limits of which petroleum must be developed and produced. Upon the first commercial discovery an Egyptian joint stock company is established to act as operating company of commercial petroleum. This operating company is jointly owned by EGPC and the contractor and has a board of directors of eight members, four of whom are designated by EGPC and four by the contractor. Its object is to act as the agency through which EGPC and the contractor carry out the development operations under the contract. The joint company also acts as the agency to carry out all aspects of the exploration work, including the preparation of the exploration work programmes and budgets, but does not own any right or title to petroleum produced or in any other asset or property obtained or used in connection with the contract or any development lease. The joint company does not carry any responsibility or liability in respect of the operations carried out by it. Any of the company's decisions shall be understood to be the result of the decisions of EGPC and the contractor or of the contractor alone as may be required under the contract. Under this organizational set-up the contractor has no operational task anymore (after the first commercial discovery) but remains nevertheless fully responsible for the funding of the operations. The contractor is obliged to share the profit production with EGPC, as the sole owner of the concession. The contractor is liable to pay income tax, but EGPC is responsible for paying the tax on behalf of the contractor (see Chapter 8).

Since their inception in 1973, the Egyptian style production sharing contracts have not much changed, except as regards the percentage shares of cost recovery and the sharing of the profit production. Cost production percentages range around 40 per cent; the contractor's profit production share may range between 25 per cent and 12.5 per cent. Actual percentages depend on the petroleum potential of the exploration acreage on offer and on market conditions, i.e., on the prevailing or expected producer prices for exported oil.

A major change came about when the need arose to become more specific about the contractor's position regarding the development of natural gas reserves. Contracts concluded before 1988 contained a 'standard' provision regarding natural gas, also known from the Indonesian contracts. This standard provision simply stated that, if natural gas would be produced or would be capable of being produced, EGPC and its contractor should study all possible economic alternatives for its use. The standard clause further provided for the production sharing principles being applicable to natural gas, if and when sold for export. Natural gas not exported as LNG, and not used in the petroleum operations, would remain the property of the State and at the free disposal of EGPC. In 1988, the standard clause was made more specific by the introduction of an integrated treatment of crude oil and natural gas development as far as cost recovery out of production is concerned.

7.3 SYRIA

As from the time production sharing contracts became generally applied in Egypt, the concept was adopted in Syria in an almost identical form. The Syrian government was in this matter assisted by officials from EGPC's concession department. A difference however was introduced. This concerned a variation in the payment of royalty. Under the Egyptian style contract EGPC as the licence holder had been made responsible for paying this royalty out of its own resources. Under the Syrian style contracts the royalty took the form of setting aside a part of the production in favour of the State, thereby reducing the production on which the contractor and state party could lay a claim by way of cost production and profit production.

7.4 PERU

In 1971 a simple form of production sharing was introduced in Peru. It had been the international petroleum industry itself (in this case Occidental Petroleum

Corporation) which acquainted the country with the concept. Under the terms of the Peruvian style contract the production as made available by the contractor was divided in two equal shares between Petroperu, the Peruvian national state oil company and the contractor. The latter was liable to pay income tax. As of 1985, production sharing contracts were replaced by risk service agreements. Under the rules of this type of agreement the contractor is paid a fee expressed in US dollars per barrel produced.

7.5 THE PHILIPPINES

By Presidential Decree of 22 December 1972, 'on the promotion of petroleum exploration and production in the Philippines', new conditions for service contracts were published. The proposed contracts are concluded between the Petroleum Board (the state party) and a foreign oil company (the contractor) and appear to be similar to the Indonesian style production sharing contracts but with this difference that the sales proceeds of the production are shared between the parties instead of the production itself.

Note 1. The main features of this type of contract are the following:

(i) the contractor undertakes, manages and executes petroleum operations, and meets all costs thereof;
(ii) up to a maximum of seventy per cent of the gross proceeds from production may be used by contractor in repayment of costs incurred; and
(iii) the Petroleum Board receives sixty per cent of the net proceeds defined as gross proceeds less cost incurred within the aforesaid seventy per cent limit and less the Philippine Participation Incentive Allowance (PPIA), if applicable.

The PPIA is expressed as a discount off sales price and is granted to the contractor if the latter agrees to set up a joint venture with a Philippine partner. This contrasts with Indonesian contracts, where the contractor has to make an offer for five per cent or ten per cent national participation without receiving compensation therefore. The contractor receives forty per cent of the net proceeds plus the PPIA, if applicable. Further terms and conditions include:

(i) valuation of crude oil for determining gross proceeds to be based as to exported crude oil on the posted price, which is the FOB price as established by the contractor in consultation with the Petroleum Board, and as to local sales on the price realized in arm's length deals;
(ii) all foreign currency earned by export sales may be kept abroad except foreign currency needed for payments to the Petroleum Board and for covering costs incurred in local currency; and
(iii) the contractor is to be subject only to income tax, but all income tax, including any tax on distribution of income to shareholders, to be assumed and paid by the Petroleum Board on behalf of the contractor.

Note 2. Sales proceeds sharing instead of production sharing comes very close, at least substantially, to the holding of an exclusive production licence.

7.6 LIBYA

In Libya production sharing contracts were adopted and applied as of 1974. The same oil company, that had been instrumental in introducing the concept in Peru, signed on 7 February 1974 the first production sharing contract with the Libyan National Oil Corporation, NOC. This state corporation had been established in 1970. Up to that time exclusive petroleum rights had been granted in the form of concessions under the terms of the Petroleum Law of 1955 (see Chapter 6). Upon the introduction of production sharing contracts, the Petroleum Law was not abolished but its scope and application was restricted to the petroleum activities of NOC. Exclusive oil rights were reserved for and only granted to NOC. The latter became authorized to enter into production sharing contracts with interested foreign oil companies, provided any proposed and negotiated agreement was submitted to the governmental authorities (the General People's Committee) for approval.

The pattern of the Libyan style contract was rather similar to that of the Peruvian style contract but there were some important differences. For instance, the production was similarly divided into two parts but not in equal proportions. The applied proportions were much more favourable to NOC, such as for instance 81/19 or 85/15. Because of this disproportionate division the Libyan contractor was exempted from having to pay any income taxes (nor royalties for that matter). There was also a difference in the matter of funding of development work. After a commercial discovery was made (in a block) NOC would contribute to the operating costs and development costs to be incurred up to the date of export. These costs would be shared between NOC and the contractor in the same proportion as the production would be shared, say 81/19 or 85/15. Under the earlier contracts, NOC's contributions to development expenditure were treated as loans which had to be paid back by the contractor over a number of years, repayment to start when a certain level of production had been reached or a certain quantity of petroleum had been produced. For instance, as soon as the cumulative production, all blocks taken together, would have reached the level of one hundred million barrels, the contractor concerned was obliged to start repayment of NOC's contributions at a rate of five per cent per year. In other contracts, in particular those concerning non-associated natural gas, no repayment are required.

A peculiar feature of the early Libyan contracts concerned the supervision of the operations. This supervision was not done by NOC but by a management committee consisting of two members representing the Libyan State and the contractor respectively. In fact, the management committee established by the nationalization decree of 1 September 1973 acted as management committee for the purposes of the production sharing contracts then introduced. Later on this was changed in the sense that the two Libyan members were appointed by NOC. Decisions are taken by simple majority; hence NOC was placed in firm control as if it would have been itself and acted as the supervisory body.

Later on, without changing the pattern of the original contracts dating from 1974, the terms concerning the production sharing proportions, NOC's contributions to development expenditure and repayment obligations, if any, were varied to make contracts economically favourable or less favourable, depending on whether non-associated gas or crude oil was involved, on the prospects of finding commercial petroleum in a particular contract area and on the level or expected level of the international oil price. So introduced the Libyan model contract of 15 April 1989 a production sharing mechanism whereby a relation is made between the sharing proportions and the status or completeness of the cost recovery. Under this mechanism the sharing proportion becomes more in favour of the state party depending on whether 'pay-out' has been achieved. The completeness of the cost recovery is indicated by the continuously changing value of a fraction representing the ratio between the contractor's cumulative cash-in (i.e., the proceeds from the sale of cost production and the contractor's share of profit production) and the contractor's cumulative cash-out. It will be clear that until production starts the value of this fraction is zero and that at payout time the fraction equals one.

7.7 MALAYSIA

Up to the signing on 14 February 1971 of Teheran Agreement by the governments of the OPEC Gulf States (see Chapter 8) the Malaysian government had over the years incorporated in its own licensing regime the results of the continuing negotiations on government take that took place with the governments of the OPEC Gulf States and their concessionaires, Malaysia's position in these matters being protected by the presence of a 'most favoured nation' clause in its licences and leases. Thereafter, however, the Malaysian government developed and pursued its own concession policies, which in this case meant that the government followed the example given years earlier (in 1960) by Indonesia and decided to nationalize the then existing petroleum licences and agreements and to make only production sharing contracts available to the foreign petroleum industry.

The new policy was introduced by the Petroleum Development Act 1974, which came into force on 1 October 1974 and by virtue of which:

(i) petroleum agreements were cancelled with effect from 1 April 1975; and
(ii) ownership in and the exclusive right of exploring for, exploiting, winning and obtaining petroleum, whether onshore or offshore, was vested into a state-owned corporation, which was named Petroliam Nasional Berhad (Petronas).

Note. In vesting the ownership of petroleum into Petronas the new Act constituted a radical departure from the Continental Shelf Act 1966, since the latter had vested all rights with respect to the continental shelf and its natural resources into the Federation to be exercisable by the Government of the Federation.

Since then petroleum operations are regulated by the Act of 1974, as amended by the Petroleum Development (Amendment) Acts of 1975 and 1977, and the

Regulations made thereunder. The prime minister is authorized to make regulations to give effect to the provisions of the Act.

In consideration for having received the ownership of petroleum resources and these exclusive rights Petronas is liable to make payments to the Federal Government or any State Government as agreed. Petronas had been granted complete freedom to set the form, terms and conditions of any new agreements, including those that were meant to replace the petroleum agreements that were cancelled. Petronas was and is under no obligation to submit finally agreed terms to the government for approval. Since on the side of the State only Petronas is involved in the making of the new agreements, changes and amendments to the latter can be arranged by Petronas and its contractor in mutual agreement without the intervention of any other state authority or of the legislature.

The new Act of 1974 was preceded by the Petroleum Mining Act 1966, which was not repealed, merely made non-applicable to Petronas, with exception of the provisions regarding getting permission to enter upon lands for carrying out petroleum operations. The Petroleum Mining Act 1966 provided for an Appropriate Petroleum Authority authorized to grant licences or to enter into petroleum agreements. Such licences or petroleum agreements, as were in existence at the time, were 'saved' by the Act of 1974 but for only a limited period. The Act declared that such licences and agreements would continue for six months from the coming into force of the Act or for such extended period as the prime minister would allow. Considering that the Act of 1974 had left it to the same prime minister to decide when the Act should come into force, the continued validity of existing licences or agreements was entirely put within the discretionary powers of the prime minister.

Note. As originally written the Act of 1974 provided for a compulsory acquisition of licences and agreements without offering the dispossessed titleholders any compensation at all, let alone 'adequate' compensation as demanded under the Constitution, and in this respect was clearly in conflict with the latter. The discrepancy was recognized and the Act of 1974 was retroactively amended by the Petroleum Development (Amendment) Act 1977. After this amendment the Act of 1974 provided for adequate compensation to be paid to the person who would have lost its licence or petroleum agreement as a result of the termination of these rights under the rules of the Act. This adequate compensation was allowed to take the form of a single sum; periodical payments or such form as would be agreed under any arrangement agreed between the former holder of the licence or petroleum agreement and another person designated by the Federal Government. With this amendment, which was made effective as of 1 October 1974, the effective date of the Act, the latter was reconciled with the Constitution.

The time during which existing licences and petroleum agreements were suspended was meant to be used for the negotiation of new arrangements with Petronas. Such arrangements were considered to qualify as adequate compensation for the rights that were lost under the Act of 1974. Since Petronas had been given the exclusive rights to explore and exploit, the new arrangements could only take the form of contracts of work. As a matter of fact Petronas decided to adopt for these contracts the production sharing principle. In the end all licences and agreements were terminated with effect from 1 April 1975. Temporary contractual arrangements were made to bridge the period between the termination date and the coming into force of the new arrangements. The negotiation of the first production sharing contracts took a long time. They were finally signed on 30 November 1976.

From the outset it was Petronas' policy that these contracts should have as their objective and be so designed as to serve the national interest of Malaysia and to contribute to the economic development of the country. This policy is reflected in various provisions of the contracts.

Note. In the contracts special attention is paid to the modalities of natural gas development. If non-associated gas is found in substantial quantities the contractor is allowed to hold the gas field, after agreeing with Petronas the extent thereof, for a period not exceeding five years. Before the expiry of the holding period the contractor has to submit for the approval of Petronas a development plan. Failure to do so shall render the gas field to be relinquished to Petronas, unless Petronas agrees to extend the holding period. Any such extension may not exceed five years. If during the holding period contractor would come to the conclusion that exploitation of the gas field is not commercially justified the gas field must be surrendered to Petronas. Contractors must develop the gas field within four years from the date on which the development plan had been approved by Petronas, albeit a development period longer than four years could also be approved by Petronas. Failing to develop within the agreed development period renders the gas field surrendered to Petronas, except in a situation of *force majeure*.

Under the terms of the contact Petronas is allowed to request an increase or decrease of the production rate of any oil field located in the production area for the following purposes and reasons: for optimizing oil and gas recovery; to minimize wastage of natural gas; for safety or operational reasons; or in the national interest. Similarly in the national interest Petronas may request its contractor to develop for Petronas' account and benefit any petroleum reservoir considered by the contractor to be sub-economical or to execute on this basis such works or build such facilities as Petronas may desire.

The production sharing mechanism is composed of the following elements. A maximum of ten per cent of crude oil won and saved (royalty oil) and a maximum of ten per cent of the natural gas available for commercial sale (royalty gas) is taken by Petronas to enable the latter to settle cash payments to the Federal and State Governments due to these governments under the Act of 1974.

A maximum of fifty per cent of the crude won and saved (cost oil) and a maximum of sixty per cent of the proceeds of the sale of the natural gas (cost gas) is reserved for the purpose of recovering by the contractor of all costs relating to crude oil operations and to natural gas operations separately. It is important to note that costs are rccoverable on a cash-out, expensed basis, no depreciation of amortization of capital expenditure being required. The cost recovery procedures with respect to oil operations is kept separate and independent from those with respect to natural gas operations, no 'integration' is permitted, except that the cost of using associated gas for oil operations and the cost of extracting NGL's from the natural gas produced is considered part of the crude oil related cost and as such recoverable from cost oil.

The remaining portion of the crude oil won and saved is called profit oil and the remaining portion of the proceeds of the natural gas sold is called profit gas. The profit oil is divided between Petronas and its contractor on the basis of successive production tranches, coupled to the level of cumulative production. To give an example: incremental sharing proportions may range from 50/50 to 70/30 in favour of Petronas. As soon as cumulative production exceeds eight million kilolitres (about fifty million barrels) profit oil is shared in a 70/30 proportion. The proceeds of profit gas may be divided between Petronas and contractors in a 50/50 proportion, but as soon as the cumulative quantity sold exceeds a certain value, say sixty billion cubic metres the proportion changes to 70/30 in favour of Petronas. Actual numbers and figures are a matter of negotiation.

In contrast with the Indonesian example a contract does not provide for a domestic supply obligation, but instead Petronas has been given the right to purchase from its contractor up to fifty per cent of the quarterly cost oil at a price equal to the cost oil value.

Note. Of interest is the attention paid to the problem caused by the production of associated natural gas. This natural gas must be used by priority for optimizing oil production. Such use includes re-injection for pressure maintenance. Associated gas that cannot be used for these purposes must be re-injected in suitable underground reservoirs, but if there are no such reservoirs the gas may be flared. Before any gas is flared any NGL's should be extracted from the associated gas if commercially feasible.

Contractors are not allowed to sell their cost gas and share of profit gas save on a joint dedicated basis with Petronas to an outlet for natural gas which can be developed on terms that render the production and delivery of the gas economically acceptable to both parties.

Another characteristic feature of the Malaysian contracts is the provision for state participation. The applicable model contract provides for an unspecified sharing of contractor's rights and obligations as between the foreign company and Carigali, a wholly owned subsidiary of Petronas. The foreign company is obliged to enter into an agreement of cooperation with Carigali and this agreement must be approved by Petronas.

Note. The principles of such agreement are set out in the model contract. The percentage participation interest is a matter of negotiation and will be specified in the agreement. An important principle to be incorporated into the actual agreement concerns Carigali's exemption from having to contribute to the cost of exploration work and from being responsible for fulfilling the minimum exploration work obligations. In case of a commercial discovery Carigali is free to decide whether or not to participate in the development thereof. Under the terms of the contract Carigali and its foreign partner are treated collectively as one party, which also applies to arbitration. Disputes between Carigali and its foreign partner are settled as provided for in their joint operating agreement.

A Malaysian contract is further characterized by containing a number of 'national interest' provisions. In this context, the contract contains elaborate rules in the matter of procurement of equipment, materials, services and supplies, the employment of expatriates and the training of contractors' and Petronas' personnel.

Note. In the matter of procurement contractors have by priority to purchase from domestic manufacturers and suppliers. If the required products and services are not available in Malaysia contractors may order, after having received Petronas' approval therefore, these goods and services from outside Malaysia. All procurement of equipment, etc. shall be on an arm's length basis and shall, unless otherwise approved by Petronas, be obtained as a result of competitive bidding. Before employing foreign employees contractors have first to obtain the written approval of Petronas for such employment. Foreign employees may not be employed in positions which can be fulfilled by nationals. Contractors' national employees should be trained to take over the positions held by foreign employees. Contractors are obliged annually to submit, the details of all salaries, benefits and privileges accorded to each classified category of contractors' personnel (both foreign and national).

Contractors are subject to the Petroleum (Income Tax) Act 1967, as amended with effect from 1 April 1975 by the Petroleum (Income Tax) (Amendment) Act 1976 (No. 1) and (No. 2). In contrast with the Indonesian example, the production sharing mechanism that was introduced by the new contracts was not seen as an income tax replacing or absorbing arrangement and the contracts were not meant to exhaust the methods to determine government take. The aforesaid Petroleum (Income Tax) Act 1967, designed and meant to impose a tax on income from petroleum operations, was maintained, albeit the Act had to be amended (with effect from 1 April 1975) in some respects in order to cater for the peculiarities of the production sharing contracts (see Chapter 8).

7.8 ANGOLA

7.8.1 Background

The State of Angola became independent in November 1975. In petroleum policy matters the government was guided by and followed the Malaysian example. In 1976 a national oil company was formed under Decree No. 57 of 9 June 1976. This company was named Sociedade Nacional de Combustivel de Angola, Empresa Pablica (Sonangol EP). Two years later, a new petroleum law was enacted (Law No. 13 on Petroleum Activities of 26 August 1978). The Law of 1978, which is a framework law leaving details to be filled in by subsidiary legislation (government regulations), declared all onshore and offshore petroleum deposits to be the property of the People of Angola and all the exploration and exploitation rights to be transferred to Sonangol. The latter was not authorized to alienate its mining rights, neither partially nor totally. As from the coming into force of the Law all then existing petroleum rights were cancelled and transferred to Sonangol. Any foreign company possessing the necessary technical capabilities and financial capacities, and wishing to explore within the national territory, was permitted to do so but only in association with Sonangol and within the areas comprised in the licences granted to Sonangol. Such association could be in the form of a 51/49 joint venture (this possibility was restricted to the onshore) or in that of a production sharing contract (this possibility was available for both the onshore and the offshore).

At the end of 1979 the terms of the envisaged production sharing contracts were made public. The terms appeared to follow the pattern of Egyptian contracts, except that an operating committee fulfils the role of the board of directors of the joint operating company envisaged under the Egyptian contracts and except that the payment was required of an excess profit tax per barrel profit oil lifted. The levy equalled the difference between the barrel's market value and a certain indexed base price (having as base date the third quarter of 1978). The excess profit tax testified to the Angolan government's preoccupation with creaming off excess profits generated by the high price level prevailing in the early 1980s. In 1984 the price cap amounted to USD 22.70/b. In the 1989 Model contract the base price was specified as USD 20/b adjusted quarterly in accordance with the UN's value index for exports of manufactured goods from developed market economies.

In 2004 the Petroleum Law of 1978 was replaced by the Law No. 10 on Petroleum Activities of 12 November 2004. The new Petroleum Law confirmed the basic principles of its predecessor, such as establishing Sonangol, which is referred to as the National Concessionaire, as the sole holder of mining rights (prospecting licences and concessions) and authorizing the latter to enter into association contracts with Angolan or foreign entities. The said association contract may take the form of joint ventures or production sharing contracts. The power to issue prospecting licences is vested in the supervising Minister. The government is responsible for granting concessions. For each area of an association contract a concession decree is issued. In joint ventures Sonangol must have majority participation. The Law further contains provisions, conditions and operational rules of a general nature which are applicable to all types of association contracts which Sonangol is authorized to enter into, including production sharing agreements. Among these conditions there are those which are typical for petroleum contracts concluded with a state entity established in a developing country and which are usually found in the contract itself and not in the governing Petroleum Law as is here the case. The subjects concern the use of national products and services; domestic supply obligation; the government's right of requisition in the event of a national emergency; the contractor's right to dispose freely of his share of production; the contractor's obligation to employ only Angolan citizens in all categories and functions, except if there are no Angolan citizens in the national market with the required qualifications and experience.

7.8.2 The 2006 Model Production Sharing Agreement

On 31 March 2006, a Model Production Sharing Agreement was issued giving effect to and complementing the provisions of the new Petroleum Law. Most of its provisions are more or less 'standard' and conform to the conditions described in Chapter 5. To this category of more or less standard provisions but nevertheless of essential importance belongs the provision laid down in Article 7 of the Model concerning the establishment of development areas. In said Article 7 it is stipulated that following each commercial discovery the extent of the whole area within the contract area capable of production from the deposit or deposits identified in the well that that originated the commercial discovery and its related appraisal wells, if any, shall be agreed upon by Sonangol and contractor. Each agreed area shall then be converted automatically into a development area effective from the date of commercial discovery. To each development area a production period is assigned lasting (. . .) years. The production period may be extended. Any development area is considered automatically terminated if within (. . .) years from the date of commercial discovery in said development area the first lifting of oil from the said development area as part of a regular program of lifting in accordance with the lifting schedule has not taken place. Development areas as established under said Article 7 play an important role in the oil allocation procedures. The latter procedures form the typical elements of the Angolan contract. Said distinctive elements concern:

(i) the absence of the obligation to deliver royalty oil to the State; and
(ii) the ring fencing applied in the determination of cost recovery oil (around the contract area for exploration costs, around development areas for development costs); and
(iii) the ring fencing applied to the sharing of profit oil (around development areas).

7.8.2.1 Determination of Cost Recovery Oil (Article 11)

Up to a maximum of (. . .) per cent of all oil produced and saved from development areas hereunder may be taken by contractor for recovery of all exploration, development, production and administration and services expenditures incurred by him under the contract. If in any year recoverable cost is less than the maximum value of the cost recovery oil, the difference shall become part of and included in the 'development area profit oil'. Development expenditures in each development area shall be uplifted by (. . .) per cent. In the event that in any year recoverable costs exceed the value of cost recovery oil from the relevant development area for such year, the excess shall be carried forward for recovery in the next succeeding year or years, but in no case after the termination of the agreement. In the case that development expenditures for a development area are not fully recovered within five years after commencement of commercial production or within five years after the year in which development expenditures are incurred whichever later occurs, then contractor's share of cost recovery oil shall be increased from year six, based on a method agreed upon by Sonangol and contractor but not exceeding sixty-five per cent. There is a ring fence around the contract area for recovery of exploration costs and around the development area for recovery of development costs.

Note. According to the latest Terms of Reference (2005/2006) the maximum cost recovery percentage is fifty per cent, and the percentage of the uplift around 120 or 130 per cent.

7.8.2.2 Determination of the Sharing of Profit Oil (Article 12)

The total oil produced in a quarter from each commercial discovery and its development area less the cost recovery oil from the same development area as provided in Article 11 shall be referred to as 'development area profit oil' and shall be shared between Sonangol and the contractor according to the after tax, nominal rate of return achieved by the contractor at the end of the preceding quarter in the corresponding development area. The sharing percentages are linked to the level of earning power achieved, in as much as the higher this level has become the smaller the contractor's percentage share of development area profit oil in the current quarter will be. The tax paid cash flow consists of the sum of contractor's proceeds of sale of his cost recovery oil and of his share of profit oil minus the fifty per cent petroleum income tax and minus development expenditures and production expenses. The cash flow is set up for each development area separately, starting from the date on which the first commercial discovery was made in such area. For the calculation of the cash flow neither any expenditure incurred prior to the date of commercial discovery for the particular development area nor any exploration expenditure shall be included.

Note. According to the 2005/2006 Terms of Reference contractor's profit oil share at the lowest level of profitability is indicatively seventy per cent, falling to twenty per cent to ten per cent as profitability increases

The 2004 Petroleum Law was accompanied by the Law No. 13 on Taxation of Petroleum Activities of 24 December 2004. This Law applies to all entities, whether Angolan or foreign, performing petroleum operations in Angolan territory, as well as in other territorial or international areas within the tax levying jurisdiction of the Republic of Angola, as recognized by international law or agreements. The main tax charges introduced by the Petroleum Tax Law (Article 4) are a petroleum production tax and a petroleum income tax. The petroleum production tax (the equivalent of a royalty) is not applicable to production sharing contracts. The petroleum income tax is levied on the taxable income which consists of revenues generated by the petroleum operations less operating costs and less depreciation charges at the rate of 16.666 per cent in respect of costs incurred for exploration operations, the drilling of development wells and for production, transportation and storage facilities. For production sharing agreements the tax rate is fifty per cent (Section III, Article 41).

Note. Sources and References: Petroleum Legislation, Angola, Revision Date 1 July 2006, 2006 Barrows Company Inc., New York. *PEPS*, IHS Inc.; Energy, Country: Angola, Date 13 July 2006, National Oil Company, Fiscal Terms, Gas, Licence Terms, Licensing Authority, Laws, Contract Type, State Participation.

7.9 THE PEOPLE'S REPUBLIC OF CHINA

In August 1982, China invited foreign oil enterprises to apply for production sharing contracts with respect to the Chinese part of the continental shelf. The basis for the contracts was laid down in the *Regulations of the People's Republic of*

China on Exploitation of Offshore Petroleum Resources in Cooperation with Foreign Enterprises, which had been made public on 30 January 1982.

The new Regulations followed the by now classic Indonesian approach: the ownership of petroleum in situ on the continental shelf was vested in the State. The Chinese National Offshore Oil Corporation (CNOOC), which had been established at the same time, was given the exclusive right to search for and exploit such petroleum. CNOOC should undertake the licensed activities in cooperation with foreign oil enterprises. The Regulations contained the main principles on the basis of which the cooperation should take place. The standard chosen appeared to be the application of a production sharing mechanism characterized by two special features: the production sharing concept is applied to each individual field and CNOOC is given the option to participate in the development of any such field with a share of up to fifty-one per cent. Development cost oil as generated by a particular field is divided between the contractor and CNOOC in proportion to both parties' individual contributions to the costs of development. Remaining oil (which means oil remaining after royalty oil [12.5 per cent], tax oil [five per cent] and cost oil [up to fifty per cent of production]) is divided between the contractor and CNOOC as participant on the one hand and CNOOC as state party on the other in the proportion (X)/(l-X). The X-factor is determined on the basis of successive tranches of levels of daily production, each tranche having its own X-factor. Actual X-factors are fixed in negotiations with applicants.

7.10 NIGERIA

As one of the Member States of OPEC, Nigeria's interest was primarily focused on the introduction of OPEC style state participation with respect to the oil mining leases under its jurisdiction. With respect to new exploration acreage the Nigerian government had shown in one instance to be prepared to experiment with the concept of production sharing. In 1972 a production sharing agreement was concluded with a US oil company, but this agreement remained an isolated event.

Note. In May 1970 the Egyptian NOSODECO contract was signed. Undoubtedly, this contract had stood model for the Nigerian 1972 contract. For instance the cost oil allocation (a fixed forty per cent) and the division of the profit oil was very similar to the Egyptian example. Nevertheless, in three important aspects the Nigerian contract deviated from the Egyptian example. First, the Nigerian contract did not have an autonomous contract area. Instead, the contract area coincided with the area of two oil prospecting licences granted to and held by the state party (in this case the Nigerian National Petroleum Corporation, NNPC). Being the licence holder made NNPC under the applicable petroleum legislation responsible for maintaining the licences in good standing, for fulfilling any area relinquishment obligations and also for the conversion of the prospecting licences into oil mining leases in the event of a commercial discovery. Secondly, the Nigerian contract provided for setting aside a portion of the production (a portion referred to as tax oil) for the payment of the special Petroleum Profits Tax, whereas the Egyptian contract did not provide for the delivery of tax oil, since the contractor's income taxes were paid by

EGPC on the former's behalf out of EGPC's share of profit oil. Under the Nigerian contract profit oil amounted to what remained of the production after deduction of cost oil (a fixed forty per cent) and tax oil (fifty-five per cent of the production remaining after cost oil, the percentage figure representing the then prevailing rate of the Petroleum Profits Tax). The so remaining profit oil portion is divided between NNPC and its contractor in proportions of 65/35 (lower tranche, up to and including 50,000 b/d) and 70/30 (upper tranche, in excess of 50,000 b/d). Thirdly, under the Nigerian contract the contractor remained throughout the duration of the contract responsible for carrying out the operations, no joint operating company was established (after making a commercial discovery) as was and still is the practice under Egyptian contracts.

Not before the early 1990s did the Nigerian government again offer oil companies the opportunity to enter into production sharing agreements with NNPC. This change in policy took place in connection with opening up for exploration the Nigerian continental shelf. In fact prospecting licences were on offer to be formally acquired by NNPC. Oil companies interested in starting an exploration venture could apply or make a bid for a production sharing contract with respect to NNPC's licences covering the exploration acreage of their choice. As the licensee, NNPC is responsible for keeping the licences in good standing and, in case of commercial discovery, for timely converting the prospecting licences in oil mining leases. The contractor has to work in accordance with the rules provided for in the licences, possible future oil mining leases and the applicable petroleum legislation (see Chapter 6).

The production sharing mechanism is composed of the following elements. A portion of oil (royalty oil) is reserved for NNPC in order to allow the latter to pay the royalty due to the State under the terms of any offshore oil mining lease (rates depend on water depth and are ranging between twenty per cent (shallow water) and 0.0 per cent (water depth in excess of one thousand metres). A further portion of production is allocated to the contractor for recovery of the operating and capital costs incurred by him in accordance with rules of accounting incorporated in the agreement. The next portion, referred to as tax oil, is allocated to NNPC in order to allow the latter to pay on behalf of itself and contractor the petroleum income tax due under the Petroleum Profits Tax Act (PPT). In this case, taxable income consists of the difference between the proceeds and deductible costs of the total venture, as calculated in accordance with the rules of the PPT. However this may be, the contractor must be seen (vis-à-vis the tax authorities of the home country of its parent company) as having been subject to the PPT or as having actually paid its share of the PPT imposed. For this reason the PPT paid by NNPC to the government is allocated between the parties on the basis of actual proceeds received and actual costs incurred. The remaining production is divided between NNPC and its contractor in proportions depending on levels of cumulative production.

7.11 OTHER NON-WESTERN COUNTRIES

Many more non-Western countries, when opening up their territory to foreign oil companies for petroleum operations, have opted for adopting the concept of production sharing as the vehicle to engage and cooperate with these companies. To mention a few: Yemen, Myanmar, Vietnam, Ivory Coast, Ghana, and Tanzania. Since the 1990s several East European countries, the Russian Federation, Kazakhstan, Azerbaijan and Turkmenistan have followed their example. Furthermore, following the example of Nigeria, a few other non-Western, major petroleum exporting countries, such as Oman (1975) and other Gulf States, experimented with the production sharing concept but in those States its application remained restricted to new exploration ventures situated outside their main producing areas. The latter remain state owned and controlled or governed by concessions or licences with majority state participation (see Chapter 6).

7.12 THE RUSSIAN FEDERATION

7.12.1 An Overview

In the Russian Federation the concept of the production sharing contract for petroleum operations was introduced by Presidential Decree 2285 of 24 December 1993. The production sharing contract was meant to offer an attractive alternative for the licensing system provided for at the time by the Federal Law No. 2395 *On Underground Resources* of 21 February 1992 (the Subsoil Law) with the aim to promote foreign investment in exploration and production of petroleum.

Note. Under the terms of the aforesaid Decree 3 production sharing contracts (viz. Kharyaga, Sakhalin I and Sakhalin II) were signed. These contracts enjoy a protected status unless otherwise agreed.

The production sharing contract was formally adopted by the Federal Law No. 225 *On Production Sharing Agreements* of 31 December 1995, which entered in force on 11 January 1996 (PSA Law).

In 1999 some subsidiary legislation (Federal Decisions) for implementing the PSA Law was issued, for instance Federal Decision No. 740 of 3 July 1999 containing rules concerning cost recovery. Reconciliation of the PSA Law with the Tax Code was achieved with the Federal Law No. 65-FZ of 6 June 2003 (PSA Tax Law) but on that occasion the PSA Law itself was also amended.

Operations in offshore areas beyond the twelve-mile territorial sea limit are separately governed by the Federal Law No. 187 *On the Continental Shelf of the Russian Federation* of 30 November 1995. Applicable provisions of the Subsoil Law (see below) apply in so far as consistent with said Law

7.12.2 The 1992/1995 Subsoil Law

The Federal Law No. 2395 *On Underground Resources* of 21 February 1992 as restated by the Federal Law No. 27 of 3 March 1995 (Subsoil Law), is a general Mining Law addressing the exploration and exploitation of all underground natural resources including oil and natural gas. (*Note*. The general scope of this Law should be borne in mind when studying its procedures and provisions.)

The Subsoil Law declares the subsoil within the Russian Federation including underground natural resources to be state property. The State Subsoil Fund shall comprise the plots being used in the form of geometric blocks of the subsoil and the subsoil not in use within the Russian Federation and its continental shelf. The possession, use and disposition of the State Subsoil Fund shall be effected jointly by the Russian Federation and the Russian Federation subdivisions. Subsoil rights may be granted for regional geologic study, prospecting for and assessment of deposits of underground natural resources, exploration and production of these resources. Subsoil rights may be granted simultaneously for geologic study (prospecting, exploration) and for the production of underground natural resources and associated activities. In such cases production operations may be conducted both during the period of geological study and directly after its completion. The subsoil rights for exploration only shall be granted for up to five years; for production of underground natural resources up to twenty years and for both above-mentioned uses combined up to twenty-five years. (*Note*. These terms have later been amended).

Subsoil rights shall be granted by a licence which shall consist of a special certificate which shall define the main terms and conditions related to the use of the subsoil. An agreement may be signed between the competent authority and the licensee to establish concrete terms and conditions of operations under the licence. A licence shall contain the 'standard' conditions of a licence for minerals but shall also include an agreement on production sharing, a (risk or non-risk) service contract, and may be supplemented by other terms and conditions. (*Note*. In practice joint ventures have and are being used and also some contracts of work. The reference to production sharing agreements has lost its relevance since the introduction of the PSA Law.)

A licence shall be granted by holding bid contests and auctions as well as other grounds set forth in this Law. The procedures and conditions for any bid or auction shall be determined by the competent authority. Information on forthcoming bid contests, auctions, and the result thereof as well as on the awards of licences shall be published in the mass media.

The Subsoil Law was amended in January 2000 to include measures relative to licence assignment, licence term, development rights and the retention of uneconomic deposits, in August 2004 to introduce measures diminishing the role of regional authorities in the issuing of licences, the functions of regional and local bodies, revenue distribution and procedures for conducting auctions and tenders. And again in early 2007 to extend the exploration period of selected areas such as inland bodies of water, territorial seas and the continental shelf from five years to eight years.

7.12.3 THE 1996 PSA LAW

7.12.3.1 Scope

The PSA Law establishes a legal basis for foreign investment in exploration, development and production of mineral raw materials within the territory of the Russian Federation as well as on the continental shelf and/or within the boundaries of the exclusive economic zone under production sharing agreements (referred to hereinafter as 'contracts').

7.12.3.2 Signing of the Contract

The contract shall be signed by the Federal government and the executive government body of the appropriate Russian Federation subdivision. If the subsoil area granted is on the continental shelf or within the exclusive economic zone the contract shall be signed by the Russian Federation government by agreement with the executive government body of the appropriate Russian Federation subdivision on whose territory operations under the contract shall be carried out.

7.12.3.3 Precedence

The obligations of the parties to the production sharing contracts shall be established pursuant to this Federal Law and the civil legislation of the Russian Federation. If the Russian Federation statutory acts contain provisions other than those made by this Federal Law, the provisions contained in this Federal Law shall apply.

7.12.3.4 Rights Granted by the Contract

The contracts shall grant the contractor exclusive rights for the exploration, development and production for a certain period. Contractor shall perform operations at his own risk. Terms and conditions for the use of the subsoil provided by the contract shall not be in conflict with the Subsoil Law. (*Note*. In 1996/1999 the attempted full-scale integration of the production sharing agreement into the Subsoil Law and other Laws with exception of the Tax Code has not yet been followed through). Within the limits of their power Russian Federation subdivisions shall issue legislative regulation of production sharing contracts for the use of subsoil areas within their respective territories.

7.12.3.5 Parties

Parties to the contract are on the one part the Russian Federation (hereinafter 'the State'), which is represented by the Federal Ministry of Natural Resources and the executive body of the Russian Federation subdivision, on whose territory the subsoil area granted for use is located, and on the other contractors such as nationals of the Russian Federation, foreign nationals, legal persons or consortia of legal persons established under joint venturc contracts without the status of a legal person (*Note*. Meant here are participants in a non-incorporated joint venture). If a consortium of legal persons acts as contractor under the contract, members of such consortium shall have joint and several rights and liabilities under the contract.

7.12.3.6 Licence

A subsoil area shall be granted for use to the contractor by a licence which certifies the right to use the subsoil area provided for in the agreement by the executive government body of the appropriate Russian Federation subdivision and the federal subsoil fund management body or its territorial subdivision within thirty days from the date on which the contract is signed. Said licence is granted for the whole period of the contract, and shall be subject to extension or re-issuance or shall lose its validity as provided for in the contract. If a consortium of legal persons without the status of a legal person acts as contractor, the licence shall be issued to one of the members of such consortium and indicate that such member acts on behalf of the consortium.

7.12.3.7 Duration

The duration of the contract shall be defined by the parties in accordance with legislation of the Russian Federation applicable as of the date of the contract. The duration of the contract shall be extended at the request of the contractor provided he fulfils his obligations, for a period sufficient to complete commercial production. Terms of and procedure for such extension shall be set forth in the contract. In the case of extension the licence for the use of the subsoil shall be re-issued.

7.12.3.8 Awarding of Contracts

Contracts may only be awarded by auction.

Note. This condition has been introduced after amendment of the PSA Law by the Federal Law No. 65-FZ of 6 June 2003 (referred to as the PSA Tax Law). Further details are given hereinafter.

7.12.3.9 Drafting of Conditions by Government Commission

Drafting of conditions for use of the subsoil and the contract shall be the task of a commission established by the federal government in agreement with the executive government body of the appropriate Russian Federation subdivision.

7.12.3.10 Programmes and Budgets

Operations under the contract (contract operations) shall be carried out in accordance with programmes and budgets subject to approval and be carried out in accordance with the law.

7.12.3.11 National Interest Provisions

Investors shall grant Russian legal entities preferential right to participate in the contract operations as subcontractors, suppliers, transport agencies or in any other capacity under contracts with the contractors with all other conditions being equal. The contract must provide that a certain amount of equipment for production and processing subject to cost recovery shall be produced on the territory of the Russian Federation.

7.12.3.12 Protection of the Natural Environment

Contractors have to prevent harmful impact of operations upon the natural environment, and to remedy the consequences of such impact. The contractor shall take all necessary measures to protect the original environment and the traditional way of life of small ethnic minority groups.

7.12.3.13 Operator

Operations shall be performed by the contractor's operator. Operator may be an affiliate or legal entity established by the contractor or foreign legal entities. Contractor is liable to the Russian Federation for the operator's actions with its assets as if such actions were the investor's own.

7.12.3.14 Area Relinquishment Schedule

Upon completion of certain stages of exploration the contractor shall relinquish areas granted, in accordance with an area relinquishment schedule included in the contract.

7.12.3.15 Management Committee

The parties may establish a management committee, made up of an equal number of representatives of each party. The number of members, rights and duties, working procedures shall be provided for by the contract.

7.12.3.16 Production Sharing Mechanism

Production shall be shared by the State and the contractor under the contract which shall define total volume of production and its value; define the portion of production for recovery of costs (hereinafter compensatory production), specify sharing of profit production by the State and the contractor which shall mean production less a portion for payment for the right to use the subsoil.

Note 1. Payments for the right to use the subsoil include bonuses, royalty (now the mineral production tax (MPT)), and surface tax.

Note 2. Which type of costs qualify for recovery from production, the manner in which these qualifying costs may be recovered (immediately recoverable or subject to annual depreciation) and the maximum percentage of the production that is available for cost recovery will be settled per individual contract, subject to a normative act (regulations) on cost recovery, such as the Federal Decision No. 740 of 3 July 1999 and to the cost recovery rules contained in Chapter 26.4 of the Tax Code (discussed hereinafter). In line with past contracts maximum percentages for cost recovery oil have been indicated: viz. seventy-five per cent for fields on land and ninety per cent for fields on the continental shelf. The sharing of profit production is also settled on an individual contract basis, subject to any conditions established by the licensing authority in the context of a bidding round. Past practice has been to link profit production sharing to the profitability of contractor's operations under the contract. Said profitability to be measured either by calculating the 'R-Factor' (i.e., the ratio of the cumulative contract revenues earned by the contractor to the cumulative contract expenses incurred by him) or the internal rate of return of the tax-paid cash flow realized by contractor in the period preceding the period in respect of which the profit production sharing must be established (see Chapter 5 and elsewhere in this chapter).

Cost recovery production and profit production to which the contractor is entitled shall be the property of the contractor and may be exported under procedures specified in the contract without quantitative restrictions on export subject to the Federal Law 'on the State Regulation of Foreign Trade Activity'.

Production received by the State as a result of production sharing under the contract shall be distributed between the Russian Federation and the Russian Federation subdivision on whose territory the subsoil area is located under agreements concluded by the appropriate government bodies.

7.12.3.17 Title to Assets

Assets acquired by the contractor for the operations shall be the contractor's property unless otherwise provided by the contract. Title to the aforesaid property may be transferred from the contractor to the State as of the date the cost of said property has been fully recovered, or as of the date of termination of the contract, or any other date as may be agreed. During the period of the contract, the contractor shall be granted an exclusive right to use such property for free to conduct the operations. Contractor shall bear responsibility for maintenance of the property and the risk of its accidental loss or damage.

7.12.3.18 Taxation (Profits Tax and Other Payments)

With the exception of the profits tax and payment for the use of the subsoil, the contractor is exempt from taxes, charges, excise levies and other obligatory payments. Collection of said taxes, charges and other obligatory payments shall be replaced by production sharing under the contract. If laws of the Russian Federation subdivisions and statutory acts of local self-government bodies do not grant the contractor an exemption from taxes and other payments to the budgets of the Russian Federation subdivisions and local budgets, the profit production the contractor is entitled to under his contract shall be increased at the expense of a corresponding cut in the State's share by the amount equal to the taxes, and other payments actually paid to the Russian Federation subdivisions and local budgets.

The profits tax shall be paid by the contractor in accordance with legislation. Subject to the profits tax shall be the value of profit production as provided in the contract and owned by the contractor. Such value shall be reduced by the contractor's payments for the use of borrowed funds and onetime payments by the contractor as well as other unrecoverable costs incurred by the contractor. Accounting procedures for the profits tax shall be established by the contract. If costs exceed the value of the contractor's share of profit production the profit tax base in future periods shall be reduced by the amount of the difference until fully recovered.

Note. The provisions regarding taxation have been restated as well as amended by the Federal Law No. 65-FZ of 6 June 2003 (referred to as the PSA Tax Law) by means of introducing into the Tax Code a new Chapter 26.4. This chapter contains a special tax regime for production sharing agreements concluded in accordance with the PSA Law. Said tax regime is not much different from the one described above and will be summarized hereinafter.

7.12.3.19 Bonus

Bonuses are established in the contract and upon achievement of a certain result.

7.12.3.20 Surface Tax

Annual payments for exploration per unit of subsoil area are subject to economic and geographical conditions, the size of the area, the type of mineral resources, duration, geological study of the subsoil and degree of risk involved.

7.12.3.21 Royalty

Regular payments (royalty) are established as a percentage of production or of the value of production and are payable by the contractor in cash or kind.

Note. On 1 January 2002 the royalty was abolished for new PSAs and effectively replaced by the mineral production tax (MPT). This tax was meant for royalty/ tax concession under Subsoil Law terms and was introduced by Federal Law No. 126-FZ of 8 August 2001, by means of adding Chapter 26 (Articles 334 to 346) to the Tax Code. The royalty value is defined as gross sales revenue less tariffs, VAT and export duty. The PSA Tax Law (discussed hereinafter) confirms that under a production sharing agreement MPT is payable on oil and gas condensate at a rate of 340 roubles per tonne multiplied by a coefficient reflecting world prices as described above. This tax rate shall be applied with a coefficient of 0.5 until the maximum level of commercial extraction of oil and gas condensate, which may be established in any given contract, is reached. The royalty value (the tax base of the MPT) is different since contractors are not paying export duty.

7.12.3.22 Bank Accounts

For the operations the contractor shall have bank accounts in rubles and/or foreign currency in banks located in Russia and/or in foreign states which shall be used solely for the purpose of the operations. The contractor shall have the right to receive and keep foreign currency proceeds from the sale of production owned by him under the contract in the said bank accounts, pay any expenditures incurred with respect to the operations as well as the profits tax and payments for the use of the subsoil from these accounts and freely dispose of the balance of funds in the accounts.

7.12.3.23 Stability-of-Contract Understanding

The terms and conditions of the contract shall remain in force during the whole period of the contract. Changes may be made only by consent of the parties, as well as at the request of only one of the parties in cases foreseen under civil law. If during the validity of the contract, the legislation of the Russian Federation, or that of the Russian Federation subdivisions or statutory acts of local self-government bodies establish regulations adversely impacting on investor's commercial results under the contract, the contract shall be amended to ensure that the investor obtains the commercial results which he would have obtained if the legislation existing at the time of signing the contract had applied.

7.12.3.24 Amendment of PSA Related Laws

The President of the Russian Federation and government bodies of Russian Federation subdivisions are requested to reconcile their statutory acts with this federal law within three months.

7.12.3.25 Blocks Available for PSAs

In the first quarter of 1996, the Russian Federation Government shall submit a draft federal law on approval of fields to be covered by this federal law to the Russian Federation State Duma.

Note. In response to the above mentioned requests for legislation the government introduced in 1996 two draft federal laws. The one draft concerned a draft federal law, *On Making Changes and Amendments to Legislative Acts of the Russian Federation Arising from the Adoption of the Federal Law On Production Sharing Agreements*. In particular the Subsoil Law had to be amended in order to clarify the legal status of the production sharing agreement as a right to use the subsoil. Apart from the Subsoil Law the draft Law introduced amendments to eleven other PSA related laws, among which the Tax Code and the aforementioned federal law *On the Continental Shelf*. The other draft federal law concerned a draft federal Law *On Modifications to the PSA Law*. This draft law contained rules for deciding which blocks should qualify for being put on the federal list of PSA-blocks. The whole PSA-package was signed by the Russian President in January 1999 but never became law.

7.12.4 The 2003 PSA Tax Law

The Federal Law No. 65-FZ of 6 June 2003 which entered into force on 9 June 2003 (usually referred to as the PSA Tax Law) added a new Chapter 26.4 to the Tax Code containing a special tax regime applicable to production sharing agreements concluded in accordance with the PSA Law (i.e., not applicable to the contracts that were concluded prior to the date the PSA Law entered into force). In fact said Chapter 26.4 amounted to a restatement of the profits tax and other taxes for which contractors are liable.

Under the restated/amended situation contractors shall pay the following taxes and levies: mineral production tax (MPT); profits tax (PT); bonuses; VAT, unified social tax, environmental and water charges, state duties, custom levies, land tax; excise duty. The amounts paid by the contractor in respect of all of the above listed taxes and levies with exception of MPT and PT are recoverable by the contractor under cost recovery provisions except for bonus payments which are tax deductible. Contractors are exempt from the payment of regional and local taxes and levies (the remedy provisions of the PSA Law (see above) have been upheld) and also from transport tax. The contractor is exempt from duty on exports from operations under his contract.

Taxable income for PT is the value of contractor's share of profit production less the following: financing costs, bonuses and other costs that are not recoverable for production sharing purposes as defined by each contract subject to regulations concerning cost recovery. If in any tax period the amount of allowable deductions exceeds the value of the contractor's share of profit production, then the amount of the difference may be carried forward to the next tax period until fully deducted. The rate of profits tax current at the time of signing the contract is the rate that will apply for the duration of the contract. The current rate of the profits tax is 24 per cent.

The MPT was introduced by Federal Law No. 126-FZ of 8 August 2001, by means of adding Chapter 26 (Articles 334 to 346) to the Tax Code. The tax was meant for royalty/tax concession under Subsoil Law terms. The royalty value is defined as gross sales revenue less tariffs, VAT and export duty. The PSA Tax Law confirms that under a production sharing agreement MPT is payable on oil and gas condensate at a rate of 340 roubles per tonne multiplied by a coefficient reflecting world prices as described above. This tax rate shall be applied with a coefficient of 0.5 until the maximum level of commercial extraction of oil and gas condensate, which may be established in any given contract, is reached. The royalty value, the tax base of the MPT, is different since contractors do not have to pay the export duty.

The PSA Tax Law did not only amend the tax provisions of the PSA Law (by introducing a special PSA Chapter into the Tax Code) but also amended other provisions of the PSA Law. The amendment involved inter alia Articles 2.4 and 6 of the PSA Law concerning the award of production sharing agreements. The new situation is now that in order for a field/project to qualify for development under a production sharing agreement, the government must have failed to attract, in an auction organized under the terms of the Subsoil Law, a contractor to undertake the development under Subsoil Law royalty/tax terms. Only in these circumstances and subject to certain additional criteria being met may the field/project be designated for development under a production sharing agreement. Said additional criteria to be met are:

(i) that development of the field/project is required to maintain employment in the area concerned, while closing down work on the field would lead to negative social consequences;
(ii) that the field/project is located on the continental shelf or in a remote area lacking in population, transportation or other infrastructure; and
(iii) that development of the field/project requires special cost-intensive technologies (for difficult to produce large reservoirs located in a geologically difficult environment).

It is further clarified that production sharing agreements can only be awarded by way of auction.

Note. Sources and References: *The Russian Investor*; International Taxation Series, Russia, Revision Date 30 October 2005, 2005 Barrows Company Inc., New York; Petroleum Legislation, Russia, Revision Date 1 January 2005, 2005 Barrows Company Inc., New York. PEPS, IHS Inc.; Energy, Country: Russia, Date 27 April 2006, Fiscal Terms, Licence Terms, Licensing Authority, Laws, Contract Type.

Chapter 8
Petroleum Taxation

8.1 GENERAL REVIEW

8.1.1 Introduction and Background

In common with any other commercial industrial activity petroleum activities have from the very beginning been the object of taxation. The oldest charge with which the oil and gas producer, as the licensee, became confronted was a payment in kind or cash to the owner of the petroleum in situ, which in most cases would be the State, exceptionally the landowner. This payment is known as royalty and has always been expressed as a percentage of the oil and/or natural gas volumes that were produced by the licensee. In addition, the licensee had to pay area rentals and various cash bonuses. Apart from these licence-related obligatory payments, the oil and gas producer who sells his production for profit and behaves in every aspect as a commercial enterprise is liable to pay taxes (income taxes) on the net result from the sales proceeds. Up to the 1970s, sales proceeds (after payment of royalties) were treated from a fiscal point of view no differently than sales proceeds derived from other industrial activities, meaning that only the income tax for incorporated commercial enterprises (corporation tax) was imposed. After this period producer prices, due to and triggered by political events, started to rise in an unprecedented manner and a spectacular difference opened up between the cost of production and the producer price (of exported oil). In reaction thereto every host government applying a licensing regime immediately seized the opportunity to increase the government take (i.e., the sum of royalty and income tax). Higher royalty rates were introduced as well as completely new special petroleum income taxes or profit sharing arrangements, partly supplementing, partly replacing the corporation tax. At the same time more and more non-Western countries adopted production sharing contracts, which enabled them to raise government take without having to adjust the prevailing corporation tax.

8.1.2 Taxation under a Licensing Regime

Holders of an exclusive production licence are obliged to make the following payments to the State by way of tax:

(i) licence-related taxes (obligatory payments) to be separated into royalty, area rentals, cash bonuses, net or gross profit shares and excess profit taxes.

Note 1. The royalty is a payment expressed as a percentage of the oil and/or natural gas volumes that are produced by the licensee and is payable in cash or kind. The royalty must be paid to the owner of the petroleum in situ and has only to be paid to the State if the State is such an owner which happens to be the case in almost all States. If exceptionally the landowner is the owner of the petroleum in situ the royalty has to be paid to him. The royalty is the most basic and oldest known petroleum tax. Royalty is a deductible item for the purposes of the income tax.

Important elements in the calculation of the royalty are:

(a) the rate expressed at a percentage of the value chosen as the base value;
(b) the place, where the value is assessed; and
(c) the base value.

As a rule, a theoretical arm's length market value is taken as the base value. If the wellhead is selected as the place of assessment of the value, all capital and operating costs involved in treating, handling and transporting the oil or natural gas production from the wellhead to the market place may be deducted from the theoretical arm's length market value.

Note 2. Area rentals are expressed as an annual amount per square km. Usually the amount increases the longer the licence is kept. Area rentals are deductible for the income tax, similarly as the royalty.

Note 3. Cash bonuses are cash payments which are due to be paid on certain occasions, e.g., at the day the licence is granted or the contract is signed or when the production for the first time reached a certain level. It may also be the (huge) cash payment due to be paid by the successful applicant if the authorization is put up for sale. Most bonuses are not deductible for the income tax.

Note 4. State net profit share is a levy which amounts to a percentage share of the net profit. The net profit is the income remaining after payment of income tax. The rules of calculation of the state net profit share will be laid down in regulations or in the licence. The payment of this share will be an allowable deduction (in the same way as royalty or rentals are) in calculating the taxable income under the rules of the income tax. A well known example of a net profit share is the net profit share that may be demanded in the bidding procedure for competitive leases to be granted with respect to the US outer continental shelf under the terms of the US Outer Continental Shelf Lands Act of 7 August 1953.

Note 5. State gross profit share is a levy which amounts to a percentage share of the gross profit. The gross profit is the income before payment of income tax, i.e., gross proceeds less operating costs and depreciation. The sharing of the gross profit takes the place of a special petroleum income tax. The rules of calculation of the gross profit and the State's share thereof will be laid down in the regulations or in the licence. The gross profit as defined does not have to be identical to taxable income as defined for the purposes of the common income tax. As a matter of fact the sharing of gross profit allows for flexibility and adjustment of the fiscal burden when needed in the light of changing circumstances. For instance, it offers the opportunity to change the State's percentage share and/or to introduce incentives in the form of special cost allowances reducing the amount of the gross profit that has to be shared with the State. Gross profit sharing under a licensing regime had its origin in the 50/50 profit sharing principle that was introduced in December 1950 the Middle Eastern petroleum agreements. This principle meant that the total sum of income tax and all other payments to the State including royalty should not exceed fifty per cent of the gross profits of the concessionaire concerned (this matter is further discussed hereinafter).

Note 6. State share of excess profits is a levy which is calculated on a unit basis as a percentage share of the difference between a certain base price, which is indexed for inflation, and the actually realized sales price. The cash payment is in itself a deductible item in calculating and determining taxable income under the common income tax (in the same way as royalty, rentals and state net profit share are). A well-known example is the US windfall profit tax, which was in effect from 1980 to 1988:

(ii) income tax, either the corporation tax or a special petroleum income tax or a combination of the two. (*Note.* By the 'corporation tax' is meant the tax which is levied on income realized by incorporated commercial enterprises generally.)

8.1.3 Taxation under Production Sharing Contracts

The contractor is liable to pay a special petroleum income tax, except that:

(i) the contractor may have to pay an excess profit tax, if producer prices exceed a predetermined level;
(ii) the contractor may be exempted from paying any tax at all (as in the Libyan type of contract);
(iii) the state enterprise concerned shall pay to the State on the contractor's behalf the income tax, for which the latter formally is liable (as in the Egyptian type of contract).

8.1.4 Review Income Taxes

8.1.4.1 Corporation Tax (Licensees)

This tax is levied on income realized by incorporated commercial enterprises generally. It is calculated as a percentage of the taxable income. The impact of the tax is determined by the manner in which the taxable income (revenues, including the value of stock changes, less allowable deductions and losses) is described and calculated, and by the applicable percentage rate.

Important factors for the calculations are:

(i) the definition of 'income' and in this context the use of transfer prices instead of prices actually realized in a sales transaction;
(ii) the treatment of and accounting for stock changes;
(iii) the description of cost items and type and scope of expenditure allowed to be deducted from sales revenues in calculating taxable income;

(iv) the modalities (rates, progression and timing) of the writing off of the cost of acquisition of capital assets;
(v) the treatment of exploration expenditure;
(vi) the treatment of losses incurred in other operations (i.e., in non-petroleum operations or in operations carried out in other licences);
(vii) the carry forward of losses suffered in previous periods, etc.; and
(viii) additions to the general reserve to meet future liabilities, in particular making deductions to cover the future costs of abandonment of offshore installations or fields generally.

8.1.4.2 Special Petroleum Income Tax (Licensees and Contractors)

A special petroleum income tax is an income tax which is specifically designed to be levied on income realized by a licensee or a contractor from selling his oil and/or natural gas production entitlement for a profit. Under a licence the production entitlement that can be sold by the licensee consists of total production from the licence area, under production sharing contracts the production entitlement that can be sold by the contractor consists of his cost recovery production and his share of profit production. The tax payable is calculated along the same lines as the corporation tax and is expressed as a percentage of taxable petroleum income. The impact of this tax depends on the applicable rate and on the manner in which taxable income is calculated and determined (see hereinafter).

8.1.5 Interrelation and Concurrency of Taxes

If the licensee is liable to pay the corporation tax, any other payment to the State, such as the royalty, area rentals, the state net profit share and excess profit tax, if applicable, is an item allowed to be deducted from taxable income for the purposes of calculating and determining the corporation tax, in the same way as any recognized and undisputed costs may be deducted from taxable income. Where the corporation tax is applied concurrently with a special petroleum income tax, or concurrently with a gross profit sharing arrangement, special provisions will have to be made to mitigate the impact of these payments. To this purpose, a special petroleum income tax may be allowed as a deduction in calculating the taxable income under the rules of the corporation tax (as is the case in the United Kingdom) or it may be the other way around (as in the case of the special hydrocarbon tax in Denmark, where said tax is levied on the net income (income after payment of the common corporation tax) of the licensee). The special petroleum tax may also have to be paid in addition to and independently from the corporation tax (as in Norway). In other States and fiscal legislations, the corporation tax is declared not to be applicable to petroleum income and is entirely replaced by a special petroleum tax. This has the effect that only a single tax on petroleum income is imposed and payable (as in Nigeria, Oman, Abu Dhabi and Malaysia). A state gross profit share (which has for instance to be paid by the licensee in the Netherlands) has to all intent and purposes the same effect as a special petroleum income tax. In the Netherlands, where petroleum income is also subject to the corporation tax, the latter is considered as a prepayment of the state gross profit share.

Note 1. Where a licensee is liable to pay a special petroleum income tax the applicable fiscal regime may in some instances allow the making of a financial arrangement under the terms of which the petroleum tax paying licensee is guaranteed in fiscal terms a minimum profit or a fee after imposition of income tax, the so-called fiscal margin (as in Nigeria, Oman and Abu Dhabi).

Note 2. A particular form of special petroleum income tax is the cash flow tax. This tax is based on the cash flow generated by the petroleum activities and is calculated as a percentage of the annual cash surplus after royalty and corporation tax. It is payable after the after tax cash flow has passed the payout point (i.e., after the cash flow has passed beyond the point in time when cumulative cash-in for the first time equals cumulative cash-out). Where applicable, the cash flow tax is levied in addition to the corporation tax and royalty (and rentals). In the period proceeding the moment of payout any annual cash deficit is increased by an amount of interest and thereafter carried forward to the next year and added to the cash-out items of that year. The adding of an amount of interest to the annual cash deficits postpones the moment that payout will be reached and hence postpones the time that the cash flow tax will have to be paid. In periods of low oil price levels and modest cash flows, there is obviously no scope for cash flow taxes.

8.1.6 Other Taxes

8.1.6.1 Export Duty

This type of levy is calculated as a percentage of the value of exported petroleum and assessed at the point of export. Export duties are rarely imposed on oil or natural gas production. If imposed, it would not change the producer price obtainable for the exports on the world market. Hence its effect would be to lower the price to be received by the producer. Export duties are levied under the taxation regime applicable in the Russian Federation. The duties are levied on the basis of a sliding scale with reference to the average price of Urals Blend in the European export markets. No export duty is levied on production originating from production sharing agreements. (*Note.* Reference PEPs IHS Inc.; Energy, Country: Russia, Fiscal Terms, Date: 27 April 2006.)

8.1.6.2 Import Duty

This type of levy is calculated as a percentage of the value of imported petroleum or as a fixed sum per unit of petroleum (weight or volume) and assessed at the point of import. The duty has to be paid by the buyer of the imported production. Import duties make imports artificially more expensive. Such duties are levied if a government of an importing country wishes to protect its domestic oil producers against imports offered at a price with which the domestic producers cannot compete.

8.1.7 Taxation as an Instrument for Responding to Changes in Oil Price Levels and Profitability

Under a licensing regime the manipulation of the modalities of the royalty (rates, place of assessment, royalty value) and those of a special petroleum income tax (tariffs, allowable deductions, special incentives in one form or another, etc.) allows a host government to pursue a flexible fiscal policy with respect to the economic outcome of petroleum production operations. For instance the rates and tariffs at which respectively a royalty or any special petroleum income tax is levied could be lowered during a period of low oil prices on the world market and vice versa. If the special tax is imposed in respect to income realized from individual fields, the impact can be regulated by classifying fields as to size and geographical location and modifying the rate of tax in accordance with the class to which the field concerned belongs or with its geographical location. As far as royalty is concerned this type of tax allows differentiating between oil and natural gas, in the sense that the rates for natural gas may be fixed at lower levels than those for oil if and when needed to stimulate the utilization of associated natural gas.

Note. Pursuing a flexible fiscal policy by means of adjusting the tariffs of the corporation tax is not possible because this tax affects all incorporated commercial enterprises in the country concerned and does not allow singling out one type of industrial activity such as the production of petroleum for special fiscal treatment.

Under a production sharing regime an inherent flexibility in adapting government take to changed circumstances is offered by the manner in which the terms applicable to the recovery of costs and the sharing of 'profit production' are formulated and the tariffs of the special petroleum income tax are adjusted.

8.2 PAST AND CURRENT PETROLEUM TAXATION (LICENCES)

8.2.1 The Middle East Petroleum Agreements of the 1930s

8.2.1.1 Obligatory Payments

The Middle East petroleum agreements of the 1930s (also referred to as concessions if the context so allows) initially only envisaged obligatory payment consisting of signature bonuses, discovery bonuses, certain annual or monthly payments and royalties tied to minimum sums. The concessionaires were exempted from paying any taxes or other impost of every kind, including income taxes, customs duties, import or export taxes (see Chapter 6).

Note. Exceptionally, the 1901 D'Arcy concession provided for the payment of 16 per cent of the annual net profits; whereas the 1933 revised Anglo-Persian concession required a net profit sharing in addition to the payment of a royalty of four gold shillings per ton. The net profit sharing consisted of a twenty per cent tax on dividends distributed to the shareholders of the concessionaire in excess of GBP 671,250. Payments on account of royalty and dividend tax were subject to a minimum annual payment of GBP 750,000.

8.2.1.2 50/50 Profit Sharing

At the end of 1950 the first revision of the fiscal regime (whereby Saudi Arabia took the lead) took place. By Royal Decree of 27 December 1950 Saudi Arabia introduced and imposed an income tax at a rate of fifty per cent, which was imposed on income earned by petroleum companies. The Aramco concession providing for exemption from paying income tax had to be revised in this respect. At the time the revision was agreed (30 December 1950), it was also agreed that the total sum of income tax and all other payments to the State should not exceed fifty per cent of the net income of the concessionaire, the so called 50/50 profit sharing principle. Initially, the net income had to be calculated after deduction of foreign income taxes, but in October 1951 it was decided that the net income to be shared had to be the net income before foreign tax.

In 1955 the concessions were revised in two respects:

(i) the concessionaires were obliged to publish the price at which they were ready to sell their production FOB port of export to all comers, which price was referred to as the posted price; and
(ii) the rate of the royalty was increased to 12.5 per cent.

Note 1. The royalty was calculated by applying the newly agreed rate to the posted price reduced by a temporary percentage discount and a marketing allowance.

Note 2. In 1958 the government of Venezuela increased the rate of its income tax from twenty-six per cent to forty-five per cent. Up to that time, actually since 1943, concessionaires in that State had been subject to an income tax at a rate of twenty-six per cent, a royalty at a rate of 16.67 per cent and an additional charge meant as a make-up payment and calculated to increase the total sum of the payments to the State to fifty per cent of the concessionaires' net oil income. It is important to note that the 16.67 per cent royalty had to be expensed and was not considered to be a pre-payment of the income tax as was the case in the Gulf States at that time. The Venezuelan system amounted to a 50/50 profit sharing arrangement and as such it had been adopted by the Gulf States, albeit with a different treatment of the royalty necessitated by the fact that in the Gulf States the rate of the applicable income tax was fifty per cent. After having increased the rate of the income tax to forty-five per cent, while concomitantly repealing the make-up payment, the Venezuelan State received more than fifty per cent of the net income of a concessionaire, hence more than any Gulf State received from its concessionaire.

After years of unchanged posted prices, the concessionaires lowered in February 1959 their prices by USD 0.18/b and for a second time, in August 1960, by USD 0.10/b. The total reduction amounted therefore to USD 0.28/b.

Table 8.1. Posted Prices (USD/b) 1953-1960

	1953	*1954*	*1955*	*1956*	*1957*	*1958*	*1959*	*1960*
Arabian Light ex Ras Tanura (34°)	1.82	1.93	1.93	1.93	2.00	2.08	1.90	1.80
Irani crude ex Bandar Mashur (34°)		1.91	1.91	1.91	1.98	2.04	1.86	1.77
Iraqi crude ex Fao (34°)	1.79	1.90	1.90	1.85	1.92	1.97	1.80	1.72
Kuwaiti crude ex Mina-al-Ahmadi (34°)	1.64	1.72	1.72	1.72	1.78	1.85	1.67	1.59

The concessionaires had been forced to reduce their posted prices in the struggle for outlets in the West-European market. In that region the concessionaires were confronted with each other's exports, with oil exported from Venezuela and, as from 1955, with oil exported from the Soviet Union. To counter the competition the concessionaires were forced to sell their crude oils below the applicable posted prices. Since the 50/50 profit sharing system was calculated on the basis of posted prices, selling below the posted price caused the concessionaires to lose the corresponding fiscal relief. Unable to continue with this practice, the concessionaires decided, as mentioned above, in February 1959 and again in August 1960, to adjust their posted prices to market realities. As mentioned before the price adjustments triggered and caused the establishment of OPEC on 1 September 1960 by the governments of Saudi Arabia, Iran, Iraq, Kuwait and Venezuela. After this event the 50/50 profit sharing arrangement based on posted prices was further developed in continuing negotiations between the governments of the OPEC Gulf States on the one hand and their respective concessionaires on the other. These practically permanent negotiations were focused on two issues, namely:

(i) the rate and fiscal treatment of royalties;
(ii) the level of the posted prices and the rate of the income tax.

With reference to (i). For fiscal purposes, royalties were no longer considered to form a part of the State's 50 per cent profit share and as such a pre-payment on account of income tax. Instead royalties were treated as an expense deductible from gross income in calculating and determining taxable income. Over the period 1964-1966 the concessions were revised to the effect that as of 1964 royalties had to be expensed. At the same time a new system of discounts in respect of the posted price (i.e., 8.5 per cent in 1964, 7.5 per cent in 1965, 6.5 per cent in 1966 and to be reconsidered in the years thereafter in the light of the then prevailing market situation) was agreed as quid pro quo for the acceptance of the increase of the government take and the deviation from the 50/50 profit sharing principle. Not surprisingly, the fact that in Venezuela (and in all Western countries) the royalty had always been expensed for the purpose of the corporation tax was used by OPEC to its advantage. As part of the total overhaul of the OPEC fiscal system set in motion by the 1971 Teheran Agreement, the rate of the royalty was on 1 July 1974 increased from 12.5 per cent to 14.25 per cent. On 1 October 1974 this rate was further increased to 16.67 per cent, to be raised again on 1 November 1974 to 20 per cent. By way of small compensation the posted price of 34° Arabian Light was lowered with effect from that date from USD 11.651/b to USD 11.251/b.

With reference to (ii). In 1966 the government of Saudi Arabia concluded an agreement with its concessionaire to the effect that the posted price would also be applicable for fiscal purposes where sales to third parties were concerned. Hitherto posted prices were only applied for fiscal purposes in connection with sales to affiliated companies.

In October 1970 the Gulf States raised the rate of the income tax from fifty per cent to fifty-five per cent. Following adoption of OPEC Resolution XXI 120 on the 21st meeting of the Conference (December 1970) calling for a settlement and conclusion of all ongoing negotiations on fiscal matters and issues, governments and concessionaires negotiated and concluded on 14 February 1971 in Teheran an agreement on posted price levels and the rate of income tax (the 1971 Teheran Agreement). Under the terms of the Teheran Agreement the income tax rate was raised to fifty-five per cent, and the concessionaires undertook to increase their posted prices in various ways and at various dates. All existing allowances were abolished. The 1971 Teheran Agreement was meant to settle in a binding manner the matter of government take and financial obligations of the concessionaires for a period from 15 February 1971 to 31 December 1975. As part of the agreement the OPEC Gulf States declared they would not take any action in the Gulf to support any OPEC Member State which might seek or demand any increase in government take above the terms then agreed.

Note. With Libya, which was not a party to the 1971 Teheran Agreement, a separate agreement (the Tripoli Agreement, 2 April 1972) along the lines of the former was made. The 1971 Teheran Agreement was immediately adopted by other oil exporting States, among which were the OPEC Member State Nigeria as well as non-OPEC Member States, such as Oman, Malaysia and Brunei. It should be recalled that the licences and concession agreements then current in those oil exporting countries contained a 'most favoured nation' clause.

The next increase (by 8.49 per cent) of the posted prices took place on 20 January 1972. It was not the result of the 1971 Teheran Agreement, but that of the Geneva Agreement signed on that date. The latter was supplemental to the 1971 Teheran Agreement and provided for a further increase of the posted price in compensation for the devaluation of the US dollar that occurred soon after the 1971 Teheran Agreement had become effective. In February 1973 the US dollar devalued by ten per cent. OPEC concluded that the 1972 Geneva Agreement (Geneva I) did not adequately respond to these drastic changes. Negotiations were resumed and led to a revision of Geneva I subsequently called Geneva II. The latter contains a calculation of a new posted price as of 1 June 1973.

On 16 October 1973, after the outbreak of the Yom Kippur war (also referred to as the Ramadan war), the OPEC Gulf States decided unilaterally to increase the posted prices, considering themselves no longer bound by the terms of the 1971 Teheran Agreement and Geneva II. The posted prices were raised by no less than seventy per cent. Soon thereafter, with effect from 1 January 1974, the posted prices were raised again (by the respective governments), on this occasion by almost 130 per cent. The Table 8.2 indicates the movements of the posted price (USD/b) for 34° Arabian Light resulting from the implementation of the 1971 Teheran Agreement and Geneva II and as of 16 October 1973 and 1 January 1974 from the unilateral actions of the governments.

Table 8.2. Posted Prices Arabian Light (USD/b) 1971-1974

15 February 1971 Teheran	*20 January 1972 Geneva I*	*1 January 1973 Teheran*	*1 June 1973 Geneva II*	*16 October 1973*	*1 January 1974*
2.180	2.479	2.591	2.898	5.119	11.651

Until the end of 1974 the posted prices were maintained at their level as fixed on 1 January 1974. Actually, on 1 November 1974 a small downwards (from US SI 1.651 to USD 11,251) adjustment took place, in compensation for the sharply increased royalty and income tax rates. The latter had on 1 October 1974 been increased to 16.67 per cent (as to the royalty) and 65.65 per cent (as to the income tax). On 1 November 1974 the rates were further increased: the royalty rate to twenty per cent and the income tax rate to eighty-five per cent. In respect of the production that could be bought back by the concessionaires under the terms of the state participation agreements a government selling price was introduced, which was set at ninety-three per cent of the posted price, now called the fiscal price. As of 1 January 1975 the official government selling price became the only price that was published, this time by the governments themselves. The fiscal price remained in use for the calculation of royalty and income tax. Application of the new rates on the basis of these fiscal prices resulted in an average government take of ninety-one per cent of the difference between the fiscal price and the cost of production.

At the 42nd meeting of the Conference (Vienna, 12–13 December 1974) it was decided to fix the average government take per barrel calculated with respect to 34° Arabian Light, the marker crude, at USD 10.12, valid for the period 1 January 1975 to 30 September 1975. Taking into account a fiscal price of USD 11.251/b the calculations must have assumed a cost of production (for Arabian Light) of USD 0.13/b. However this may be, after the aforesaid meeting of the Conference the developments of the OPEC fiscal regime lost their meaning since by that time in most Gulf States the concessions had completely been taken over by the State.

Note with reference to Oman. One of the few Gulf States where state participation went no further than sixty per cent was Oman. In that State further developments were restricted to the fiscal system. With effect from 1 May 1977 the financial provisions of the Main Agreement which up to that date were those as had been agreed between the OPEC Member States at subsequent meetings of the Conference beginning with the Teheran Agreement of 1971, were amended by an agreement, referred to as the 'Long Term Financial Agreement' dated 11 April 1977. Pursuant to this financial agreement the Sultan's Share (in this agreement confirmed to be eighty per cent of the Profit) would be increased or decreased so as to allow Petroleum Development (Oman) Ltd (since the effective date of the Transfer Agreement of 15 May 1980 named Private Oil Holdings Oman Ltd [POHOL]) to receive a certain net reward (also referred to as a production fee). As from 1977 and with the introduction of the net reward arrangement the petroleum agreement was further developed in bilateral negotiations between the government and POHOL without reference to events taken place elsewhere in the Middle East Gulf Area.

8.2.2 The United Kingdom

8.2.2.1 Obligatory Payments under the Licence

The licensee is obliged to make the following payments:

(i) an initial payment, which is the payment of a prescribed sum per block or per square km comprised in the licensed area;
(ii) annual payments, which are payments to be made annually of a prescribed sum per square km comprised in the licensed area; and
(iii) royalty payments (abolished with effect from 1 January 2003).

With reference to (i) and (ii). The rates of respectively the initial payment and the annual payments are set out in Schedule 2 to the licence.

With reference to (iii). With effect from 1 January 2003 royalty payment was abolished. Prior to said date the modalities of the payment of a royalty, such as the royalty value and the payment schedule, were described in the model clauses incorporated in the licence. The royalty rate was specified in Schedule 3 to the licence. During the period that royalty was payable, the appropriate rates have always been specified to be 12.5 per cent. The first exemption from paying royalty was granted under the Petroleum Royalties (Relief) Act 1983. This exemption was applicable to petroleum produced from certain 'new fields'. A 'new field' was defined as a field which is situated in the seaward areas north of 55 degrees latitude and the development programme of which received approval after 1 April 1982. Further exemption from royalty was granted with effect from 1 July 1988, under the Petroleum Royalties (Relief) and Continental Shelf Act 1989 with respect to petroleum produced from all 'new fields', including fields in the Southern Basin (the area between latitudes fifty-two and fifty-five) and onshore. The royalty value was defined either as the wellhead value (pre-1976 licences) or the so-called 'landed value', in each case derived from the price realized in sales made at arm's length or, in respect of other sales, on the theoretical market value as defined in the Oil Taxation Act 1975 (i.e., a deemed arm's lengths sales price at a point of delivery in the United Kingdom), and adjusted by the market value of any stock changes over the chargeable period.

8.2.2.2 Payments under Income Tax Laws

Currently, a licensee is liable to pay tax on its income from its operations under the licence under the rules of two different income taxes, on the one hand the corporation tax (CT) and the supplementary corporation tax (SCT) and on the other a special petroleum income tax. CT and SCT are imposed under the rules of the Income and Corporation Taxes Act 1970 as amended. SCT was introduced in the 2002 Finance Act. The special petroleum income tax is imposed under the rules of the Oil Taxation Act 1975, as amended, and is referred to as the Petroleum Revenue Tax (PRT).

Whereas CT and SCT are imposed on the aggregate income realized by the licensee from petroleum produced under all and any of his licences, PRT is imposed on petroleum income realized by each person participating in a separate field ('oil field') except that under the 1993 Finance Act oil fields, in respect of which the development programme has been approved after 16 March 1993, are exempt from paying PRT. Stating it differently, any oil field income is after aggregation with income realized by the same person from other oil fields in which he is participating, whether produced under the same or different licences also taxable under CT and SCT.

For the purposes of CT and SCT the carrying on of the authorized petroleum activities must be treated as a separate trade, distinct from all other activities carried on as part of the trade (Oil Taxation Act 1975). This separation of activities for tax purposes is called the 'ring fencing' of activities. The existence of the ring fence around petroleum activities means that losses, which would qualify for being claimed under the corporation tax, cannot be claimed against income arising from petroleum activities, unless the losses arose from these activities. On the other hand losses incurred in carrying on petroleum activities can be offset against income arising from other activities.

For the purposes of the PRT taxable income (assessable profit) is calculated separately with respect to the income and expenditures attributable to each person participating in any oil field and with respect to such person and such oil field consists in the difference, calculated with respect to a six months' chargeable period, between gross profit plus tariff and disposal receipts on the one hand and the royalty (prior to its abolition), allowable expenditures (i.e., expenditures that can be claimed by the participant as a deduction from gross profit to arrive at the assessable profit), the supplementary amount (i.e., an 'uplift' on expenditures qualifying to be uplifted) and the cash equivalent of the oil allowance (a proportional part of the gross profit) on the other. Gross profit is the value of the petroleum disposed of during a six months' chargeable period calculated on the basis of prices realized in arm's lengths sales or, where other sales are concerned, on the monthly market value (which is the price deemed to be realized in arm's lengths sales at a point of delivery in the United Kingdom) plus fifty per cent of the market value of the change in stocks over the chargeable period. The oil allowance is fixed for each field separately and expressed in long tons per chargeable period with a cumulative limit. The oil allowance was initially fixed for all offshore fields at 250,000 tonnes with a cumulative limit of five million tonnes. For offshore fields situated north of 55 degrees latitude, and approved for development between 31 March 1982 and 16 March 1993 the allowance was doubled to 500,000 tonnes per chargeable period with a cumulative limit of ten million tonnes. With effect from 1 July 1988 the oil allowance for fields in the Southern Basin (i.e., fields situated between 52 and 55 degrees latitude) was halved to 125,000 tonnes with a cumulative limit of 2.5 million tonnes. Allowable expenditure includes all expenditure incurred for exploration and exploitation of the field, abandonment costs, but only insofar as they are related to safety and the prevention of pollution, the cost of transport to the United Kingdom and any unrelievable field losses. The supplementary deductible amount or 'uplift', as it is called, is available in the form of an amount of thirty-five per cent of the expenditure (hereinafter referred to as qualifying capital expenditure) made in respect of field evaluation, bringing the field into production, commencing the transportation of the production from the field to the United Kingdom and providing any installation for the initial treatment and storage of oil (in short the pre-production expenditure) and further some production enhancing expenditure. The uplift may only be applied as a deduction until the payout time of the field has been reached. The total tax payable with respect to a field for a six calendar month chargeable period is limited to eighty per cent of the amount by which the adjusted profit for that period exceeds fifteen per cent of the accumulated qualifying capital expenditure at the end of that period. This so called 'safeguard' is available for a period, which is half as long again as the payback period. The assessable profit may be further reduced by the allowable losses carried forward from the previous chargeable period. The adjusted profit for a chargeable period is a greater amount than the assessable profit for that period and amounts to the difference between gross profit on the one hand and royalty (prior to its abolition) and non-qualifying expenditure on the other.

The rate at which the PRT is levied is determined in the annual Finance Acts on the basis of the estimated profitability of the petroleum production operations. At the introduction of the tax under the terms of the 1975 Oil Taxation Act the rate was set at forty-five per cent. With effect from 1 January 1979 the rate was increased to sixty per cent and with effect from 1 January further increased to seventy per cent. On 1 January 1983 the rate reached its highest level: seventy-five per cent. With effect from 30 June 1993 the rate was lowered to fifty per cent. The frequent changes were all due to and dictated by the fluctuations of the price of oil on the world market.

For the purposes of CT and SCT gross profit is assessed in the same manner as for the PRT, except that one hundred per cent, instead of fifty per cent, of the market value of stock changes is included (similar as is the case when the royalty value is determined). All amounts of PRT paid by a participant in a field may be deducted by him from his taxable petroleum income. Other allowable expenditures include exploration and appraisal costs, intangible development drilling costs, operating expenditure and the cost of closing down a field, including the cost of demolishing an offshore installation. In the April 2002 Budget expenditure qualifying as 'plant and machinery' which since November 1996 comprise all development costs received a one hundred per cent first year capital allowance and could therefore be expensed and did longer have to be depreciated at twenty-five per cent per year on a declining balance basis as till then had been the rule. In the same Budget assets with expected lives of twenty-five years or more and which have to be depreciated at a rate of six per cent per year were given a first year capital allowance of twenty-four per cent. Abandonment costs may be carried back for three years. The applicable rates of CT and SCT (in a similar manner as the rate of the PRT) are established in the annual Finance Acts and may and will change from year to year.

Note. Currently, the applicable tax rates are: CT: thirty per cent; SCT: twenty per cent. Reference: *PEPS*, IHS Inc.; Energy, Country: United Kingdom, Fiscal Terms, Date: 28 March 2006.

8.2.3 NORWAY

8.2.3.1 Obligatory Payments under the Licence

The licensee is required to make the following payments:

(i) cash bonuses;
(ii) annual surface rentals, and
(iii) a production fee (royalty).

With reference to (i). When granting a production licence a non-recurring fee (cash bonus) may be levied and a fee may be stipulated which is calculated on the basis of production volume (production bonus).

With reference to (ii). The licensee shall pay a fee (area fee) for a production licence, after expiry of the initial period, calculated per square km.

With reference to (iii). The licensee shall pay a fee calculated on the basis of the quantity and value of petroleum produced at the shipment point of the production area (production fee). With regard to petroleum which is injected, exchanged or stored prior to being delivered to be taken ashore or used for consumption, the production fee shall be calculated on the basis of the quantity and value of the petroleum at the shipment point for the original production area at the time when the petroleum according to contract is delivered to be taken ashore or used for consumption. Nevertheless, a production fee shall not be paid for petroleum produced from deposits where the development plan is approved or where the requirement to submit a plan for development and operation is waived after 1 January 1986.

The Ministry may with six months' notice decide that the production fee shall be paid wholly or partly in the form of produced petroleum. The Ministry may in such cases require the licensee to make sure that such petroleum is transported, processed, stored and made available at prices, priority and other conditions which are no less favourable than the terms applicable to the licensee's own petroleum from the relevant production area. The Ministry may, with six months' notice, decide that the fee shall again be paid in cash.

The King may issue regulations relating to the size of the fees and bonuses and on the method of calculation, including provisions on stipulation of the value which shall form the basis for the calculation, on the metering of the petroleum, e.g., for fiscal purposes, and on information which the licensee shall provide about the production. The King may determine that the fees mentioned shall not be paid wholly or partly, or that the duty to pay such fees shall be postponed.

Until the issue of a new regulation pursuant to the Act of 29 November 1996 No. 72, the modalities of the payment of royalty are as stipulated in the Royal Decree of 14 June 1985. The royalty rate for 1965 licences was maintained at ten per cent. For all other licences, in respect of which a royalty is payable, (i.e., production licences under which petroleum is produced on the basis of a development plan approved or where the requirement to submit a development plan was waived before 1 January 1986), the royalty rate for oil is derived with respect to each production area from a sliding scale setting royalty rate against daily oil production. On this scale royalty rates increase in steps of two per cent from a minimum eight per cent to a maximum sixteen per cent, corresponding to daily average oil production likewise increasing in steps from up to 6,500 to 55,000 cubic metres. If a royalty rate has been higher than twelve per cent, it will not be reduced below this number when production declines. The rate for natural gas was 12.5 per cent (till its abolition in 1992). However, the Ministry may stipulate lower royalty rates upon the granting of a licence or thereafter, if the depth of the sea, the capacity of the wells or other circumstances so dictate. The value of petroleum for the purposes of the royalty is stipulated by the Ministry, on the basis of the norm price, which is the price established in accordance with the Norm Price Decree (Royal Decree of 25 June 1976, as amended), with any reasonable adjustment to account for any difference between the price reference point and the shipment point.

Note. Royalty on natural gas for all fields was abolished in 1992. With effect from 1 January 2000 royalty on crude oil from a series of important royalty paying oil fields were either abolished or phased out over a three- or six-year period. Reference: *PEPS*, IHS Inc.; Energy, Country: Norway, Fiscal Terms, Date: 9 January 2006.

8.2.3.2 Payments under Income Tax Laws

The licensee deriving income from the exploitation of sub-sea petroleum deposits and from pipeline transportation is liable to pay tax on such income in accordance with the Act of 13 June 1975 No. 35 (the Petroleum Tax Act). The latter Act provides for the payment of the corporation taxes, including municipal and state tax, and for the payment of a special (petroleum income) tax.

Taxable income for the purpose of paying the corporation taxes is determined in the same manner as provided under the Income Tax Act, with such adjustments as are stipulated in the Petroleum Tax Act and consists of the revenues from the aforesaid exploitation reduced by the allowable deductions. The main deductions are the royalty, if any, the deductions allowed on account of capital expenditure, expensed costs and losses carried forward. Revenues are evaluated on the basis of norm prices, which are prices determined in accordance with the Norm Price Decree (Royal Decree of 25 June 1976, as amended). Expenditure incurred on account of the purchase of production assets may be written off at a rate of 16.67 per cent per annum. Such capital expenditure incurred prior to 1 January 1987, may be written off as from the year in which the production asset in question is brought into regular use; for capital expenditure incurred after 1 January 1987, the writing off starts in the year in which the expenditure is incurred. Losses may be carried forwards for fifteen years. One-half of the losses sustained in non-petroleum activities inside Norway may also be deducted.

Taxable income for the purpose of paying the special petroleum tax is determined in the same manner as for the purposes of the corporation taxes, except that losses incurred in non-petroleum activities are not deductible and except that special deductions are available. With effect from 1 January 1992 two special deductions were introduced in replacement of a special deduction and a production allowance existing before that date. The new deductions consist firstly in an 'uplift' of 1.75 per cent per annum of the purchase price of operating assets, to be applied during fifteen successive years as from the year the asset concerned is brought into regular use, or from the year in which the expenditure was made depending on whether the purchase price was incurred before or after 1 January 1987 (the same uplift existed before 1 January 1992 but was applied with a rate of 6.67 per cent), and secondly in an uplift of five per cent of such purchase prices incurred after 1 January 1992 for a period of six years. The second uplift came in replacement of the hitherto existing production allowance. This now abolished production allowance amounted to fifteen per cent of the revenues generated by fields included in development plans which were submitted to the Ministry for approval after 1 January 1986 (and which concern the same fields in respect of which no royalty has to be paid).

Note. Beginning 1 January 2005 under the Act of 10 December 2004 No. 78 the uplift was increased to 7.5 per cent per annum for the first four years.

Taxable income for the purpose of the special tax cannot be reduced by costs made by the licensee in non-petroleum activities.

On 1 January 1992 and with effect from that date the rate of the corporation taxes was fixed in the aggregate at twenty-eight per cent (reduced from previous 50.8 per cent) and the rate of the special petroleum tax was fixed at fifty per cent (increased from previous thirty per cent). The rate of the special petroleum tax is fixed annually by the *Storting*. (*Note*. Reference: *PEPS*, IHS Inc.; Energy, Country: Norway, Fiscal Terms, Date: 9 January 2006.)

When installations utilized in exploitation or pipeline transportation are required to be removed, the expenses are not deductible for income tax purposes. On the other hand, government contributions for removal are not treated as taxable income. No deductions shall be allowed to cover future expenses for removal of installations used in exploitation or pipeline transport.

8.3 PAST AND CURRENT PETROLEUM TAXATION (PS CONTRACTS)

8.3.1 INDONESIA

In the 1980s the Indonesian government was persuaded to change the income tax paying procedure and to accept direct payment of income taxes by the contractor, due to the difficulties which US oil companies experienced (in their dealings with the US Internal Revenue Service [the IRS]) in getting the tax payments made by Pertamina on behalf of contractor accepted and recognized for the purpose of the US foreign tax credit. The IRS considered and treated the Indonesian production sharing agreements as royalty paying agreements in respect of which no income tax in the sense as understood under US fiscal rules was paid. Under the terms of the revised or new contracts the contractor paid itself the income taxes (at a rate of forty-five per cent) for which it was liable. The contractor had also to pay a dividend withholding tax at a rate of twenty per cent, resulting in a combined rate of fifty-six per cent in respect of its profits (i.e., the value of the contractor's share of profit oil). Furthermore the amended contracts provided for a revised share of profit oil in favour of the contractor and brought cost recovery rules in line with cost deduction schedules as applicable under customary western style company income tax legislation.

8.3.2 EGYPT

In addition to and separate from the share of profit production assigned to the state party, the contractor is subject to all Egyptian income tax laws.

Note. As the licensee, EGPC is obliged to pay a royalty, in cash or kind, of ten per cent of the total quantity of petroleum produced. The royalty must be paid out of EGPC's share of profit production.

Contractor's annual taxable income is in the contract described as consisting of:

(i) the proceeds of the sale of contractor's contractual entitlement to petroleum; less

(ii) the costs incurred by contractor, as allowed and qualifying to be recovered from cost recovery petroleum under the terms of the contract but without regard to the volume percentage limitation applicable to such cost recovery petroleum; and less

(iii) the value of EGPC's share of excess cost recovery petroleum as paid to EGPC; but plus
(iv) an amount equal to contractor's Egyptian income taxes.

Consequently, income realized under the contract is kept separate from any other income (and losses) realized by contractor in other business. The contractor's Egyptian income taxes are paid on its behalf by EGPC out of the value of the latter's share of oil or gas production and the tax so paid by EGPC is considered to be income to the contractor (and consequently is added to contractor's sales proceeds in calculating contractor's taxable income). In calculating its Egyptian income taxes, EGPC may deduct all royalties paid by EGPC to the State as well as the contractor's Egyptian income taxes paid by EGPC to the State on behalf of the contractor. EGPC shall furnish to the contractor the proper official receipts evidencing the payment of the contractor's Egyptian income tax within ninety days following the receipt by EGPC of the contractor's tax declaration for the preceding tax year.

Note. The accounting procedure of the contract contains the following clarification: The contractor's taxable income equals the contractor's provisional income (i.e., sales proceeds less recoverable costs and less payment for EGPC's share of excess cost recovery petroleum) plus the contractor's Egyptian income tax; hence: the contractor's tax equals tax rate/(1 - tax rate) multiplied by the contractor's provisional income. The value of the fraction: tax rate/(1-tax rate) depends on the value of the applicable tax rate and determines whether the contractor's Egyptian income tax can realistically be paid out of the value of EGPC's share of the oil or gas production. The sample calculation included in the said accounting procedure assumes an income tax rate of forty per cent, which results in a value of the said fraction of 0.667 and therefore in the contractor's income tax becoming equal to 66.7 per cent of the contractor's provisional income.

8.3.3 Malaysia

In addition to and separate from the share of profit production assigned to the state party (Petronas), the Contractor is obliged to pay:

(i) income tax under the rules of the Petroleum (Income Tax) Act 1967, as amended with effect from 1 April 1975 by the Petroleum (Income Tax) (Amendment) Act 1976 (No. 1) and (No. 2);
(ii) cash payment to Petronas per kilolitre of contractors' share of profit oil;
(iii) export duty at a rate of twenty-five per cent of the value of his profit oil barrels in so far as these are exported; and
(iv) so-named 'research cess' at a rate of 0.5 per cent of the cost oil value of the contractor's share of production.

Note 1. In contrast with the Indonesian example, the production sharing mechanism that was introduced by the new contracts was not seen as an income tax replacing or absorbing arrangement and the contracts were not meant to exhaust the methods to determine government take. The aforesaid Petroleum (Income Tax) Act 1967, designed and meant to impose a tax on income from petroleum operations, was maintained, albeit the Act had to be amended (with effect from 1 April 1975) in some respects in order to cater for the peculiarities of the production sharing contracts.

Note 2. A maximum of ten per cent of the crude oil production and a maximum of ten per cent of the total proceeds from the joint sale of natural gas is taken by Petronas in order to enable the latter to make to the Federal and State governments the cash payments (royalty) required under the Petroleum Development Act 1974.

Note 3. The cash payment amounts to seventy per cent of the difference between the cost oil value and a certain base price. The latter is annually increased by five per cent, calculated with respect to the last base price. It depends on the developments on the oil market whether or not any cash payment is required.

Note 4. In April 1980 an export duty was introduced at a rate of twenty-five per cent of the value of all crude oil which is exported. As far as the contractor is concerned the duty is only payable with respect to its profit oil barrels in so far these are exported.

With reference to (i). Gross income realized by the same chargeable person (defined as 'Petronas or any other person carrying on petroleum operations') is calculated on realized price or on market value, whichever is the higher. Certain expenditures, such as intangible drilling expenditures incurred with respect to contract areas that are not contiguous, are 'ring fenced' and can only be deducted from the gross income derived under the same production sharing contract, in respect of which the expenditures were made. Import and export duties were no longer allowed as tax credits. These duties are allowed as deductions. The cash payment due on profit oil, if any, as well as the research cess became recognized as a deduction from gross income. By Amendment No. 2 the tax rate was reduced from fifty per cent to forty-five per cent.

The model contract contains a mild form of a government take stability guarantee. It is declared that the terms of the contract have been negotiated against the background of the provisions of the 1967 Act as amended with effect from 1 April 1975 and the export duty as applicable on the effective date of the contract. It is then stated that, in case the income tax would be changed or new taxes specially meant for the petroleum industry would be introduced in such a manner that the burden of taxation would significantly be increased or decreased, Petronas and its contractors shall mutually agree on an arrangement which should restore the original economic position of the parties. Apparently, the stability provision is meant to work both ways: if the burden of taxation is decreased, e.g., when a lower rate for the income tax is introduced, contractors are not be allowed to benefit from such relief.

8.3.4 Angola

Contractors are liable to pay a special petroleum income tax under the terms of the Law No. 13 on Taxation of Petroleum Activities of 24 December 2004. This Law applies to all entities, whether Angolan or foreign, performing petroleum operations in Angolan territory, as well as in other territorial or international areas within the tax levying jurisdiction of the Republic of Angola, as recognized by international law or agreements. The main tax charges introduced by the Petroleum Tax Law (Article 4) are a petroleum production tax and a petroleum income tax. The petroleum production tax (the equivalent of a royalty), however, is not applicable to production sharing contracts. The petroleum income tax is levied on the taxable petroleum income of contractor which consists of revenues generated by the petroleum operations less operating costs and less depreciation charges at the rate of 16.666 per cent in respect of costs incurred for exploration operations, the drilling of development wells and for production, transportation and storage facilities. For production sharing agreements the tax rate is fifty per cent (Article 41) (see Chapter 7).

8.3.5 The Russian Federation

Under Chapter 26.4 to the Tax Code containing a special tax regime applicable to production sharing agreements contractors shall pay the following taxes and levies: mineral production tax (MPT); profits tax (PT); bonuses; VAT, unified social tax, environmental and water charges, state duties, custom levies, land tax; excise duty. The amounts paid by the contractor in respect of all of the above listed taxes and levies with exception of MPT and PT are recoverable by the contractor under cost recovery provisions except for bonus payments which are tax deductible. Contractors are exempt from the payment of regional and local taxes and levies and also from transport tax. The contractor is exempt from duty on exports from operations under his contract.

Note. Aforesaid Chapter 26.4 of the Tax Code had been introduced into said Code by the Federal Law No. 65-FZ of 6 June 2003 (usually referred to as the PSA Tax Law). Further details can be found in Chapter 7.

Taxable income for PT is the value of contractor's share of profit production less the following: financing costs, bonuses and other costs that are not recoverable for production sharing purposes as defined by each contract subject to regulations concerning cost recovery. If in any tax period the amount of allowable deductions exceeds the value of the contractor's share of profit production, then the amount of the difference may be carried forward to the next tax period until fully deducted. The rate of profits tax current at the time of signing the contract is the rate that will apply for the duration of the contract. The current rate of the profits tax is twenty-four per cent.

Chapter 9

International Regulation of Offshore Petroleum Operations

9.1 THE LAW OF THE SEA

9.1.1 ORIGIN AND INITIAL DEVELOPMENT

After the end of the Second World War drilling technology and techniques had progressed to the point where drilling offshore beyond the shallow coastal waters became feasible. This meant that in a technical sense offshore petroleum operations were ready to move beyond the limits of the territorial sea and coastal State jurisdiction. At that moment the breadth of the territorial sea and hence the seaward extent of the territorial sovereignty of the coastal State was not internationally agreed. Maritime nations, among them the US, adhered to the three-mile limit which was historically derived from the range of shore-based cannon. The importance and significance of this range followed from the doctrine brought forward by Hugo Grotius in the seventeenth century, stating that claims to sovereignty over the contiguous sea was justified on the basis of defending the coastal zone against aggressors but that the coastal State should be able, if challenged, to support the claimed sovereignty by military force. In practical terms Grotius' doctrine led to the theory that the seaward limits of territorial sovereignty should coincide with and could not exceed the range of shore-based cannon. In 1728 the Italian jurist Galiani proposed to fix this range at three miles, but this limit was only supported and put in practice by the seafaring nations such as the US. Under the auspices of the League of Nations an attempt to achieve uniformity in this matter was made at the 1930 Hague Codification Conference, but no agreement could be reached. Beyond the seaward limits of the territorial sea the domain of the high seas begins. In respect of the high seas four freedoms were internationally recognized: the freedom to navigate; the freedom to fish; the freedom to lay cables and pipelines;

and the freedom of overflight. The possibility of exploring for and of exploitation of mineral resources of the seabed and the subsoil of the area of the high seas had never been considered. Attention had always been focused on the superjacent waters and on the commercial use that could be made of these waters, in practice amounting only to navigation and fishing. The question of where the jurisdiction lay over the natural resources underneath the high seas had to be solved before the industry could move with confidence into the domain of the high seas.

9.1.2 The 1945 Truman Proclamation

In 1945 the US tried to solve the problem by simply unilaterally claiming the sovereignty over the resources of the subsoil and the seabed of the continental shelf beneath the high seas but contiguous to the coasts of the US. The claim is stated in the Proclamation issued on 28 September 1945 by the US President Truman as follows:

> Now, therefore, I, Harry S. Truman, President of the USA do hereby proclaim the following policy of the USA with respect to the natural resources of the subsoil and seabed of the Continental Shelf. Having concern for the urgency of conserving and prudently utilizing its natural resources, the Government of the USA regards the natural resources of the subsoil and seabed of the Continental Shelf beneath the high seas, but contiguous to the coasts of the US as appertaining to the US, subject to its jurisdiction and control. In cases where the Continental Shelf extends to the shores of another State, or is shared with an adjacent State, the boundary shall be determined by the US and the State concerned in accordance with equitable principles. The character as high seas of the waters above the Continental shelf and the right to their free and unimpeded navigation are in no way thus affected.

The Proclamation restricts the natural resources jurisdiction of the US to its continental shelf. Remarkably, the Proclamation does not contain a definition of continental shelf. However, in the accompanying press release the continental shelf is described as submerged land which is contiguous to the continent and which is covered by no more than one hundred fathoms (six hundred feet) of water. The Proclamation was a declaratory statement, on the basis of which natural resources in unknown quantities were acquired at no costs at all. In the preamble of the Proclamation practical and economic arguments and reasons are listed meant to justify this acquisition. In this connection reference is made to the long-range worldwide need for new sources of petroleum and other minerals; the expectation that such resources underlie many parts of the Continental Shelf off the coasts of the USA and that with modern technological progress their utilization is already practicable, or will become so at an early date; the requirement of recognized jurisdiction over these resources in the interest of their conservation and prudent utilization, when and as development is undertaken; and the view of the US government that the exercise of jurisdiction over the natural jurisdiction over said natural resources by the contiguous nation is reasonable and just. In support of this view the following arguments were brought forward:

(i) the effectiveness of measures to utilize or conserve these resources is contingent upon cooperation and protection from the shore;
(ii) the continental shelf may be regarded as an extension of the land mass of the coastal nation and thus is naturally appurtenant to it;
(iii) the said resources frequently form a seaward extension of a pool or deposit lying within the territory; and
(iv) self-protection compels the coastal nation to keep close watch over activities off its shores, which are of the nature necessary for the utilization of the said resources.

Note. The prolongation argument, i.e., the argument that the continental shelf should be regarded as an extension of the land mass of the coastal State and thus naturally appurtenant to it, carries enough weight to support and justify the proclaimed jurisdiction. All other arguments that are mentioned in the preamble are in fact superfluous. They only serve to make clear why in the first place the US was keen and found it necessary to proclaim such jurisdiction. The prolongation-argument has more than twenty years later been confirmed by the International Court of Justice at The Hague in its judgment in the *North Sea Continental Shelf Cases* (20 February 1969). In this judgment, the Court held that the rights of the coastal State in respect of the area of continental shelf constituting a natural prolongation of its land territory under the sea existed *ipso facto* and *ab initio*, by virtue of its sovereignty over the land.

The jurisdiction claimed by the Truman Proclamation became federal law on the occasion of the enactment of the Act of 7 August 1953, known as the Outer Continental Shelf Lands Act. In the accompanying Congressional declaration of policy; jurisdiction; construction it is stated (after partial amendment in 1978) that:

(i) it is the policy of the US that the subsoil and seabed of the outer continental shelf appertain to the US and are subject to its jurisdiction, control, and power of disposition as provided in the Act;
(ii) the Act shall be construed in such manner that the character as high seas of the waters above the outer continental shelf and the right to navigation and fishing therein shall not be affected;
(iii) the outer continental shelf is a vital national resource held by the federal government for the public, which should be made available for expeditious and orderly development, subject to environmental safeguards, in a manner which is consistent with the maintenance of competition and other national needs . . .
(iv) operations in the outer continental shelf should be conducted in a safe manner by well-trained personnel using technology, precautions, and techniques sufficient to prevent or minimize the likelihood of blow-outs, loss of well control, fires, spillages, physical obstruction to other users of the waters or subsoil and seabed, or other occurrences which may cause damage to the environment or to property, or endanger life or health.

Note. The 1958 Convention on the Continental Shelf became effective as law in the United States in 1964.

9.1.3 Preparatory Work on Conventions

The Truman Proclamation triggered a series of similar unilateral statements, proclamations and legislative actions of other coastal States, in particular in Latin America and the Middle East Gulf area and the uncertainties about the scope and geographical extent of the jurisdictions claimed indicated that the matter of sovereignty over the natural resources of the continental shelf urgently needed regulation by an international convention.

Note. Mexico (29 October 1945), Argentina (11 October 1946), Chile and Peru in 1947, did not hesitate to issue similar proclamations. In these Latin-American proclamations jurisdiction was claimed over both epicontinental sea and continental shelf albeit the freedom of navigation was explicitly declared to remain unaffected. Whereas Mexico and Argentina, the latter in particular, were in possession of an extensive continental shelf, Peru and Chile were both not so fortunate. In fact the continental shelf adjacent to the coast of the two countries (and the shelf adjacent to the coast of Ecuador for that matter) is quite small and narrow. Since the coastal States were primarily interested in protecting their fisheries, i.e., the living natural resources of the superjacent waters as opposed to the non-living natural resources of the seabed and subsoil, they needed a larger area of jurisdiction, in which to create a protected fishing zone. This persuaded those countries, joined for the occasion by Ecuador, to claim in 1952 sovereignty and jurisdiction over the contiguous sea up to a distance of two hundred nautical miles as measured from their respective coasts. Incidentally, thirty years later, this distance of two hundred nautical miles would be adopted as the outer limit of the Exclusive Economic Zone, one of the main features of the 1982 Convention on the Law of the Sea. The example given by the Truman Proclamation was also followed in the Middle East Gulf area. On 28 May 1949 Saudi Arabia issued a similar proclamation and soon thereafter, all in June of that year, Bahrain, Qatar, Kuwait and the Trucial sheikdoms, among them Abu Dhabi and Dubai, did the same. The Saudi proclamation claimed jurisdiction over the subsoil and seabed of those areas of the Gulf seaward from the coastal sea of Saudi Arabia, but contiguous to its coasts. A Saudi royal decree of the same date fixed the breadth of the coastal sea at six miles, three miles more than the international standard supported and recognized by the seafaring nations at that time. Interestingly, the stated jurisdiction was not restricted to the natural resources of the seabed and subsoil. This left open the possibility that the Saudi government could claim jurisdiction over cables and pipelines crossing its seabed. No. statement was made with respect to the living resources of the superjacent waters as had been the case in Latin America. Overlapping claims would be settled in agreement with the State concerned in accordance with equitable principles. Iran enacted legislation concerning its continental shelf on 19 June 1955. The areas as well as the natural riches of the seabed and subsoil up to the limits of the continental shelf, which extends from the coasts of Iran and the Iranian islands in the Persian Gulf and the sea of Oman, were declared to belong to the Iranian government and to fall under its sovereignty. On 27 November 1957 the government of Iraq issued a statement claiming seabed and subsoil jurisdiction in respect of waters contiguous to Iraqi territorial waters along the pattern established by the earlier declarations of Saudi Arabia and the other Gulf countries.

By around 1948 the UN International Law Commission had already started work on a codification of the law of the sea. By 1957 the Commission had made such progress with its work that the UN General Assembly could decide (21 February 1957) to convene an international conference of plenipotentiaries in order to prepare and establish the text of one or more conventions on matters regarding the law of the sea. The intended conference convened in Geneva between 24 February and 27 April 1958. The Conference succeeded in adopting (in the Final Act) the text of four separate conventions, namely the Convention on the Continental Shelf; the Convention on the Territorial Sea and the Contiguous Zone; the Convention on the High Seas; and the Convention on Fishing and Conservation of the Living Resources. As happened before in 1930, the Conference failed to reach agreement on the breadth of the territorial sea. On the other hand, agreement was reached on a seaward limit of twelve nautical miles in respect of the contiguous zone (i.e., a zone in the high seas contiguous to the territorial waters, in which zone the coastal State may exercise certain rights for protection of its interests). By limiting the contiguous zone to twelve miles it is implied that no coastal State should claim a seaward limit of its territorial waters exceeding this distance. In order to bring the matter to a conclusion the UN General Assembly called for a second conference on the law of the sea. The Second UN Conference on the Law of the Sea took place again in Geneva from 17 March 1960 to 27 April 1960. At the Conference it proved impossible to get the required majority of votes for any proposed limit. The matter had to await the Third UN Conference on the Law of the Sea, which took place between December 1973 and April 1982. At this Conference the seaward limit of the territorial sea was finally settled at twelve nautical miles, the same distance earlier agreed as the seaward limit of the contiguous zone, now no longer existing as a separate zone.

9.1.4 The 1958 Convention on the Continental Shelf

The 1958 Convention on the Continental Shelf provided the legal basis for undertaking offshore petroleum operations beyond the territorial waters of a coastal State, but within the limits of its continental shelf.

In Article 1 of the Convention the term 'continental shelf' is defined for the purpose of the Convention. It has turned out to be a legal and not a geomorphological or geological definition. The expression is used as referring to:

(i) the seabed and the subsoil of the submarine areas adjacent to the coast but outside the area of the territorial sea, to a depth of 200 metres or, beyond that limit, to where the depth of the superjacent waters admits of the exploitation of the natural resources of the said areas; and
(ii) the seabed and subsoil of similar submarine areas adjacent to the coasts of islands.

Although the Convention assigns any island a continental shelf, it assigns sovereign rights over the continental shelf to coastal States and not to islands. It is clear that the Convention intended to say that sovereign rights over the continental shelf adjacent to an island may be exercised by the coastal State which exercises sovereign rights over that island. The limit of the seaward boundary of the continental shelf is described in the Convention as a water depth of two hundred metres or, beyond that limit, to where the depth of the superjacent waters admits of the exploitation of the natural resources of the said area. From the given description of the limit it appears that, although a depth of two hundred metres forms in many places the geomorphological limit of the continental shelf, it was not intended that the coastal State should be restricted to this water depth if it would prove technically possible to develop and produce natural resources beyond that depth. On the other hand, it was equally not intended that under the Convention coastal States could ever claim jurisdiction over the natural resources of the deep seabed area. To whom such sovereignty should be accorded was regarded, in absence of any technical possibility to make use of such rights, as a problem that did not have to be addressed by the Conference and could be left for regulation in the future.

Note. From the description of the continental shelf as an area adjacent to the coast (but outside the territorial waters) it follows that strictly speaking between the coast and the seaward boundary of the shelf there may not be areas in which the water depth exceeds the water depth designated as the seaward boundary (i.e., two hundred metres or more, as the technical case may be). However, it was generally considered, as appears from state practice, e.g., the UK/Norway North Sea Boundary Agreement of 10 March 1965 that such areas of greater water depth should not be allowed to limit the continental shelf but should be treated as local features of the continental shelf. Nonetheless, the status of local feature was not considered to mean that such local areas of greater water depths would fall under the jurisdiction of the coastal State concerned. The latter could not exercise any sovereign rights over deeper water area until it would have become technically feasible to undertake exploitation work in such areas. The whole problem of the presence on the continental shelf of local areas of greater water depth has now disappeared as a result of the different approach followed by the 1982 Convention on the Law of the Sea in accordance to which the continental shelf extends at least to two hundred nautical miles as measured from the shoreline, regardless of the geomorphological features of the sea bottom.

The Convention assigns to the coastal State sovereign rights over the continental shelf for the purpose of exploring it and exploiting its natural resources. These rights are exclusive in the sense that if the coastal State does not explore the continental shelf or exploit its natural resources, no one may undertake these activities or make a claim to the continental shelf without the express consent of the coastal State. Said natural resources consist of the mineral and other non-living resources of the seabed and subsoil together with living organisms belonging to sedentary species, that is to say, organisms which at the harvestable stage, either are immobile on or under the seabed or are unable to move except in constant physical contact with the seabed or the subsoil (Article 2).

In conformity with the Truman Proclamation the sovereign rights of the coastal State over the continental shelf do not affect the legal status of the superjacent waters as high seas, or that of the airspace above those waters (Article 3).

Subject to its right to take reasonable measures for the exploration of the continental shelf and the exploitation of its natural resources the coastal State may not impede the laying or maintenance of submarine cables or pipelines on the continental shelf (Article 4).

Note. It is not immediately clear what these 'reasonable measures' might involve and whether for instance the coastal State could have asked for guarantees with regard to matters of safety and the protection of the marine environment. It must be assumed that by 'reasonable measures', to which the Convention referred, such measures were meant as needed to be taken by the coastal State to protect its own exploration and exploitation operations from being damaged or hindered by the submarine pipeline or cable laying operations of another State. As discussed hereinafter this matter has been clarified under the 1982 Convention on the Law of the Sea.

The exploration of the continental shelf and the exploitation of its natural resources must not result in any unjustifiable interference with navigation, fishing or the conservation of the living resources of the sea, nor result in any interference with fundamental oceanographic or other scientific research carried out with the intention of open publication (Article 5.1). Subject to the provisions of Articles 5.1 and 5.6 the coastal State is entitled to construct and maintain or operate on the continental shelf installations and other devices necessary for its exploration and the exploitation of its natural resources, and to establish safety zones around such installations and devices, and to take in those zones measures necessary for their protection (Article 5.2). The safety zones referred to in Article 5.2 may extend to a distance of five hundred metres around the installations and other devices which have been erected, measured from each point of their outer edge. Ships of all nationalities must respect these safety zones (Article 5.3). Such installations and devices, though under the jurisdiction of the coastal State, do not possess the status of islands. They have no territorial sea of their own, and their presence does not affect the delimitation of the territorial sea of the coastal State (Article 5.4). Due notice must be given of the construction of any such installations, and permanent means for giving warning of their presence must be maintained. Any installations, which are abandoned or disused, must be entirely removed (Article 5.5). Neither the installations or devices nor the safety zones around them may be established where interference may be caused to the use of recognized sea lanes essential to international navigation (Article 5.6). The coastal State is obliged to undertake, in the safety zones, all appropriate measures for the protection of the living resources of the sea from harmful agents (Article 5.7).

Note. From its wording it can be concluded that Article 5 is only applicable to installations and devices, which have been erected and extend above the surface of the sea. Consequently this article does not apply to or concern submarine pipelines. This in turn meant that no safety zones could be established around such pipelines and that abandoned pipelines did not have to be entirely removed. As a matter of fact submarine pipelines are only referred to in this Convention in connection with the right of other States to lay submarine cables or pipelines on the continental shelf. Article 5 does also not apply to submarine constructions other than submarine pipelines such as submarine production installations which are fixed to the seabed. With respect to this Convention no conclusions may be drawn from this because these submarine devices were not in use at the time the Convention was made. The matter of establishing safety zones around installations is further discussed in the context of Articles 60 (EEZ) and Article 80 (continental shelf) of the 1982 Convention on the Law of the Sea.

Where the same continental shelf is adjacent to the territories of two or more States whose coasts are opposite each other, the boundary of the continental shelf appertaining to such States shall be determined by agreement between them. In the absence of agreement and, unless another boundary line is justified by special circumstances, the boundary is the median line, every point of which is equidistant from the nearest points of the baselines from which the breadth of the territorial sea of each State is measured (Article 6.1). Where the same continental shelf is adjacent to the territories of two adjacent States the boundary of the continental shelf shall be determined by agreement between them. In the absence of agreement and, unless another boundary line is justified by special circumstances, the boundary shall be determined by application of the principle of equidistance from the nearest points of the baselines from which the breadth of the territorial sea of each State is measured (Article 6.2).

Note. The median line or another line drawn on the basis of the principle of equidistance is not always the boundary line of the last resort. Apart from the possibility that special circumstances exist which make the application of the principle of equidistance difficult or impossible, the Convention (Article 12) allows any State at the time of signature, ratification or accession to make reservations with respect to this article (as well as for that matter to any other article of the Convention, except Articles 1 to 3 inclusive). At the time, a number of States, realizing that due to the configuration of their coast the application of the equidistance principle would leave them with only a small portion of the adjacent continental shelf, made use of the possibility offered by the aforesaid Article 12 and submitted a reservation with respect to Article 6. The demarcation of the continental shelf between opposite or adjacent coastal States has frequently given rise to disputes between the States concerned. In most cases coastal States involved in such a demarcation dispute agreed to submit their dispute to international arbitration or to the International Court of Justice at The Hague. Well-known examples of such a settlement are the respective boundary agreements reached between the Netherlands, Denmark and the German Federal Republic on the division of their part of the North Sea continental shelf. Said agreements were based on a judgment of the International Court of Justice at The Hague to which the States involved had agreed to submit their dispute (the *North Sea Continental Shelf Cases*, 20 February 1969).

9.1.5 The 1970 UN General Assembly Resolution 2749 (XXV)

The 1958 Convention was meant to regulate natural resources operations carried out on the continental shelf. The subject of deep sea mining fell outside its scope. On the other hand deep sea mining was not listed in the Convention on the High Seas under the rights and freedoms of the high seas (viz. the freedom of navigation, fisheries, the laying of submarine cables and pipelines, the freedom of scientific research and the use of the airspace above the waters). In fact deep sea mining was completely ignored by the 1958 Conventions. In the 1960s further developments in drilling technology and bottom sampling made clear that in the near future deep sea mining could become technically feasible. To make this a reality it was felt that a mining regime had to be developed that was suitable to regulate the exploration for and exploitation of deep sea mineral resources. The absence of such specific regime, so it was rightly feared, would have an inhibiting effect on the preparedness of the mining industry to embark on deep sea mining.

For creating the desired mining regime consideration was given to:

(i) an appropriate adaptation of the 1958 Convention ensuring that this Convention would also become applicable to natural resources beyond the continental shelf;
(ii) treating deep sea mining as one of the freedoms of the high seas; and
(iii) using the flag-state approach.

With reference to (i). Under such an approach the two hundred metres water depth criterion would have to be left out, retaining the technological water depth (i.e., the water depth where exploitation of minerals is still possible), as the only yardstick for determining the limits of the jurisdiction of the coastal State. In other words the seaward boundary of the continental shelf would move gradually to greater water depth as more sophisticated technology would be developed and become available. However, application of this system yields different results for different coastal States depending on their technological capabilities and local circumstances. This was perceived as a practical obstacle for applying this solution because it was not clear if one coastal State would be able to exploit minerals at say a depth of 2,500 metres and was entitled in this way to extend its continental shelf to that depth whether all other coastal States would automatically be similarly entitled to extend their continental shelf to that depth or only if they would have actually proven the exploitability at that depth on their coast. Apart from that a more serious objection was that the presence of mid-oceanic ridges and plateaus, separated from the continental shelf by greater water depths, where no exploitation is possible, could not be accommodated within this system. The conclusion that was reached at the time was that the rules of the Convention could only be used if water depth as limiting factor for the jurisdiction of the coastal State would be completely removed therefrom. However it was also realized that, if water depth would not be the limiting factor, application of the Convention would have the result that the oceans and other sea areas of the world would be divided by and between oceanic coastal States, an equally unacceptable result considering that there does not exist a natural, geomorphological link between any coastal State and the seabed and subsoil of the oceans. The latter area can not be seen as the natural prolongation of the land mass of the coastal State, the basis for the latter's claim (under international law, since the 1969 Judgment of the International Court of Justice in the *North Sea Continental Shelf Cases*) to jurisdiction over any adjacent sea area.

Moreover, dividing the natural resources of the deep sea area between the oceanic coastal States only was not considered appropriate or equitable with respect to the landlocked States and other coastal States, i.e., coastal States bordering on narrow dependent seas or gulfs.

With reference to (ii). To treat deep sea mining as one of the freedoms of the high seas would have as consequence that any person would be entitled to explore the area and exploit its deep sea mineral resources. However without a mining regime being in place a prospector making a discovery would have no means to protect its discovery and mining site against other prospectors also looking for exploitation possibilities. A chaotic situation would certainly develop, proving that the undertaking of mining operations, being tied to specific areas of the ocean bed, is completely different from that of exercising any of the other freedoms of the high seas. By their nature, mining operations can not be undertaken within the framework of exercising a freedom, but need a proper mining regime, be it of national or international origin, providing for administration and protection of rights. Without such regime there would be no incentive for the industry to start searching for the resources of the deep sea area.

With reference to (iii). The flag-state approach would give any State the right to use its vessels to explore and exploit a certain area of the deep sea and to acquire the jurisdiction over such an area by occupation. This solution was also considered not to be a very practical one. A flag-state mining regime would certainly require international registration of claims, rules for settling disputes and overlapping claims, and international supervision over the operations, if only to prevent that harm is done to the marine environment.

Meanwhile, fearing that the deep sea resources in one way or other would come under the control of the Western States, given the progress made by those States and their enterprises in deep seabed mining technology, developing countries supported in 1967 a proposal made by Malta, to the effect that the UN General Assembly should declare that the deep seabed and its mineral resources should be reserved for peaceful purposes in the interest of mankind. GA Resolution 2749 (XXV) of 17 December 1970 laid down the following principles with respect to the international area (the Area) as envisaged in Malta's proposal:

(i) the Area and its natural resources are the common heritage of mankind;
(ii) no occupation by States or other natural or legal persons is permitted;
(iii) no rights may be granted or claims be made which contradict the future international mining regime and the principles enunciated in this declaration;
(iv) all mining operations have to fall under the scope of the aforesaid international mining regime;
(v) the Area must be open for peaceful use by all States without discrimination in accordance with the international mining regime;
(vi) all mining operations must be carried out in the interests and to the benefit of mankind, taking into account the interests and needs of the developing countries;

(vii) the international regime must be incorporated in an universal Law of the Sea Convention to which every State may accede;
(viii) the status of the waters of the Area and the air space above it may not be affected.

9.1.6 The 1982 Convention on the Law of the Sea

9.1.6.1 Background

The Law of the Sea Convention of 1982 was the outcome of the Third UN Conference on the Law of the Sea (UNCLOS III). This Conference had been convened for the purpose of giving effect to the principles contained in UNGA Resolution 2749 (XXV) of 17 December 1970. The Convention took place between December 1973 and April 1982. It became effective on 16 November 1994 and as of that date superseded the four Geneva Conventions, without impairing or changing any interstate agreements that in the meantime were made under the rules of these Conventions. In particular bilateral agreements relating to the delimitation of the continental shelf are respected as well as all other international agreements or conventions regarding the use of the sea and the prevention of pollution of the marine environment.

The main objective of the 1982 Convention had been to internationalize the deep seabed area which in the UNGA Resolution 2749 (XXV) of 17 December 1970 had been referred to as the Area, and to proclaim that this Area and its mineral resources in situ (which include petroleum) are the common heritage of mankind and that all rights in these mineral resources are vested in mankind as a whole, on whose behalf the International Seabed Authority shall act. This Authority is established by the Convention as the organization through which Member States shall organize and control activities in the Area, particularly with a view to administering the mineral resources.

9.1.6.2 Contents of the Convention (Parts V, VI, VIII, XI and XII)

9.1.6.2.1 Part V: The Exclusive Economic Zone (EEZ)

Every coastal State is assigned an Exclusive Economic Zone (EEZ). The EEZ is described as an area beyond and adjacent to the territorial sea which shall not extend beyond two hundred nautical miles from the baselines from which the breadth of the territorial sea is measured (Articles 55 and 57).

Note: Depending on local geomorphological circumstances the EEZ of a particular coastal State may well extend beyond the outer edge of the continental margin and could therefore comprise part of the deep seabed area.

As described in Article 56.1(a), (b), (c) a coastal State has in its EEZ:

- sovereign rights for the purpose of exploring and exploiting, conserving and managing the natural resources, whether living or non-living, of the waters

superjacent to the seabed and of the seabed and its subsoil, and with regard to other activities for the economic exploitation and exploration of the zone, such as the production of energy from the water, currents and winds;
- jurisdiction with regard to:
  - (i) the establishment and use of artificial islands, installations and structures;
  - (ii) marine scientific research; and
  - (iii) the protection and preservation of the marine environment; and
- other rights and duties provided for in this Convention.

In exercising its rights with respect to the EEZ a coastal State must have due regard to the rights and duties of other States and shall act in a manner compatible with the provisions of this Convention (Article 56.2).

The rights of other States in the EEZ include the freedoms of navigation and overflight and of the laying of submarine cables and pipelines, and other internationally lawful uses of the sea related to these freedoms, such as those associated with the operation of ships, aircraft and submarine cables and pipelines, and compatible with the other provisions of this Convention. In exercising their rights and performing their duties under this Convention in the EEZ, States shall have due regard to the rights and duties of the coastal State and shall comply with the laws and regulations adopted by the coastal State in accordance with the provisions of this Convention and other rules of international law in so far as they are not incompatible with this Part V (Article 58).

In its EEZ the coastal State shall have the exclusive right to construct and to authorize and regulate the construction, operation and use of:

(i) artificial islands;
(ii) installations and structures for the purposes provided for in Article 56 and other economic purposes;
(iii) installations and structures which may interfere with the exercise of the coastal State in the zone.

The coastal State shall have exclusive jurisdiction over such artificial islands, installations and structures, including jurisdiction with regard to customs, fiscal, health, safety and immigration laws and regulations (Article 60.2).

Due notice must be given of the construction of such artificial islands, installations or structures, and permanent means for giving warning of their presence must be maintained. Any installations or structures, which are abandoned or disused, must be removed to ensure safety of navigation, taking into account any generally accepted international standards established in this regard by the competent international organization (currently the International Maritime Organization [IMO]). Such removal must also have due regard to fishing, the protection of the marine environment and the rights and duties of other States (Article 60.3).

Note. In contrast with the 1958 Convention (Article 5), no absolute obligation exists to remove installations or structures in their entirety.

The coastal State may, where necessary, establish reasonable safety zones around such artificial islands, installations and structures in which it may take appropriate measures to ensure the safety both of navigation and of the artificial islands, installations and structures (Article 60.4).

The breadth of the safety zones shall be determined by the coastal State, taking into account applicable international standards. Such zones shall be designed to ensure that they are reasonably related to the nature and function of the artificial islands, installations or structures, and shall not exceed a distance of five hundred metres around them, measured from each point of the outer edge, except as authorized by generally accepted international standards or as recommended by the competent international organization. Due notice shall be given of the extent of safety zones (Article 60.5).

All ships must respect these safety zones and shall comply with generally accepted international standards regarding navigation in the vicinity of artificial islands, installations, structures and safety zones (Article 60.6).

Artificial islands, installations and structures and the safety zones around them may not be established where interference may be caused to the use of recognized sea lanes essential to international navigation (Article 60.7).

Artificial islands, installations and structures do not possess the status of islands. They have no territorial sea of their own, and their presence does not affect the delimitation of the territorial sea, the EEZ or the continental shelf.

Note. Like Article 5 of the 1958 Convention, Article 60 (which also applies to the continental shelf through Article 80) appears exclusively to address installations and structures that – just like artificial islands which are specifically mentioned – extend above the surface of the sea. Submarine constructions such as submarine pipelines and submarine production installations which are fixed to the seabed are excluded. This has two consequences. In the first place a coastal State may not establish a reasonable safety zone around any submarine construction. In the second place a coastal State appears not to have been given jurisdiction over submarine constructions. For submarine pipelines this problem is solved by Article 79 (continental shelf). Said article confirms the jurisdiction of the coastal State over cables and submarine pipelines constructed or used in connection with the exploration of its continental shelf or the exploitation of its resources. The jurisdiction over submarine constructions other than submarine pipelines must instead be based on Article 56.1 (a) (EEZ) or Article 77.1 (continental shelf).

The delimitation of the EEZ between States with opposite or adjacent coasts must be effected by agreement on the basis of international law in order to achieve an equitable solution (Article 74.1). If no agreement can be reached within a reasonable period of time, the States concerned shall resort to the procedures provided for in Part XV (Settlement of Disputes) (Article 74.2).

Note. In contrast to the 1958 Convention (Article 6) no reference is made to the principle of equidistance (median line) as the method of delimitation of the last resort. On the other hand the reference to international law brings the whole body of judgments delivered by the International Court of Justice in continental shelf demarcation cases within focus.

Pending agreement on a delimitation the States concerned, in a spirit of understanding and cooperation, shall make every effort to enter into provisional arrangements of a practical nature and, during this transitional period, not to jeopardize or hamper the reaching of the final agreement. Such arrangement shall be without prejudice to the final delimitation (Article 74.3).

Note. An important example of such a provisional arrangement of a practical nature is the 1989 Australia/Republic of Indonesia Agreement creating the Timor Gap Zone of Cooperation, an area measuring 62,000 square km. This interstate agreement concerning joint development is described hereinafter in Chapter 10.

9.1.6.2.2 Part VI: The Continental Shelf

Apart from the EEZ, the Convention (Part VI) accords every coastal State a continental shelf. The continental shelf is defined as comprising the seabed and subsoil of the submarine areas that extend beyond its territorial sea throughout the natural prolongation of the land territory to the outer edge of the continental margin, or to a distance of two hundred nautical miles from the baselines from which the breadth of the territorial sea is measured where the outer edge of the continental margin does not extend up to that distance (Article 76.1). The continental margin is described (Article 76.3) as comprising the submerged prolongation of the land mass of the coastal State and consisting of the seabed and subsoil of the shelf, the slope and the rise, but excluding the deep ocean floor with its oceanic ridges or the subsoil thereof.

Note. As defined in Article 76.1, 'continental shelf' includes the continental slope and rise and therefore comprises a generally much larger submarine area than the continental shelf as perceived and described by the 1958 Convention (i.e., the seabed and the subsoil of the submarine areas adjacent to the coast, but outside the area of the territorial sea, to a depth of two hundred metres). In the latter case the continental shelf is described in geomorphological terms (in a geomorphological sense the two hundred metres water depth marks the transition from the continental shelf into the continental slope). In fact the definition given in Article 76.1 builds on the judgment delivered on 20 February 1969 by the International Court of Justice at The Hague in the *North Sea Continental Shelf Cases.* In its judgment the Court held that any coastal State had an original right to those areas of the continental shelf which constituted the natural prolongation of its land territory into and under the sea. But Article 76 went further: it describes 'continental shelf' as 'continental margin', i.e., as consisting of the seabed and subsoil of the shelf, the slope and the rise, without replacing the term 'continental shelf' itself by the term 'continental margin' throughout Article 76. Thus there has been introduced a discrepancy between this definition and the geomorphological definition of continental shelf.

In terms of the Convention the continental shelf, irrespective of the geomorphological conditions of the submarine areas, at least coincides with the EEZ, but may extend beyond this Zone. In quite a few instances the outer edge of the continental margin lies beyond two hundred nautical miles from the baselines. On the other hand the outer edge of the continental margin may fall short of the two hundred nautical miles distance. Where this situation occurs and bearing in mind that the deep seabed area is geologically speaking the area that is contiguous to said outer edge the 'continental shelf' as defined in Article 76 will include part of the deep seabed area.

Note. The attribution of a 'minimum' continental shelf has solved the problem caused by the presence on the continental shelf of deep troughs. The latter's presence raise the question of whether they are to be treated as a local geomorphological feature of an otherwise continuous continental shelf or must be treated as a basic geological feature that forms the outer edge of the continental margin. By according a minimum continental shelf a deep trough situated within the two hundred nautical miles distance from shore will present no jurisdictional problems. Examples of such near-by deep troughs are the Okinawa trough in the Southern East China Sea and the Timor trough situated between East Timor (formerly part of Indonesia) and Northern Australia.

The outer edge of the continental margin must be established by a coastal State, where ever the margin extends beyond the EEZ (Article 76.4). In delineating the outer edge a coastal State shall make use of fixed points to be defined by geographical coordinates and connect these points by straight lines not exceeding sixty nautical miles in length (Article 76.7). A coastal State may choose between two types of fixed points:

(i) fixed points, at each of which the thickness of sedimentary rocks is at least one per cent of the shortest distance from such point to the foot of the continental slope; or
(ii) fixed points, which are located no more than sixty nautical miles from the foot of the continental slope.

The fixed points selected by the coastal State must be located either within 350 nautical miles from the baselines from which the breadth of the territorial sea is measured, or within one hundred nautical miles from the 2,500 metres isobath (Article 76.5). The coastal State is obliged to provide information on the limits of the continental shelf, wherever it is extending beyond the EEZ, to the Commission on Limits of the Continental Shelf set up under the Convention on the basis of equitable geographical representation. The Commission shall make recommendations to coastal States on matters related to the establishment of the outer limits of their continental shelf. These recommendations on their own are not binding, but if a coastal State establishes the limits of the continental shelf in accordance with these recommendations, such limits shall be final and binding (Article 76.8). Through this provision the Commission's recommendations carry great weight. The delimitation of the continental shelf between States with opposite or adjacent coasts is provided for (Article 83) in the same manner and in accordance with the same procedures and principles as set out in Article 74 with respect to such delimitation of the EEZ.

Note. When considering the division of the North Pole area between the surrounding States among which Canada and the Russian Federation feature prominently, it appears that first the extent of each State's share of the Arctic continental shelf must be determined in accordance with Article 76. Only where such shares prove to overlap the provisions of Article 83 shall apply.

The sovereign rights of a coastal State over the continental shelf are identical to the rights granted under the 1958 Convention (Article 77) but with respect to submarine pipelines, which include therefore pipelines laid by other States, a coastal State has been given the power to take reasonable measures for the prevention, reduction and control of pollution therefrom (Article 79.2). Furthermore, the delineation of the course for the laying of other States' pipelines is subject to the consent of the coastal State (Article 79.3). The coastal State has the right to establish conditions for submarine cables or pipelines entering its territory or territorial sea and has jurisdiction over cables and pipelines constructed or used in connection with its natural resources operations (Article 79.4). States laying submarine cables or pipelines must respect cables or pipelines already in position and must not thereby jeopardize the possibilities of repairing these cables or pipelines (Article 79.5). With respect to artificial islands, installations and structures on the continental shelf the rights and jurisdiction of the coastal State are declared to be the same as those described in Article 60 with respect to these constructions within its EEZ (Article 80).

Note. It follows from the reference made to Article 60, that submarine constructions on the continental shelf cannot be protected by establishing safety zones around them. As to jurisdiction over these constructions there is a difference with the EEZ: jurisdiction over submarine pipelines can be based on Article 79.4, whereas jurisdiction over other submarine constructions such as submarine production installations shall have to be based on the basic rights set out in Article 77.1.

Jurisdiction over marine scientific research and the protection and preservation of the marine environment is exclusively reserved for the EEZ (Article 56 1 (b) (ii) and (iii)). As far as the protection of the environment is concerned, leaving out the continental shelf has no consequences in view of Article 208. The latter Article imposes on any coastal State the duty to prevent, reduce and control pollution of the marine environment arising from seabed activities under their jurisdiction (which by definition include any mining activity on the continental shelf). On the continental shelf the coastal State has been given the exclusive right to authorize and regulate drilling for all purposes (Article 81).

Article 82 stipulates that a coastal State should make payments in cash or kind in respect of the exploitation of the non-living resources of that part of the continental shelf that extends beyond the two hundred nautical miles limit. The payments or contributions consist in the payment of an annual royalty commencing after the first five years of production.

9.1.6.2.3 Part VIII: The Regime of Islands

The 1982 Convention has made important contribution to the development of the law of the sea in that it contains a definition of what constitutes an island. Under the given definition not every piece of dry land qualifies for the application of the Convention in the matter of islands.

According to Article 121 of the Convention:

1. An island is a naturally formed area of land, surrounded by water, which is above water at high tide.
2. Except as provided for in paragraph 3, the territorial sea, the contiguous zone, the exclusive economic zone and the continental shelf of an island are determined in accordance with the provisions of this Convention applicable to other land territory.
3. Rocks which cannot sustain human habitation or economic life of their own shall have no exclusive economic zone or continental shelf.

If politically independent, islands are coastal States in their own right (e.g., United Kingdom, Ireland, Iceland, Madagascar, Cuba, etc.), otherwise islands fall under the jurisdiction of a coastal State (e.g., Andaman Islands, Corsica, Gotland, etc.) or constitute an archipelagic State (i.e., a State consisting of an agglomeration of islands (e.g., Indonesia, the Philippines). If not being rocks as described in Article 121.3 islands can claim an EEZ and a continental shelf. This claim may make the territorial jurisdiction over an island (if not being a coastal State in its own right) very valuable and important if the area surrounding the island is thought to be a prospective area. If so, jurisdictional disputes about an island will be difficult to settle because much more is at stake than the territorial jurisdiction over the island itself. On the other hand jurisdictional disputes about reefs and rocks which do not qualify as islands in terms of Article 121 will be easier to settle.

9.1.6.2.4 Part XI: The Area

The deep seabed and its subsoil in so far not forming part of the EEZ or the continental shelf (as defined in Article 76) constitute the Area. Petroleum belongs to the mineral resources of the Area, but from a geological point of view it is not expected that petroleum reservoirs will actually be found in the deep seabed, except for gas hydrates, a chemical compound consisting of methane and water. Since at the moment little can be said about the technical possibilities of producing these hydrates it can safely be assumed that petroleum operations will exclusively take place in areas which are placed by the Convention under national jurisdiction.

Note. Having been granted the exclusive jurisdiction over installations and operations within the EEZ and the continental shelf the coastal State is responsible for preparing and establishing rules and regulations with regard to these operations and installations. However, on account of inter-state agreements and conventions, to which the coastal State is or becomes a party, the coastal State is also responsible for implementation of the rules, standards, procedures agreed in the context of these inter-state agreements and conventions. This means that the coastal State must incorporate and include the rules, etc. in its national legislation.

9.1.6.2.5 Part XII: The Protection of the Marine Environment

9.1.6.2.5.1 Introduction

Part XII of the Convention covers extensively the duties of States with regard to the protection and preservation of the marine environment. Major petroleum operations and other activities taking place within the marine environment can be summarized as follows:

(i) the placement, construction, use and operation of offshore (drilling, production and storage) installations, vessels or facilities, construction and laying of submarine pipelines, the drilling of exploration, appraisal and development wells by means of the offshore installations;
(ii) taking measures to prevent, reduce or control pollution of the marine environment;
(iii) taking measures to safeguard and protect the safety and health of personnel at work on the offshore installations;
(iv) taking measures concerning the partial or total removal of disused or abandoned offshore installations.

Note. Vessels mean waterborne craft and includes floating craft, whether self-propelled or not, excluding floating offshore installations. Offshore installation means any man-made structure placed in the offshore area of operation for the purpose of petroleum operations. Offshore installations may be floating installations (e.g., vessels converted into production barges or moored vessels) or fixed to or standing on the seabed. In conformity with Article 60 offshore installations are taken to be installations that extend above the surface of the sea and therefore do not include submarine production installations that are fixed to the seabed.

Generally, pollution of the marine environment may be caused by:

(i) natural events (in particular submarine volcanoes); and
(ii) human activities.

In the context of Conventions and other interstate agreements designed for the protection of the marine environment 'pollution' is restricted to pollution caused by human activities. Taking the 1992 Convention for the Protection of the Marine Environment of the North-East Atlantic (OSPARCON) as example, 'pollution' is described in that Convention in the following terms:

> 'Pollution' means: the introduction by man, directly or indirectly, of substances or energy into the maritime area which results, or is likely to result, in hazards to human health, harm to living resources and marine ecosystems, damage to amenities or interference with other legitimate uses of the sea.

Note 1. Where human activities are concerned distinction may be made between accidents or non-intended events on the one hand and deliberate or intentional acts or actions on the other. The latter have been summarized above as far as offshore petroleum operations are concerned. For the definition of 'Pollution' this distinction makes no difference.

Note 2. 'Introduction of energy' refers to the release of energy in the course of and for the purpose of conducting seismic surveys.

Note 3. Substances which may be intentionally or non-intentionally released or discharged into the marine environment include:

(i) substances produced by, generated by or used in the drilling and production phase of the operations, such as crude oil, condensate, natural gas and production water, drilling mud, drilling cuttings and any type of waste material;
(ii) substances accidentally escaping during the drilling or production operations or substances leaking from casing, tubing, storage tanks or submarine pipelines, such as crude oil, condensate, natural gas, production water or chemicals;
(iii) disused or abandoned offshore installations or parts thereof; and
(iv) substances generated by or used in and for the normal operation of offshore installations and vessels, such as sewage, used oils, any type of waste material.

9.1.6.2.5.2 Contents of Part XII

The main provisions of Part XII (Articles 192 to 212) are the following:

- States (whether coastal States or not) have the obligation to protect and preserve the marine environment (Article 192);
- States have the sovereign right to exploit their natural resources pursuant to their environmental policies and in accordance with their duty to protect and preserve the marine environment (Article 193);
- States shall take, individually or jointly as appropriate, all measures consistent with the Convention that are necessary to prevent, reduce and control pollution of the marine environment from any source, using for this purpose the best practicable means at their disposal and in accordance with their capabilities, and they shall endeavour to harmonize their policies in this connection. The measures to be taken shall deal with all sources of pollution of the marine environment. These measures shall include, inter alia, those designed to minimize to the fullest possible extent:

  (i) the release of toxic, harmful or noxious substances from land-based sources, from or through the atmosphere or by dumping;
  (ii) pollution from vessels or from offshore (mining) installations resulting from operational discharges or accidents (e.g., collisions at sea) and regulating the design, construction, equipment, operation and manning of vessels and offshore installations (Article 194).

- States shall cooperate on a global basis and, as appropriate, on a regional basis, directly or through competent international organizations (*Note.* Currently the International Maritime Organization [IMO]) in formulating and elaborating international rules, standards and recommended practices and procedures consistent with this Convention, for the protection and preservation of the marine environment, taking into account characteristic regional features (Article 197);
- Coastal States shall adopt laws and regulations to prevent, reduce and control pollution of the marine environment arising from or in connection

with seabed activities subject to their jurisdiction and from artificial islands, installations and structures under their jurisdiction pursuant to Articles 60 and 80. Such laws, regulations and measures shall be no less effective than international rules, standards and recommended practices and procedures. States shall endeavour to harmonize their policies in this connection at the appropriate regional level. States, acting especially through competent international organizations or diplomatic conference, shall establish global and regional rules, standards and recommended practices and procedures to prevent, reduce and control pollution of the marine environment referred to above (Article 208);

Note. Submarine pipelines (or submarine production installations, which are fixed to the seabed, for that matter) are not specifically mentioned in Part XII as a source of pollution, despite the fact that they are widely used in submarine long distance transport of oil or natural gas as well as in offshore petroleum operations for submarine transport of oil/gas production, oil products and chemicals to and from offshore installations. Submarine pipelines may become a source of pollution of the marine environment, if their condition deteriorates. It must be concluded that the matter of prevention, etc. of pollution arising from the presence and use of submarine pipelines can only been dealt with in the context of Article 208 referred to above. As mentioned before, the sole specific reference to submarine pipelines and the possibility of these pipelines being a source of pollution appear in Article 79 containing provisions regarding submarine pipelines on the continental shelf. It must be assumed that the same holds true for other submarine constructions such as submarine production installations which are fixed to the seabed, in view of the coastal State general jurisdiction over these submarine constructions and the coastal State's general obligations under the provisions of Part XII.

- States shall adopt laws and regulations to prevent, reduce and control pollution of the marine environment by dumping. Such laws, regulations and measures shall ensure that dumping is not carried out without the permission of the competent authorities of States. States, acting especially through competent international organizations or diplomatic conference, shall endeavour to establish global and regional rules, standards and recommended practices and procedures to prevent, reduce and control such pollution. Dumping within the territorial sea and the EEZ or on the continental shelf shall not be carried out without the express prior approval of the coastal State, which has the right to permit, regulate and control such dumping after due consideration of the matter with other States which by reason of their geographical situation may be adversely affected thereby. National laws, regulations and measures shall be no less effective in preventing, reducing and controlling such pollution than the global rules and standards (Article 210).

Note. Dumping is defined in the manner as this expression is defined in the earlier 1972 London Dumping Convention. Dumping means any deliberate disposal at sea of wastes or other matter from vessels, aircraft, platforms or other man-made structures at sea as well as any deliberate disposal at sea of vessels, aircraft, platforms or other man-made structures at sea. Dumping does not include the disposal at sea of wastes or other matter incidental to, or derived from the normal operations of vessels, aircraft, platforms, or other man-made structures at sea and their equipment (operational discharges) or the placement of matter for a purpose other than the mere disposal thereof. The disposal of wastes or other matter directly arising from, or related to seabed activities (which include petroleum operations) is not treated as dumping. Such discharges are addressed in the context of preventing, reducing and controlling pollution of the marine environment arising from or in connection with seabed activities subject to their jurisdiction and from artificial islands, installations and structures under their jurisdiction.

Summarizing, the Convention recognizes several sources of pollution of the marine environment and imposes on all States, whether having a coastline or not, the obligation to protect and preserve this environment. The Convention does not contain any specific rules, procedures or standards by which States should guided or which States should adopt, when trying to implement their aforesaid duties. The Convention is only concerned in assigning and allocating responsibilities between States (States generally or coastal States exclusively) for designing and promulgating such rules and standards, although with respect to some sources of pollution (dumping and vessels) States are obliged to act through the competent international organization in order to establish international rules and standards. Currently, there exists only one such competent international organization, viz. the International Maritime Organization (IMO), the successor of the Intergovernmental Maritime Consultative Organization (IMCO). Meanwhile, States have made great efforts in negotiating and agreeing global and regional rules and standards and procedures to prevent, reduce and control pollution of the marine environment.

9.2 DUMPING CONVENTIONS

9.2.1 The 1972 Oslo Convention for the Prevention of Marine Pollution by Dumping from Ships and Aircraft

The Convention for the Prevention of Marine Pollution by Dumping from Ships and Aircraft signed in Oslo on 15 February 1972, as amended by the Protocols of 2 March 1983 and 5 December 1989, has a regional scope covering the North East Atlantic and Arctic Oceans and dependent seas, such as the North Sea, but excluding the Baltic Sea. The Oslo Convention had been restricted to dumping by or from ships or aircraft, ships in this case being defined as including fixed or floating platforms. The Convention does not cover the dumping of ships (and platforms) or aircraft. The Commission established by the Convention requested in 1984 the IMO to develop rules and standards for removal of offshore installations. The IMO Guidelines and Standards became available in 1989.

The Oslo Dumping Convention was replaced by the 1992 Convention for the Protection of the Marine Environment of the North-East Atlantic (OSPARCON).

9.2.2 The 1972 London Convention on the Prevention of Marine Pollution by Dumping of Wastes and Other Matter

The Convention on the Prevention of Marine Pollution by Dumping of Wastes and Other Matter signed in London on 29 December 1972 is a convention of global scope and worldwide application. As defined, 'dumping' includes the deliberate disposal at sea of vessels, aircraft, platforms or other man-made structures at sea, but excludes operational discharges. Furthermore the Convention explicitly states not to be intended to cover discharges of substances directly arising from or related to the exploration, exploitation and associated offshore processing of seabed mineral resources. The dumping at sea of inter alia platforms or other man-made structures which include offshore mining installations requires a prior general permit, by which is meant permission granted in advance and in accordance with Annex III of the Convention. Furthermore, any permit shall be issued only after careful consideration of all the factors set forth in said Annex III, including prior studies of the characteristics of the dumping site. Permits allowing dumping may be issued by the contracting State having jurisdiction over the offshore mining installation. Each contracting State shall designate an appropriate authority to issue the required general permits. Such an authority is obliged to comply with Annex III.

Note. Said Annex III contains the factors to be considered in establishing the criteria governing the issue of permits for the dumping of inter alia installations at sea. These factors fall into two categories:

(i) characteristics and composition of the matter to be dumped, such as total amount, form (solid, sludge, liquid), chemical, physical and biological properties, toxicity, persistence, accumulation in biological materials or sediments, susceptibility to changes and interaction, probability of production of taints or other changes reducing marketability of resources (fish, shellfish, etc.); and
(ii) characteristics of the dumping site and method of deposit, such as location of the site, rate of disposal, methods of packaging and containment, water characteristics, dispersal characteristics, bottom characteristics.

As can be seen, a number of factors are not relevant for the dumping of installations as such but are only relevant for any substances that may be contained in the installations and that therefore will be included in the dumping.

9.3 CONVENTIONS ON THE PROTECTION OF THE MARINE ENVIRONMENT

9.3.1 The 1973/1978 Convention for the Prevention of Pollution from Ships (MARPOL)

The Convention for the Prevention of Pollution from Ships (MARPOL) 1973, as modified by the Protocol of 1978, which was made under the auspices of the former Intergovernmental Maritime Consultative Organization (IMCO), (currently the IMO) establishes general rules and standards concerning operational discharges from ships in general. These rules and standards have been made applicable to offshore mining installations by treating these installations as ships for the purposes of the Convention. Under other pollution prevention conventions the rules and standards of MARPOL have been adopted by reference, but these conventions may contain additional rules for specific sea areas.

Note. MARPOL also provide for rules specifically meant for oil tankers and aimed to prevent that oil is released from a tanker at sea. Furthermore, in 1993 a new IMO Convention became effective providing for design and construction standards for future oil tankers, making the latter less likely to lose their cargo in case of an accident. Before 2000 seventy per cent of the in 1993 existing fleet must be replaced by a new type of tanker or adapted to the new construction standards.

9.3.2 The 1974 Paris Convention for the Prevention of Marine Pollution from Land-Based Sources

The rules of the Convention for the Prevention of Marine Pollution from Land-based Sources signed in Paris on 4 June 1974, as amended by the Protocol of 26 March 1986, are also applicable to all discharges, whether operational or related to petroleum operations, from offshore mining installations. This effect has been achieved by treating (for the purposes of the Convention) offshore mining installations as land-based sources. Similarly as the Oslo Dumping Convention, the Paris Convention was replaced by the 1992 Convention for the Protection of the Marine Environment of the North-East Atlantic (OSPARCON).

9.3.3 The 1992 Convention for the Protection of the Marine Environment of the North-East Atlantic (OSPARCON)

9.3.3.1 Definitions and Scope

The Convention for the Protection of the Marine Environment of the North-East Atlantic (OSPARCON) was signed in Paris on 22 September 1992. OSPARCON replaces and represents a further development of both the Oslo Dumping Convention of 1972 and the Paris Convention of 1974.

The 'maritime area' in respect of which OSPARCON is applicable and which is the same as the area of the Oslo Dumping Convention, means the internal waters and the territorial seas of the Contracting Parties, the sea beyond and adjacent to the territorial sea under the jurisdiction of the coastal State to the extent recognized by international law, and the high seas, including the bed of all those waters and its sub-soil, situated within the following limits:

(i) those parts of the Atlantic and Arctic Oceans and their dependent seas which lie north of 36 degrees north latitude and between 42 degrees west longitude and 51 degrees east longitude, but excluding:

(1) the Baltic Sea and the Belts, etc.,
(2) the Mediterranean Sea, etc.;

(ii) that part of the Atlantic Ocean north of 59 degrees north latitude and between 44 degrees west longitude and 42 degrees west longitude.

'Dumping', defined similarly as in the 1972 London Dumping Convention, means:

(i) any deliberate disposal in the maritime area of wastes or other matter:

(1) from vessels or aircraft;
(2) from offshore installations;

(ii) any deliberate disposal in the maritime area of:

(1) vessels or aircraft;
(2) offshore installations and offshore pipelines.

'Dumping' does not include:

(i) the disposal in accordance with the International Convention for the Prevention of Pollution from Ships, 1973, as modified by the Protocol of 1978 relating thereto (MARPOL 1973/1978), or other applicable international law, of wastes or other matter incidental to, or derived from, the normal operations of vessels or aircraft or offshore installations other than wastes or other matter transported by or to vessels or aircraft or offshore installations for the purpose of disposal of such wastes or other matter or derived from the treatment of such wastes or other matter on such vessels or aircraft or offshore installations;
(ii) placement of matter for a purpose other than the mere disposal thereof, provided that, if the placement is for a purpose other than that for which the matter was originally designed or constructed, it is in accordance with the relevant provisions of the Convention; and
(iii) for the purposes of Annex III, the leaving wholly or partly in place of a disused offshore installation or disused offshore pipeline, provided that any such operation takes place in accordance with any relevant provision of the Convention and with other relevant international law.

'Pollution' means the introduction by man, directly or indirectly, of substances or energy into the maritime area which results, or is likely to result, in hazards to human health, harm to living resources and marine ecosystems, damage to amenities or interference with other legitimate uses of the sea. *Note.* Foregoing definition has been discussed under Part XII hereinabove.

'Offshore activities' means activities carried out in the maritime area for the purposes of the exploration, appraisal or exploitation of liquid and gaseous hydrocarbons.

'Offshore sources' means offshore installations and offshore pipelines from which substances of energy reach the maritime area.

'Offshore installation' means any man-made structure, plant or vessel or parts thereof, whether floating or fixed to the seabed, placed within the maritime area for the purpose of offshore activities. (*Note.* Without doubt a submarine production installation, which is fixed to the seabed, is considered to be an offshore installation.)

'Offshore pipeline' means any pipeline which has been placed in the maritime area for the purpose of offshore activities.

'Vessels or aircraft' means waterborne or airborne craft of any type whatsoever, their parts and other fittings. This expression includes air-cushion craft, floating craft whether self-propelled or not, and other man-made structures in the maritime area and their equipment, but excludes offshore installations and offshore pipelines.

The Contracting Parties are obliged, individually and jointly to adopt programmes and measures and to harmonize their policies and strategies in order to prevent and eliminate pollution and to protect the maritime area against the adverse effects of human activities so as to safeguard human health and to conserve marine ecosystems and, when practicable, restore marine areas which have been adversely affected. To this end Contracting Parties shall, individually and jointly, adopt programmes and measures and shall harmonize their policies and strategies.

The Contracting Parties shall apply the precautionary principle, by virtue of which preventive measures are to be taken; and the polluter pays principle, by virtue of which the costs of pollution prevention, control and reduction measures are to be borne by the polluter.

The programmes and measures to be adopted shall take full account of the use of the latest technological developments and practices designed to prevent and eliminate pollution. In this context the Contracting Parties shall define the application of 'best available techniques' and 'best environmental practice', including, where appropriate, 'clean technology' and ensure the application thereof in carrying out such programmes and measures.

No provision of the Convention shall be interpreted as preventing the Contracting Parties taking, individually or jointly, more stringent measures with respect to the prevention and elimination of pollution of the maritime area or with respect to the protection of the maritime area against the adverse effects of human activities.

Note. An example of joint action by a sub-group of Contracting Parties is provided by the North Sea Ministerial Conferences, discussed hereinafter.

9.3.3.2 Sources of Pollution

The Contracting Parties shall take, individually and jointly, all possible steps to prevent and eliminate pollution from the sources listed herein below in accordance with the provisions of the Convention, in particular as provided for in respectively Annex I, Annex II and Annex III:

(i) pollution from land-based sources (Annex I);
(ii) pollution by dumping or incineration (Annex II, except where offshore installations and offshore pipelines are involved);
(iii) pollution from offshore sources (Annex III). As far as offshore sources are concerned, the Convention does not make a distinction between operational discharges and discharges connected with the offshore petroleum operations. Instead, all pollution matters concerning the use on, or the discharge from offshore installations or offshore pipelines of substances, which may reach and affect the maritime area, are covered by Annex III. However, the rules and provisions regarding operational discharges as contained in MARPOL remain valid. The provisions of Annex III also reflect the work done on developing the aforementioned OSCOM Guidelines concerning the disposal at sea of disused offshore installations.

9.3.3.3 The Commission

For the purpose of the Convention a Commission, made up of representatives of each of the Contracting Parties, is established. It is the duty of the Commission inter alia to supervise the implementation of the Convention; to review the condition of the maritime area and the effectiveness of the measures being adopted; to draw up (in accordance with the general obligations of the Convention) programmes and measures for the prevention and elimination of pollution and for the control of activities which may, directly or indirectly, adversely affect the maritime area; and to consider and, where appropriate, adopt proposals for the amendment of the Convention, including its Annexes and Appendices. The decisions and non-binding recommendations of the Commission must be adopted by the unanimous vote of the Contracting Parties, but should unanimity not be attainable the Commission may nonetheless in a second round adopt decisions or recommendations by a three-quarters majority vote. A decision is binding for those Contracting States that voted for it and did not withdraw their acceptance within 120 days after the adoption of the decision (provided that at the expiry of that period three-quarters of the Contracting Parties have accepted the decision) and on any other Contracting Party that has notified that it is able to accept the decision.

9.3.3.4 Annex III (re. Pollution from Offshore Sources)

The main provisions of Annex III are:

(i) when adopting programmes and measures the Contracting Parties shall require, either individually or jointly, the use of best available techniques, best environmental practice and, where appropriate, clean technology. The terms 'best available techniques' and 'best environmental practice' are described and explained in Appendix 1 of the Convention. The relative value and significance of these concepts is recognized. It is stressed that what constitutes 'best available techniques' and 'best environmental practice in a particular case will change with time in the light of technological advances, economic and social factors, as well as changes in scientific knowledge and understanding. (*Note.* The expression 'Clean Technology' is not described. However in Annex II of the 1992 Baltic Sea Convention [discussed hereinafter] the concept is described as 'production systems which minimize, avoid or eliminate the generation of hazardous waste and hazardous products');

(ii) when setting priorities and in assessing the nature and extent of the programmes and measures the Contracting Parties shall use the criteria given in Appendix 2. These criteria concern persistency; toxicity or other noxious properties; tendency to bio-accumulation; radioactivity; risk of undesirable changes in the marine ecosystem; interference with harvesting of sea-foods or with other legitimate uses of the sea; effects on the taste and/or smell of products for human consumption from the sea or effects on the characteristics of the water in the marine environment. In this connection substances to be subject to programmes and measures include inter alia heavy metals and their compounds, organohalogen compounds, organic compounds of phosphorus and silicon, biocides such as pesticides and herbicides, oils and hydrocarbons of petroleum origin, radioactive substances and persistent synthetic materials which may float, remain in suspension or sink;

(iii) the use on, or the discharge or emission from, offshore sources of substances, which may reach and affect the maritime area, shall be strictly subject to authorization or regulation by the competent authorities of the contracting states. Such authorization or regulation shall, in particular, implement the relevant applicable decisions, recommendations and all other agreements adopted under the Convention;

(iv) no disused offshore installation or disused offshore pipeline shall be dumped and no disused offshore installation shall be left wholly or partly in place in the maritime area without a permit issued by the competent authority of the relevant contracting state on a case-by-case basis. The Contracting Parties shall ensure that their authorities, when granting such permits, shall implement the relevant applicable decisions, recommendations and all other agreements adopted under the Convention;

(v) no such permit shall be issued if the disused offshore installation or disused offshore pipeline contains substances which lead to pollution in the sense as defined, i.e., substances which result or are likely to result in hazards to human health, harm to living resources and marine eco-systems, damage to amenities or interference with other legitimate uses of the sea;

(vi) any Contracting Party which intends to take the decision to issue a permit for the dumping of a disused offshore installation or a disused offshore pipeline placed in the maritime area after 1 January 1998 shall, through the medium of the Commission, inform the other Contracting Parties of the reasons for accepting such dumping, in order to make consultation possible. (*Note.* It should be recalled that the aforesaid date of 1 January 1998 is the date mentioned in the 1989 IMO Standards as the date on or after which no installation should be placed on any continental shelf or within any EEZ, unless the design and construction of the installation is such that entire removal upon permanent disuse would be feasible.);

(vii) each Contracting Party shall keep, and report to the Commission, records of the disused offshore installations and disused offshore pipelines dumped and of the disused offshore installations left in place in accordance with the provisions of the Annex, and of the dates, places and methods of dumping.

For the purposes of this Annex III the Commission has been given the following duties:

(i) to collect information about substances, which are used in offshore petroleum operations, in order to agree lists of substances for the purpose of authorizations and regulations regarding their use and discharges;

(ii) to list substances, which are toxic, persistent and liable to bio-accumulate, and to draw up plans for the reduction and phasing out of their use on, or discharge from offshore installations or offshore pipelines;

(iii) to draw up criteria, guidelines and procedures for the prevention of pollution from dumping of disused offshore installations and of disused offshore pipelines, and the leaving in place of offshore installations in the maritime area.

9.3.3.5 Amendments

Any Contracting Party may propose an amendment to the Convention. The Commission shall adopt any proposed amendment by unanimous vote of the Contracting Parties. The adopted amendment shall be submitted to the Contracting Parties for ratification. The amendment shall enter into force for those Contracting Parties which have ratified it on the 30th day after receipt of notification of its ratification by at least seven Contracting Parties.

Note. OSPARCON does not prescribe any specific rules or standards on how to conduct offshore petroleum operations; it only contains procedures leading to the establishment of such rules and standards. The latter have to be developed by the Commission and to be implemented by the Contracting Party under whose jurisdiction the offshore installations and pipelines are used. The Contracting Parties are bound to fulfil the general obligations imposed by the Convention, to use best available techniques and best environmental practice, including, where appropriate, clean technology, as these terms are defined in Appendix 1, and to apply or take into account the criteria and substances listed in Appendix 2 in assessing the nature and extent of their programmes and measures, and also bound to the rules and provisions of MARPOL with regard to operational discharges from (specialized) vessels and offshore installations. Within this framework and always subject to its obligations under the 1982 Convention on the Law of the Sea, any Contracting Party (coastal State) retains its freedom to regulate the offshore petroleum operations under its jurisdiction with regard to the subject matter of the Convention as it sees fit. It is only bound to the decisions of the Commission, if it notifies in writing that it agrees with them.

9.3.4 The North Sea Ministerial Conferences

The protection of the marine environment of the North Sea has been given special attention by the governments of the North Sea coastal States and the non-coastal States having within their territory rivers discharging into the North Sea. By the North Sea is meant the body of water between latitudes 62 degrees north and 48 degrees thirty minutes north. The aforesaid governments are closely working together within the framework of successive ministerial conferences in establishing and adopting rules and standards in order to prevent or at least minimize direct or indirect pollution of the North Sea from any type of source. The first Conference was convened by the Federal Republic of Germany in 1984. Since then there Conferences have been held every three years.

The Conferences are attended by Representatives of respectively the OSPARCON-Commission and the IMO. At their meetings rules and standards are agreed to which aim to achieve significant or substantial reductions or phasing out within agreed time frames of inputs or atmospheric emissions of polluting substances that directly (from ships or offshore installations or by dumping or incineration at sea) or indirectly (from land-based sources through the major rivers and estuaries) reach the North Sea and its atmosphere. All possible sources of pollution, including offshore petroleum operations, dumping operations involving dumping of sewage sludge, industrial waste and dredged materials, but in particular land-based sources (chemical industry, agriculture and municipal sewage plants located along the coast or the major rivers) are being regulated.

The governments involved aim to implement their recommendations and agreements through the OSPARCON-Commission, while leaving open the possibility that for the North Sea area rules and standards will be formulated which differ from those established generally with regard to the OSPARCON maritime area.

9.3.5 The 1992 Convention on the Protection of the Marine Environment of the Baltic Sea Area (Helsinki Convention [HELCON])

9.3.5.1 Definitions and Scope

The Convention on the Protection of the Marine Environment of the Baltic Sea Area (Helsinki Convention [HELCON]) was signed on 9 April 1992 in Helsinki. It entered into force on 17 January 2000, replacing thereby the Convention on the Protection of the Marine Environment of the Baltic Sea Area signed in Helsinki on 22 March 1974 as amended. The present Contracting Parties comprise all States bordering on the Baltic Sea Area as further defined in the Convention as well as the European Community. For the purpose of the Convention, the aforesaid Baltic Sea Area is described as the Baltic Sea and the entrance to the Baltic Sea bounded by the parallel of 57 44 43 degrees North in the Skagerrak (thus excluding Norway which initially had signed the Convention). It includes the internal waters, i.e., for the purpose of this Convention waters on the landward side of the base lines from which the breadth of the territorial sea is measured up to the landward limit according to the designation by the Contracting Parties. The Convention is rather similar to OSPARCON, but there are important differences, to name a few: the absence of any reference to offshore or submarine pipelines, the definition of 'Ships' comprising fixed or floating platforms; and the absolute prohibition to engage in 'Dumping', except for the disposal at sea of dredged material.

In the main text of the Convention the key expressions 'Ship', 'Pollution' and 'Dumping' have been defined as follows:

'Pollution' means the introduction by man, directly or indirectly, of substances or energy into the sea, including estuaries, which are liable to create hazards to human health, to harm living resources and marine ecosystems, hindrance to legitimate uses of the sea including fishing, impairment of the quality for use of sea water, and reduction of amenities.

'Pollution from land-based sources' means pollution of the sea by point or diffuse inputs from all sources on land reaching the sea waterborne, airborne or directly from the coast. It includes pollution from any deliberate disposal under the seabed with access from land by tunnel, pipeline or other means.

'Ship' means a vessel of any type whatsoever operating in the marine environment and include hydrofoil boats, air-cushion vehicles, submersibles, floating craft and fixed or floating platforms.

'Dumping' means:

(i) any deliberate disposal at sea or into the seabed of wastes or other matter from ships, other man-made structures at sea or aircraft;
(ii) any deliberate disposal at sea of ships, other man-made structures at sea or aircraft.

'Dumping' does not include:

(i) the disposal at sea of wastes or other matter incidental to, or derived from the normal operations of ships, other man-made structures at sea or aircraft and their equipment, other than wastes or other matter transported by or to ships, other man-made structures at sea or aircraft, operating for the purpose of disposal of such matter or derived from the treatment of such wastes or other matter on such ships, structures or aircraft;
(ii) placement of matter for a purpose other than the mere disposal thereof, provided that such placement is not contrary to the aims of the present Convention.

The Convention describes the fundamental obligations of the Contracting Parties in the following terms: the Contracting Parties are obliged to take all appropriate legislative, administrative or other relevant measures to prevent and eliminate pollution in order to promote the ecological restoration of the Baltic Sea Area and the preservation of its ecological balance. In fulfilling their fundamental obligations the Contracting Parties shall apply the precautionary principle and the polluter pays principle, and promotes the use of Clean Production and Technology including the main elements of Best Environmental Practice and Best Available Technology. The criteria for the use of Best Environmental Practice and Best Available Technology are worked out in Annex II. These concepts and the concept of Pollution, Incineration and Dumping are described in the same terms as in OSPARCON.

9.3.5.2 Sources of Pollution

Several sources of pollution are specifically dealt with as follows:

(i) pollution from land-based sources. Criteria and measures concerning the prevention of pollution from this source are worked out in Annex III;
(ii) pollution from ships (as defined in the Convention 'ships' include fixed and floating platforms): the Contracting Parties are obliged to take measures as set out in Annex IV. In this connection the provisions of MARPOL 73/78 are confirmed applicable in addition to more stringent rules needed in view of the vulnerability of the Baltic Sea Area;
(iii) pollution from incineration. The Contracting Parties shall prohibit incineration in the Baltic Sea Area;
(iv) pollution by dumping. The Contracting Parties shall prohibit dumping in the Baltic Sea Area, except when the safety of human life or of a ship or aircraft at sea is at stake. Dumping of dredged materials shall be subject to a prior special permit by the appropriate national authority in accordance with the provisions of Annex V;
(v) pollution resulting from exploration or exploitation of the seabed and its subsoil or from any associated activities. In order to prevent and eliminate pollution from such activities the Contracting Parties are obliged to implement the procedures and measures as set out in Annex VI, as far applicable.

9.3.5.3 The Commission

For the purpose of the Convention the Baltic Marine Environment Protection Commission, referred to as 'the Commission' is established. The Commission is the same Commission as the one which previously had been established under the 1974 Convention on the Protection of the Marine Environment of the Baltic Sea Area.

The Commission is made up by the representatives of the Contracting Parties. Each Contracting Party shall have one vote in the Commission. Unless otherwise provided in the Convention, the Commission shall take its decisions unanimously.

It belongs to the duties of the Commission inter alia to keep the implementation of the Convention under continuous observation; to make recommendations on measures relating to the purpose of the Convention; to keep under review the contents of the Convention including its Annexes and to recommend to the Contracting Parties such amendments to the Convention including its Annexes as may be required including changes in the lists of substances and materials as well as the adoption of new Annexes.

9.3.5.4 Annex VI (re. Offshore Pollution)

Annex VI is specifically concerned with the prevention of pollution from offshore activities.

For the purposes of the Annex the following key expressions have been defined as follows:

'Offshore activities' means any exploration and exploitation of oil and gas by a fixed or floating offshore installation or structure including all associated activities thereon.

'Offshore unit' means any fixed or floating offshore installation or structure engaged in gas or oil exploration, exploitation or production and related activities, or in loading or unloading of oil.

Note. Annex VI is not concerned with non-petroleum mining activities or with operational discharges from offshore installations (since these installations have been defined as ships operational discharges are regulated in Annex IV: Prevention of pollution from ships).

The Contracting Parties undertake to prevent and eliminate pollution from offshore activities by using the principles of Best Available Technology and Best Environmental Practice as defined in Annex II.

Before an offshore operation is permitted to start, an environmental impact assessment must be made. In connection with such assessment the environmental sensitivity of the sea area around a proposed offshore unit should be assessed with respect to the following:

(i) the importance of the area for birds and marine mammals;
(ii) the importance of the area as fishing or spawning grounds for fish and shellfish, and for aquaculture;
(iii) the recreational importance of the area;

(iv) the composition of the sediment, including the heavy metals content thereof; and
(v) the abundance and diversity of benthic fauna and the content of selected aliphatic and aromatic hydrocarbons.

Annex VI provides for specific rules regarding discharges during the exploration and exploitation phases respectively.

During the exploration phase: the use of oil-based drilling mud or mud containing other hazardous or harmful substances shall be restricted to cases where it is necessary for geological, technical or safety reasons and only after a prior authorization of the appropriate national authority. In such cases appropriate measures must be taken and appropriate installations provided in order to prevent the discharges of such mud into the marine environment. Oil-based drilling mud and drilling cuttings arising from the use of oil-based drilling mud are not allowed to be discharged in the Baltic Sea Area, but must be taken ashore for final treatment or disposal in an environmentally acceptable way. The discharge of water-based mud and cuttings has to be authorized by the appropriate national authority. The content of the water-based mud must be proved to be of low toxicity before use. The discharge of cuttings arising from the use of water-based drilling mud is not permitted in specifically sensitive parts of the Baltic Sea Area, such as confined or shallow areas with limited water exchange and areas characterized by rare, valuable or particularly fragile ecosystems.

During the exploitation phase: all chemicals and materials shall be taken ashore and can be discharged only exceptionally the permit by the appropriate national authority in each individual operation. The discharge of production water and displacement water is prohibited unless the oil content is proved to be less than fifteen mg/l as measured by the method of analyzing and sampling to be developed by the Commission. The appropriate national authority may, if compliance with this limit value cannot be achieved by the use of the best available technology and best environmental practice, allow a higher limit value, which shall, however, be as low as possible but not exceed forty mg/l. The permitted discharge shall not, in any case, create any unacceptable effects on the marine environment. In order to benefit from the future development in cleaning and production technology the discharge permits shall be regularly reviewed by the appropriate national authority and the discharge limits shall be revised accordingly.

Annex VI contains furthermore the following regulations:

(i) each offshore unit shall have a pollution emergency plan approved in accordance with the procedure established by the appropriate national authority. The plan shall contain information on alarm and communication systems, organization of response measures, a list of propositioned equipment and a description of the measures to be taken at different types of pollution incidents;
(ii) the Contracting Parties shall ensure that abandoned, disused offshore units and accidentally wrecked offshore units are entirely removed and brought ashore under the responsibility of the owner and that disused drilling wells are be plugged;

(iii) the Contracting Parties shall continuously exchange information through the Commission on the location and nature of all planned or accomplished offshore activities and on the nature and amounts of discharges as well on contingency measures that are undertaken.

Note. At the moment hardly any offshore activity takes place in the area covered by the Convention, and consequently Annex VI and a few other regulations and provisions do not find practical application. However this will change as soon as the construction starts of a major offshore gas pipeline called Nordstream that will carry Russian gas to Germany. This pipeline is projected to run through the middle of the Baltic Sea. Already in the planning stage the rules of the Convention should be taken account of.

9.3.6 UNEP Conventions on the Prevention of Pollution

OSPARCON and Baltic Sea Area Convention are not the only regional convention concerning the prevention of pollution which inter alia covers the matter of pollution resulting or likely to result from offshore petroleum operations. Several of other conventions have been developed under the Regional Sea programmes of the UN Environment programme (UNEP). Any UNEP programme is aimed at the adoption of an Action Plan and a convention containing the legal framework. Any convention is accompanied and complemented by Annexes and Protocols covering specific types of pollution.

Examples of such UNEP inspired regional conventions are:

(i) the 1976 Barcelona Convention and its Protocols (Convention for the Protection of the Mediterranean Sea against Pollution, accompanied by three Protocols, viz. a Dumping Protocol (Protocol for the Prevention of Pollution of the Mediterranean Sea by Dumping from Ships and Aircraft) with three Annexes, Protocol concerning cooperation in combating Pollution of the Mediterranean Sea by Oil and other Harmful Substances in Cases of Emergency and a Protocol concerning Mediterranean Specially Protected Areas;

(ii) the 1978 Kuwait Convention (Kuwait Regional Convention for Cooperation on the Protection of the Marine Environment from Pollution) with an Action Plan and a single Protocol (Protocol concerning Regional Cooperation in Combating Pollution by Oil and other Harmful Substances in Cases of Emergency);

(iii) the 1982 Red Sea Convention (Regional Convention for the Conservation of the Red Sea and the Gulf of Aden Environment) with an Action Plan and a single Protocol (Protocol concerning Regional Cooperation in combating Pollution by Oil and other Harmful Substances in Cases of Emergency);

(iv) the 1983 Caribbean Convention currently accompanied by a single Protocol concerning cooperation in combating oil spills in the wider Caribbean region, no Action Plan is referred to in the Convention.

The aforementioned Kuwait Convention, its Protocol and Action Plan were signed on 24 April 1978 by the governments of the Gulf States. Offshore petroleum operations take place in the area of the Convention, but only on a modest scale. The real importance of the Convention rests on the fact that it concerns a sea area through which a huge tonnage of crude oil is being shipped. Therefore shipping and accidents with ships were seen as the main possible sources of pollution for the prevention and abatement of which the Convention had to make provisions. As customary in these cases, the Convention is only providing a framework within which procedures and standards can be developed for the purpose of preventing, abating and combating pollution of the marine environment in the Convention Area. The so-developed procedures and standards are to be used by the Contracting Parties in taking measures and establishing national standards, laws and regulations for this purpose. The Contracting Parties are obliged to endeavour to harmonize their national policies and to cooperate with the competent international, regional and sub-regional organizations to establish and adopt regional standards, recommended practices and procedures. In the case of pollution caused by intentional or accidental discharges from ships the Contracting Parties are obliged to ensure effective compliance in the Convention Area with applicable international rules (e.g., MARPOL).

9.3.7 The 1989 IMO Guidelines and Standards on Removal

Article 60.3 of the 1982 Convention on the Law of the Sea requires that any installations or structures which are abandoned or disused shall be removed to ensure safety of navigation, taking into account any generally accepted international standards established in this regard by the competent international organization. In 1984 the IMO was asked by the Oslo Commission to prepare such standards. In 1989 IMO published its Guidelines and Standards on Removal. These Guidelines are applicable to installations in the EEZ or on the continental shelf and to be taken into account by coastal States when considering on a case by case basis whether or not such an installation or structure or parts of them should be allowed to remain on the seabed. In accordance with the Guidelines (which are reminiscent of the provisions of the 1972 London Dumping Convention and the latter's Annex III) in each particular case the following factors should be considered:

(i) the safety of navigation;
(ii) the rate of deterioration;
(iii) the potential effect on the marine environment, including living resources;
(iv) the risk that the material will shift from its position; (v) the risk that the material will shift from its position;
(v) the costs, technical feasibility and risks of injury to personnel associated with removal; and

(vi) the determination of a new use, which would justify allowing the installation or parts thereof to remain on the seabed.

The Standards include the following directives:

(i) all abandoned or disused installations or structures standing in less than sevety-five metres of water and weighing less than four thousand metric tons in air, excluding the deck and the superstructure, should be entirely removed. This requirement will be extended from 1 January 1998 to include all installations standing in less than one hundred metres of water and weighing less than four thousand metric tons in air, excluding the deck and the superstructure, placed on the seabed on or after that date;
(ii) removal is to be performed in such a way as to cause no significant adverse effects upon navigation or the marine environment;
(iii) where an existing installation or structure, including one which should be entirely removed or a part thereof, will serve a new use, if permitted to remain wholly or partially in place on the seabed (such as enhancement of a living resource e.g., fish and shellfish) or where an existing installation or structure, other than one which should be entirely removed or part thereof, can be left there without causing unjustifiable interference with other uses of the sea, the coastal State may determine that the installation or structure may be left wholly or partially in place;
(iv) notwithstanding the above requirements for entire removal, where such removal is not technically feasible or would involve extreme cost or an unacceptable risk to personnel or the marine environment, the coastal State may determine that the installation need not be entirely removed;
(v) on or after 1 January 1998 no installation or structure is to be placed on any continental shelf or in any EEZ, unless the design and construction of the installation or structure is such that entire removal upon abandonment or permanent disuse would be feasible;
(vi) installations no longer serving their primary purpose and located in lanes used for international navigation, in deep draft sea lanes or routing systems which have been adopted by the IMO, should be entirely removed without exception;
(vii) any abandoned or disused installations or part thereof, which is left projecting above the sea, is to be adequately maintained to prevent structural failure. An unobstructed water column of no less than fifty-five metres should be provided above any partially removed installation or structure, which does not project above the sea surface, to ensure safety of navigation;
(viii) any structure which is not entirely removed from the seabed is to be clearly indicated on nautical charts and properly marked with aids to navigation;
(ix) the coastal State should ensure that legal tide to installations and structures, which have not been entirely removed from the seabed, is unambiguous and that responsibility for maintenance and the financial ability to assume liability for future damages are clearly established.

Chapter 10

Charters and Interstate Agreements Related to Petroleum

10.1 THE NEW INTERNATIONAL ECONOMIC ORDER

10.1.1 Background

In a national context and under national petroleum legislation the ownership of petroleum resources in situ within the land territory and territorial waters of a State is vested in the State concerned, except where private mineral rights are acknowledged, in which case the ownership of petroleum in situ is vested in the owner of the land on which or in the subsoil of which the petroleum reservoir lies. The issue of ownership or similar rights with respect to natural resources in situ on the continental shelf outside the territorial waters of a coastal State was for the first time raised and settled through a Proclamation of US president Truman on 28 September 1945. In his Proclamation the US president Truman stated that the government of the US regarded the natural resources of the subsoil and seabed of the continental shelf contiguous to the coasts of the US to appertain to the US subject to its jurisdiction and control. In an accompanying press release the continental shelf was described as submerged land which is contiguous to the continent and which is covered by no more than one hundred fathoms of water. The Truman Proclamation triggered a series of similar unilateral statements, proclamations and legislative actions of other coastal States, in particular in Latin America and the Middle East Gulf area. The issue was finally settled when the 1958 Convention on the Continental Shelf was adopted. Under this Convention it was determined that any coastal State has sovereign rights over natural resources located on its continental shelf. The latter was described similarly as was done in connection with the Truman Proclamation as the submarine area adjacent to the coast but outside the territorial waters up to a depth of two hundred metres.

Then, during the 1960s and 1970s the developing countries raised within the UN General Assembly the question of whether the sovereignty of the State over the country's natural resources (onshore as well as offshore) was permanent and inalienable and, if so, what the legal and economic consequences would be of such permanent sovereignty. The developing countries had raised this issue as part of their attempts to introduce a New International Economic Order (NIEO). The desire for a new ordering of the world economy arose out of disappointment with the then existing one, which was being perceived as to block the economic development of the many new States that had been formed as a result of the post-war de-colonization process. In particular the fact that those countries' natural resources were owned and exploited by foreign multinational corporations was much resented by the former. From the NIEO the developing countries expected that it would terminate the economic dominance of the western enterprises, mining enterprises in particular. At the same time it was realized that this could only be achieved if a country could nationalize the rights and other assets of such enterprise, if and when the country concerned would feel it necessary to do so. The developing countries saw in the doctrine of permanent and inalienable state sovereignty over natural resources the desired justification for taking nationalization measures and felt that this doctrine should therefore be at the core of the NIEO. Preliminary work by the United Nations on the concept of permanent sovereignty started with the establishment by GA Resolution 1314 (XIII) of 12 December 1958 of a Commission on Permanent Sovereignty over Natural Resources. On 14 December 1962 the UN General Assembly adopted, without significant changes to the text that was prepared by the Commission, a resolution on the subject (GA Resolution 1803 (XVII)). On the main subject, that of the permanent sovereignty the text did not contain anything extraordinary: the meaning of the sovereignty being permanent was not explained. However the resolution had the following to say about nationalization and its justification:

1. In cases where authorization is granted, the capital imported and the earnings on that capital shall be governed by the terms thereof, by the national legislation in force and by international law.
2. Nationalization, expropriation or requisitioning shall be based on grounds or reasons of public utility, security or the national interest which are recognised as overriding purely individual or private interests, both domestic and foreign. In such cases the owner shall be paid appropriate compensation, in accordance with the rules in force in the State taking such measures in the exercise of its sovereignty and in accordance with international law.

Although the UNGA resolution appears to exclude diplomatic protection (to which the developing countries were vehemently opposed), the resolution was nevertheless generally supported by the Member States, mainly because of the reference in both paragraphs to international law. Briefly said, customary international law, at it was at that time and still is at present, allows nationalization of property, provided the measures are taken in the public interest and under due process of law and further are accompanied by provisions for the payment of prompt and adequate compensation.

Even after having achieved the deletion of any reference to diplomatic protection, developing countries were still not satisfied. They objected to the incorporation of a reference to international law. According to these countries international law was the law of Western States and as such took no account of their different economic situation. Since the developing countries were faced with the fact that their economies were dominated by foreign multinational corporations, they felt hampered in pursuing a policy of nationalization or expropriation, if the conditions prescribed by international law would be applicable. Therefore, the developing countries wished to make their domestic law to the only guiding principle in taking nationalization and expropriation measures and in making provisions for payment of compensation. The debate around the issues of permanent sovereignty of States and of domestic versus international law in judging whether nationalization or expropriation was appropriate and whether the offered compensation was adequate reached its final stage in a number of GA Resolutions adopted in 1974. Resolution 3201 (S-VI) of 1 May 1974 contained a Declaration on the Establishment of a New International Economic Order, by which was meant a new system of international economic relations based on equity, sovereign equality and interdependence of interests of developed and developing countries. The Declaration was accompanied by a Programme of Action (GA Resolution 3201 (S-VI)). Both documents are legally non-binding but are expressing certain political intentions. Both Resolutions were adopted without voting. At the end of the year the Charter of Economic Rights and Duties of States was adopted (GA Resolution 3281 (XXIX) of 12 December 1974) in which the principles of the Declaration were worked out.

Note. The Declaration was accompanied by a Programme of Action (GA Resolution 3201 (S-VI)). Of interest to the petroleum industry are the statements in the Programme regarding raw materials.

1. Raw materials

All efforts should be made:

(a) To put an end to all forms of foreign occupation, racial discrimination, apartheid, colonial, neo-colonial and alien domination and the exploitation through the exercise of permanent sovereignty over natural resources.

(b) [T]o take measures for the recovery, exploitation, development, marketing and distribution of natural resources, particularly of developing countries, to serve their national interests, to promote collective self-reliance among them, and to strengthen mutually beneficial international economic cooperation with a view to bringing about the accelerated development of developing countries.

10.1.2 The 1974 Charter of Economic Rights and Duties of States

The Charter of Economic Rights and Duties of States was adopted by GA Resolution 3281 (XXIX) of 12 December 1974. The Charter was meant to be an effective instrument towards the establishment of the New International Economic Order as envisaged by the Declaration of 1 May 1974. Most Western States voted against or abstained from voting. The most interesting and far reaching provisions of the Charter concern the permanent sovereignty of States, the freedom of any State to choose its own political, social and cultural system, supervision of multinational corporations and the right to nationalize foreign property. These provisions are set out in Articles 1 and 2 of Chapter II: Economic Rights and Duties of States.

Article 1 of Chapter II reads:

> 1. Every State has the sovereign and inalienable right to choose its economic system as well as its political, social and cultural systems in accordance with the will of its people, without outside interference, coercion or threat in any form whatsoever.

Article 2 of Chapter II reads:

> 1. Every State has and shall freely exercise full permanent sovereignty, including possession, use and disposal, over all its wealth, natural resources and economic activities.
> 2. Each State has the right:
>   (a) To regulate and exercise authority over foreign investment within its national jurisdiction in accordance with its laws and regulations and in conformity with its national objectives and priorities. No State shall be compelled to grant preferential treatment to foreign investment;
>   (b) To regulate and supervise the activities of transnational corporations within its national jurisdiction and take measures to ensure that such activities comply with its laws, rules and regulations and conform with its economic and social policies. Transnational corporations shall not intervene in the internal affairs of a host State. Every State should, with full regard for its sovereign rights, cooperate with other States in the exercise of the right set forth in this subparagraph;
>   (c) To nationalize, expropriate or transfer ownership of foreign property, in which case appropriate compensation should be paid by the State adopting such measures, taking into account its relevant laws and regulations and all circumstances that the State considers pertinent. In any case, where the question of compensation gives rise to a controversy, it shall be settled under the domestic law of the nationalizing State and by its tribunals, unless it is freely and mutually agreed by all States concerned that other peaceful means be sought on the basis of sovereign equality of States and in accordance with the principle of free choice of means. (*Note*. The matter of compensation to be received in case of nationalization or expropriation is to be dealt with under domestic law, although it could be mutually agreed to refer the case to international arbitration. There is no explicit reference to international law but when the case in question in mutually agreement

> is submitted to international arbitration it could also be agreed to apply [the principles of] international law. Anyway the provisions of paragraph (c) must be seen as an improvement on a corresponding paragraph in the Declaration of 1 May 1974 in which no mention at all is made of compensation.)

Article 3 of Chapter II is of special interest to governments and the petroleum industry where the exploitation of a single, continuous petroleum reservoir that extends beyond and across an interstate boundary is being considered. This article reads as follows:

> In the exploitation of natural resources shared by two or more countries each State must cooperate on the basis of a system of information and prior consultation in order to achieve optimum use of such resources without causing damage to the legitimate interest of others.

When exercising their economic rights as outlined in Chapter II, States should be mindful to fulfil in good faith their international obligations as set out in the Charter of the United Nations. This principle is set out in the Charter's Chapter I, in which it is stated that economic as well as political and other relations among States shall be governed, inter alia, by the principle described in paragraph (j) viz. 'fulfilment in good faith of international obligations', and is meant to comprise all provisions of the UN Charter through the provisions of Article 33, Chapter IV, which article reads as follows:

> Article 33.1. Nothing in the present Charter shall be construed as impairing or derogating from the provisions of the Charter of the United Nations or actions taken in pursuance thereof. 2. In their interpretation and application, the provisions of the present Charter are interrelated and each provision should be construed in the context of the other provisions.

From Chapter I, paragraph (j) read in conjunction with Chapter IV Article 33, it can be concluded that a State is not completely free in pursuing its political and social policy but has to be mindful of its obligations under the Charter of the United Nations which include the obligation to protect, respect and maintain the human rights of its citizens.

Note. Since the Cold War in 1991 came to an end, Western States have got and used the political freedom to focus their attention on other States' performance of the latter's international obligations as formulated in the Charter of the United Nations with respect to protecting and respecting human rights and of maintaining and respecting international peace and security. To the extent non-Western States (with exception of course of the non-western political powers such as the Russian Federation, The People's Republic of China and the countries protected by them) are failing in these respects they run the risk that their sovereignty and territorial integrity or at least their economic situation might not be respected by the international community as a whole, represented by the Security Council of the United Nations, or in absence of unanimity in the Council by a group of interested Western States acting on their own account.

10.1.3 The Issue of Nationalization or Expropriation

Article 2.2 (c) of Chapter II reads as follows:

> (c) To nationalize, expropriate or transfer ownership of foreign property, in which case appropriate compensation should be paid by the State adopting such measures, taking into account its relevant laws and regulations and all circumstances that the State considers pertinent. In any case, where the question of compensation gives rise to a controversy, it shall be settled under the domestic law of the nationalizing State and by its tribunals, unless it is freely and mutually agreed by all States concerned that other peaceful means be sought on the basis of sovereign equality of States and in accordance with the principle of free choice of means.

This paragraph (c) has raised the question of whether customary international law is applicable to nationalization/expropriation measures. On this issue opinions differ. Within the UN system the permanent sovereignty of States is seen as a *conditio sine qua non* for national independence and economic self-sufficiency (but not as an excuse for violating human rights or disturbing international peace and security) and from this perspective compensation to be received in case of nationalization/expropriation is left to be dealt with under rules of domestic law, unless otherwise agreed. By contrast, it is felt by Western States that because the Charter of Economic Rights and Duties of States lack their support, it cannot be considered to derogate from customary international law (in nationalization matters). Without resolving this issue, many developing countries have since the adoption of this Charter entered with West-European countries into bilateral agreements for the promotion and protection of foreign investments. Those agreements incorporate a remarkable blending of the 'Western' rules of international law with respect to nationalization or expropriation of foreign investments and the principles of the Charter. In accordance with these bilateral agreements foreign investments may not be nationalized or expropriated, except for an internal public or social purpose, against prompt, adequate and effective compensation (fair value, payment of interest, freely transferable), and in conformity with the internal law. The legality of the expropriation and the amount of compensation must be established by due process of law in the territory of the nationalizing State.

When drafting the text of the European Energy Charter and the Charter Treaty implementing this Charter, the matter of how to define state sovereignty over natural resources and to provide for non-discriminatory treatment and protection of foreign investments had to be addressed again. Interestingly, in these documents state sovereignty is no longer described as being 'permanent' or 'inalienable'. In contrast therewith, an attempt has been made to find a balance between such sovereignty and giving non-discriminatory access to the natural resources of the host State. As far as nationalization, the compensation payable therefore and the further treatment of foreign investment were concerned, the presence of the OECD Member States was instrumental in incorporating in the European Energy Charter and Charter Treaty the more or less standard 'Western' rules known from the many bilateral investment agreements referred to above. As a result the UN Charter and the European Energy Charter have turned out to be very different documents. But the older UN Charter has already exercised its influence in the world of petroleum: without any doubt, the adoption of by developing States of the production sharing contract, as pioneered by Indonesia, can be ascribed to the application of the principles enshrined in the Charter.

10.2 THE EUROPEAN ENERGY CHARTER AND CHARTER TREATY

10.2.1 The 1991 European Energy Charter

The European Energy Charter was signed at The Hague on 17 December 1991 by the European Community and forty-nine countries. Those countries included the Member States of the European Community; the other Member States of the OECD, amongst them the United States, Canada, Japan and Australia; the countries of Central and Eastern Europe; the Baltic States; and almost all Member States of the CIS, including the Russian Federation.

The European Energy Charter is a political declaration by the signatory States which sets out the objectives, principles and actions which they will undertake in the fields of trade and investment in energy products and energy production. The stated objectives of the Charter are to improve security of energy supply and of maximizing the efficiency of production, conversion, transport, distribution and use of energy, to enhance safety and to minimize environmental problems, on an acceptable economic basis. Within the framework of State sovereignty and sovereign rights over energy resources the signatory States declare to undertake to promote the development of an efficient energy market throughout Europe, and a better functioning global market, in both cases based on the principle of non-discrimination and on market-oriented price formation, taking due account of environmental concerns. Finally, the signatory States declare to be determined to create a climate favourable to the operation of enterprises and to the flow of investments and technologies by implementing market principles in the field of energy.

For the European Community that at the time was already engaged in strenuous efforts to create an internal market for natural gas the Energy Charter offered the possibility, or so it was hoped, to integrate the gas fields of Eastern Europe, the Russian Federation and the other CIS Member States (which at the time had not yet reached the size and importance they have today) into the West-European gas market. Apart from this very ambitious goal (to say nothing about the aspiration to achieve a 'better functioning of the global market') the European Community hoped by means of the Charter to get a grip on the operational safety of present and future nuclear power stations located in the eastern region. In this context it is further stated in the Charter that the signatory States will take action inter alia in the field of energy efficiency and environmental protection which will imply inter alia achieving and maintaining a high level of nuclear safety and ensuring effective cooperation in this field. Western capital was encouraged to make itself available for the further development of those eastern energy resources, for export as well as for domestic use. It was hoped that the inflow of this capital would ensure that such development would take place in an environmentally acceptable manner, specifically with respect to the operation and maintenance of nuclear power stations. The underlying reason for this emphasis on nuclear safety was to prevent another nuclear disaster as had occurred in April 1986 in the nuclear power station at Chernobyl.

The Charter is a political declaration, which does not commit its signatories, but it was the intention to negotiate and conclude an Energy Charter Treaty to give effect to its objectives and principles in a legally binding form. The intended Energy Charter Treaty was signed in December 1994. The Treaty has as its stated purpose to establish a legal framework in order to promote long-term cooperation in the energy field. The Charter Treaty and its provisions are described below.

10.2.2 The 1994 Energy Charter Treaty

The signatories of the 1991 European Energy Charter had committed themselves to negotiate in good faith and conclude a Charter Treaty in order to place the commitments contained in the Charter on a secure and binding international basis. The aimed for Treaty was established and signed in Lisbon on 17 December 1994.

The Treaty's main subjects and corresponding commitments in the way as described in its articles are set out below.

10.2.2.1 Article 1: Definitions

'Energy Materials and Products' means the items included in Annex EM.

As defined in Annex EM 'Energy Materials and Products' include among other energy items fossil fuels, such as coal, oil, oil products, natural gas and other gaseous hydrocarbons whether liquefied or in gaseous state. Also included are the basic energy materials meant for nuclear energy, amongst which uranium ores as well as uranium enriched in U235 and plutonium, and further spent fuel elements of nuclear reactors, and electricity, the energy product delivered by (nuclear) power stations.

'Returns' means the amounts derived from or associated with an investment, irrespective of the form in which paid, including profits, dividends, interest, capital gains, royalty payments, management, technical assistance or other fees, and payments in kind.

10.2.2.2 Article 3: International Markets

The Contracting Parties shall work to promote access to international markets on commercial terms and generally to develop an open and competitive market for Energy Materials and Products.

10.2.2.3 Article 6(1): Competition

Each Contracting Party shall work to alleviate market distortions and barriers to competition. In Economic Activity in the Energy Sector:

10.2.2.4 Article 7: Transit

Each Contracting Party shall take the necessary measures to facilitate the transit of energy materials and products consistent with thc principle of freedom of transit and without distinction as to origin, destination or ownership of such energy materials and products or discrimination as to the pricing on the basis of such distinctions, and without imposing any unreasonable delays, restrictions or charges. 'Transit' is defined in this article as the carriage through the area of a Contracting Party, or to or from port facilities in its area, of Energy Materials and Products originating in the area of another contracting party and destined for the area of a third Contracting Party.

10.2.2.5 Article 8: Transfer of Technology

The Contracting Parties agree to promote access to and transfer of energy technology on a commercial and non-discriminatory basis, to assist effective trade in Energy Materials and Products and investment and to implement the objectives of the Charter subject to their laws and regulations and to the protection of Intellectual Property rights.

10.2.2.6 Article 9: Access to Capital

Each Contracting Party shall endeavour to promote conditions for access to its capital market by companies and nationals of other contracting parties, for the purpose of financing trade in Energy materials and Products and for the purpose of investment in Economic Activity in the Energy sector in the Areas of those other Contacting Parties, on a basis no less favourable than that which it accords in like circumstances to its own companies and nationals or companies and nationals of any other Contracting Party or any third State, whichever is the most favourable.

10.2.2.7 Article 10: Promotion, Protection and Treatment of Investments

Each Contracting Party shall encourage and create stable, equitable, favourable and transparent conditions for investors of other Contracting Parties to Make Investments in its Area. Such conditions shall include a commitment to accord at all times to investments of investors of other Contracting Parties fair and equitable treatment. Such investments shall also enjoy the most constant protection and security and no Contracting Party shall in any way impair by unreasonable or discriminatory measures their management, maintenance, use, enjoyment or disposal. In no case shall such investments be accorded treatment less favourable than that required by international law, including treaty obligations. Each Contracting Party shall observe any obligations it has entered into with an investor or an investment of an investor of any other Contracting party. Each Contracting Party shall endeavour to accord to investors of other Contracting Parties, as regards the Making of Investments in its Area, the Treatment as defined in this article. As defined in this article 'Treatment' means treatment accorded by a Contracting Party which is no less favourable than that which it accords to its own investors or to investors of any other Contracting Party or any third State, whichever is the most favourable. Each Contracting Party shall accord to investments in its Area of investors of another Contracting Party, and their related activities, including management, maintenance, use, enjoyment or disposal, treatment no less favourable than that which it accords to investments of its own investors or of the investors of any other Contracting party or any third State and their related activities including management, maintenance, use, enjoyment or disposal, whichever is the most favourable.

10.2.2.8 Article 11: Key Personnel

A Contracting Party shall examine in good faith requests of investors of another Contracting Party and key personnel who are employed by such investors or by investments of such investors, to enter and remain temporarily in its Area to engage in activities connected with the making or the development, management, maintenance, use, enjoyment or disposal of relevant investments, including the provision of advice or key technical services.

10.2.2.9 Article 12: Compensation for Losses

An investor of any Contracting Party which suffers a loss with respect to any investment in the Area of another Contracting Party owing to war or other armed conflict, state of national emergency, civil disturbance, or other similar event in that Area, shall be accorded by the latter Contracting Party, treatment, as regards restitution, indemnification, compensation or other settlement, which is the most favourable of that which that Contracting Party accords to any other investor, whether its own investor, the investor of any other Contracting Party, or the investor of any third State. Without prejudice to the foregoing, an investor of a Contractor Party which, in any of the situations referred to above, suffers a loss in the Area of another Contracting Party resulting from (a) requisitioning of its investment or part thereof by the latter's forces or authorities, or (b) destruction of its investment by the said forces or authorities, which was not required by the necessity of the situation, shall be accorded restitution or compensation which in either case shall be prompt, adequate and effective.

10.2.2.10 Article 13: Expropriation

Investments of investors of a Contracting Party in the Area of any other Contracting Party shall not be nationalized, expropriated or subjected to a measure or measures having effect equivalent to nationalization or expropriation (hereinafter referred to as 'Expropriation'), except where such Expropriation is:

(i) for a purpose which is in the public interest;
(ii) not discriminatory;
(iii) carried out under due process of law; and
(iv) accompanied by the payment of prompt, adequate and effective compensation.

With reference to (iv). The compensation shall amount to the fair market value of the investment expropriated at the time immediately before the Expropriation or impending Expropriation became known in such a way as to affect the value of the investment (hereinafter referred to as the 'Valuation Date'). The fair market value shall at the request of the investor be expressed in a Freely Convertible Currency on the basis of the market rate of exchange existing for that currency on the Valuation Date. Compensation shall also include interest at a commercial rate established on a market basis from the date of expropriation until the date of payment. The investor affected shall have a right to prompt review under the law of the Contracting Party making the Expropriation, by a judicial or other competent and independent authority of that Contracting Party, of its case, of the valuation of its investment, and of the payment of compensation, in accordance with the principles set out in the foregoing.

10.2.2.11 Article 14: Transfers Related to Investments

Each Contracting Party shall with respect to investments in its Area of investors of any other Contracting Party guarantee the freedom of transfer into and out of its Area, including the transfer of:

(i) the initial capital plus any additional capital for the maintenance and development of an investment;
(ii) returns;
(iii) payments under a contract, including amortization of principal and accrued interest payments pursuant to a loan agreement;
(iv) unspent earnings and other remuneration of personnel engaged from abroad;
(v) proceeds from the sale or liquidation of all or any part of an investment;
(vi) payments of compensation pursuant to Articles 12 and 13.

Transfers shall be effected without delay and in a freely convertible currency, but a Contracting Party may protect the rights of creditors and ensure compliance with laws on the issuing, trading and dealing in securities and the satisfaction of judgments in civil, administrative and criminal and adjudicatory proceedings through the equitable, non-discriminatory and good faith application of its laws and regulations.

10.2.2.12 Article 18: Sovereignty over Energy Resources

The Contracting Parties recognize state sovereignty and sovereign rights over energy resources. They reaffirm that these must be exercised in accordance with and subject to the rules of international law. Without affecting the objectives of promoting access to energy resources, and exploration and development thereof on a commercial basis, the Treaty shall in no way prejudice the rules in Contracting Parties governing the system of property ownership of energy resources. Each State continues to hold in particular the rights to decide the geographical areas within its Area to be made available for exploration and development of its energy resources and the rate at which they may be depleted or otherwise exploited, to specify and enjoy any taxes, royalties or other financial payments payable by virtue of such exploration and exploitation and to regulate the resource conservation and the environmental and safety aspects of such exploration, development and reclamation within its Area, and to participate in such exploration and exploitation, inter alia, through direct participation by the government or through state enterprises. The Contracting Parties undertake to facilitate access to energy resources inter alia by allocating in a non-discriminatory manner on the basis of published criteria authorizations, licences, concessions and contracts to prospect and explore for or to exploit or extract energy resources.

Note. The principles regarding the essence and scope of state sovereignty over energy resources as set out in this article can also be found (albeit restricted to the prospecting, exploration and production of hydrocarbons in the EC Member States) in the EC Directive 94/22.EC of 30 May 1994 (see Chapter 6).

10.2.2.13 Article 19: Environmental Aspects

Taking into account its obligations under those international agreements concerning the environment to which it is a party, each Contracting Party shall strive to minimize in an economically efficient manner harmful environmental impacts occurring either within or outside its Area from all operations within the Energy Cycle in its Area, taking proper account of safety. In doing so each Contracting Party shall act in a cost-effective manner. In its policies and actions each Contracting Party shall strive to take precautionary measures to prevent or minimize environmental degradation. The Contracting Parties agree that the polluter in the Areas of Contracting Parties, should, in principle, bear the cost of pollution, including trans-boundary pollution, with due regard to the public interest and without distorting investment in the Energy Cycle or international trade. Accordingly, Contracting Parties shall inter alia:

(1) take account of environmental considerations throughout the formulation and implementation of their energy policies;
(2) promote market-orientated price formation and a fuller reflection of environmental costs and benefits throughout the Energy Cycle; and

(3) have particular regard to improving Energy Efficiency, to developing and using renewable energy sources, to promoting the use of cleaner fuels and to employing technologies and technological means that reduce pollution.

For the purposes of this article 'Energy Cycle' means the entire energy chain, including activities related to prospecting for, exploration, production, conversion, storage, transport, distribution and consumption of the various forms of energy, and the treatment and disposal of wastes, as well as the decommissioning, cessation or closure of these activities, minimizing harmful Environmental Impacts.

10.2.2.14 Article 20: Transparency

Laws, regulations, judicial decisions and administrative rulings of general application which affect trade in Energy materials and products are among the measures subject to the transparency disciplines of the GATT and relevant related instruments. Laws, regulations, judicial decisions and administrative rulings of general application made effective by any Contracting Party, and agreements in force between Contracting Parties, which affect other matters covered by this Treaty shall also be published promptly in such a manner as to enable Contracting Parties and investors to become acquainted with them.

Note. The matter of transparency of rules and regulations is also dealt with in the aforementioned EC Directive 94/22/EC of 30 May 1994.

10.2.2.15 Article 21: Taxation

Nothing in this Treaty shall apply to impose obligations with respect to Taxation Measures of the Contracting Parties, except that the provisions imposing national treatment obligations or 'most favoured nation' obligations shall apply to Taxation Measures of the Contracting Parties other than those on income or on capital. Article 13 shall apply to taxes. Whenever an issue arises under Article 13, to the extent it pertains to whether a tax constitutes an expropriation or whether a tax alleged to constitute an expropriation is discriminatory, the investor or the Contracting Party alleging expropriation shall refer the issue of whether the tax is an expropriation or whether the tax is discriminatory to the relevant Competent Tax Authority, which must strive to resolve the issues so referred.

10.2.2.16 Article 22: State and Privileged Enterprises

Each Contracting Party shall ensure that any state enterprise which it maintains or establishes shall conduct its activities in relation to the sale or provision of goods and services in its Area in a manner consistent with the Contracting Party's obligations under Part III (Articles 10 through 17) of this Treaty. No Contracting Party shall encourage or require such a state enterprise to conduct its activities in its Area in a manner inconsistent with the Contracting Party's obligations under other provisions of this Treaty.

Each Contracting Party shall ensure that if it establishes or maintains a state entity and entrusts the entity with regulatory administrative or other governmental authority such entity shall exercise that authority in a manner consistent with the Contracting Party's obligations under this Treaty. No Contracting Party shall encourage or require any entity to which it grants exclusive or special privileges to conduct its activities in its Area in a manner inconsistent with the contracting party's obligations under this Treaty.

10.2.2.17 Article 26: Settlements of Disputes between an Investor and a Contracting Party

Disputes between a Contracting Party and an investor of another Contracting Party relating to an investment of the latter in the Area of the former, which concern an alleged breach of an obligation of the former under Part III (Articles 10 through 17) shall, if possible, be settled amicably. If such dispute can not be settled amicably the investor may choose to submit it for resolution to the courts or administrative tribunals of the Contracting Party to the dispute or in accordance with any applicable, previously agreed dispute settlement procedure; or to international arbitration. In this context and for this purpose each Contracting Party gives its unconditional consent to the submission of a dispute to the International Centre for Settlement of Investment Disputes (ICSID) or a sole arbitrator or ad hoc arbitration tribunal established under the Arbitration Rules of the United Nations Commission on International Trade Law (UNCITRAL) or to arbitral proceeding under the Arbitration Institute of the Stockholm Chamber of Commerce. A tribunal set up as set out hereabove shall decide the issues in dispute in accordance with this Treaty and applicable rules and principles of international law. In some instances exceptions in this regard are allowed to Contracting Parties which are listed in Annex IA or Annex ID.

Note. The Charter Treaty is in many respects a sophisticated legal document that sought to form a common European energy market on the basis of the EC's internal market for energy. It incorporated the principles of international law and those of bi-lateral agreements concerning the protection of foreign investments. In essence the makers of the Treaty tried to link the advanced legal systems of the Western participating States with the less advanced legal systems generally prevailing in the eastern countries. This difference in sophistication made it necessary to introduce transitional arrangements for the benefit of the latter as long as and to the extent their legal systems were not brought up to Western standards. The Treaty itself was overtaken by political events: first the US left the Charter, and then in the following years almost all East European countries joined the EU having the result that the transitional arrangements with respect to these eastern countries could disappear. But a far more fundamental event which in fact diminished the value of the Treaty, at least as far as natural gas matters are concerned, was and is Russia's refusal to ratify the Treaty. Considering that Russia has become EU's major supplier of pipeline gas and will in the near future become a major supplier of LNG to Japan the Treaty would have offered a perfect legal framework for encouraging European (and Japanese) capital and technology to participate within a legally protected framework in these developments in the natural gas trade and production.

10.3 INTERSTATE JOINT DEVELOPMENT AGREEMENTS

10.3.1 Background

If with respect to a prospective area the jurisdiction over petroleum and other mineral resources in situ is claimed by two or more States, the States involved will be frustrated in their attempts to issue petroleum authorizations for that particular area. In the light of a jurisdictional dispute, no entity is likely to apply for the authorizations put on offer not wishing to run the risk that the granting of the authorization and the authorization itself is challenged by the other State.

In the matter of jurisdiction over petroleum and other mineral resources in situ distinction should be made between jurisdictional disputes concerning land based mineral resources and those concerning the mineral resources of the continental shelf and the deep-seabed. The jurisdiction with respect to land-based minerals is part and parcel of the jurisdiction with respect to land territory, on the surface or in the subsoil of which the minerals are located. Disputes between States regarding such jurisdiction have to be solved on the basis of historic rights and claims. Throughout history territorial land boundaries have been fixed after military conflict and in accordance with the stipulations of the ensuing peace treaty. In less populated or empty areas (e.g., Antarctica) boundary claims are usually asserted with less violence, but suffer mostly from a lack of documentary evidence. Offshore jurisdiction is derived from territorial jurisdiction. Only after territorial jurisdiction has been fixed, offshore jurisdiction and disputes with regard thereto can be addressed. In many instances, the matter of offshore jurisdiction has been settled in interstate boundary treaties. Such treaties have in the past been made between the coastal States involved through international arbitration or through the International Court of Justice within the framework of the 1958 Convention on the Continental Shelf. After the 1982 Convention on the Law of the Sea came into force use can be made of the rules for delimitation of the EEZ or continental shelf contained therein. In particular, the attribution of a 'minimum' continental shelf has solved the problem caused by the presence on the continental shelf of deep troughs, such as the Okinawa trough in the Southern East China Sea and the Timor trough situated between East Timor (formerly part of Indonesia) and Northern Australia.

With respect to the mineral resources of the deep seabed the 1982 Convention (Part XI) made an important contribution to preventing any conflict regarding the jurisdiction over these resources by creating the Area and by proclaiming the mineral resources thereof to be the common heritage of mankind. Furthermore by including a definition of what constitutes an island (Article 121), thereby excluding rocks which cannot sustain human habitation or economic life of their own, the Convention made another contribution to preventing or at least narrowing down jurisdictional conflicts.

Note. Because islands have been accorded an EEZ and continental shelf the jurisdiction over an island is very valuable and important if the area around the island is thought or has proven to be a prospective area. By excluding rocks which cannot sustain human habitation or economic life of their own, from this privilege jurisdictional disputes about such reefs and rocks, will be easier to settle because there is much less at stake.

States parties to a boundary dispute, be it in relation to an onshore or an offshore area, may also decide, in view of the complexities of the case or because of the paucity of the documentary evidence, not to try and fix the boundary in any definitive manner, but to conclude instead an agreement specifically dealing with the exploration for and exploitation of the mineral resources located or expected to be located on the seabed or in the subsoil of the contested area. Usually such a special agreement is designed to allow any interested person access to and the right to develop these mineral resources. As a matter of fact, this approach is recommended under the 1982 Convention with respect to disagreements between States about the delimitation of the EEZ or the continental shelf (see Articles 74.3 [EEZ] and 83.3 [continental shelf]).

Despite all solutions and settlements achieved and apart from Antarctica and the Area created by the 1982 Convention on the Law of the Sea, there still remain large areas of petroleum prospectivity, where no petroleum operations can take place due to boundary claims raised by different States. These claims and resulting disputes concern the sovereignty with respect to islands (e.g., the northern part of the Persian Gulf and the area of the South China Sea), the delimitation of EEZ's or continental shelves between adjacent or opposite coastal States (e.g., the North Pole area, involving in particular Canada and Russia; Norway/Russia; China/Vietnam; China/Japan; Greece/Turkey) or the territorial limits of two adjacent States (e.g., Yemen/Saudi Arabia; Yemen/Oman). Even after the territorial jurisdiction over an island has been settled disputes may arise concerning the extent of offshore jurisdiction attributable to the island in case the island is located very closely to the mainland of the opposite coastal State, e.g., the Anglo-Normandic islands in the bay of Granville (France) or the Greek Islands on the Aegean coast of Turkey. As far as the British Channel Islands were concerned, the United Kingdom and France, after agreeing to and setting up an international tribunal, managed to work out their own solution, which entailed the treatment of the islands as enclaves on the French continental shelf (judgment of 25 July 1977).

10.3.2 The 1959 Treaty on Antarctica and the 1991 Protocol

The continent by the name of Antarctica had since the beginning of the twentieth century been subject to territorial claims put forwards by seven States, viz. Australia, New Zealand, Norway, Argentina, Chile, France and the United Kingdom. Australia's territorial claim was the largest, covering about two fifth of the continent. France's claim was the smallest. The claims of the United Kingdom, Chile and Argentina overlap. All territorial claims are wedge-shaped. One segment remains unclaimed. Antarctica received considerable scientific attention during the International Geophysical Year of 1958. Scientists from twelve interested States got the opportunity to participate in a programme that was specifically developed for the continent. These scientific activities prompted US President Eisenhower to propose in May 1958 to the interested States, among which the seven States who had made and held territorial claims, to conclude a treaty designed to preserve the continent as an international laboratory for scientific research and to ensures that it be used only for peaceful purposes. This initiative led to the signing on 1 December 1959 in Washington of the Treaty on Antarctic which entered into force on 23 June 1961. The main provisions of the Treaty are laid down in the following articles.

Article I: 1. Antarctica shall be used for peaceful purposes only. There shall be prohibited, inter alia, any measure of a military nature, such as the establishment of military bases and fortifications, the carrying out of military maneuvers, as well as the testing of any type of weapons. 2. The present Treaty shall not prevent the use of military personnel or equipment for scientific research or for any other peaceful purpose.

Article IV: 1. Nothing contained in the present Treaty shall be interpreted as (a) a renunciation by any Contracting Party of previously asserted rights of or claims to territorial sovereignty in Antarctica; (b) a renunciation or diminution by any Contracting party of any basis of claim to territorial sovereignty in Antarctica which it may have whether as a result of its activities or those of its nationals in Antarctica, or otherwise; (c) prejudicing the position of any Contracting State as regards its recognition or non-recognition of any other State's right of or claim or basis of claim to territorial sovereignty in Antarctica. 2. No acts or activities taking place while the present Treaty is in force shall constitute a basis for asserting, supporting or denying a claim to territorial sovereignty in Antarctica or create any rights of sovereignty in Antarctica. No new claim, or enlargement of an existing claim, to territorial sovereignty in Antarctica shall be asserted while the present Treaty is in force.

Article V: 1. Any nuclear explosions in Antarctica and the disposal there of radioactive waste material shall be prohibited. 2. In the event of the conclusion of international agreements concerning the use of nuclear energy, including nuclear explosions and the disposal of radioactive waste material, to which all of the Contracting Parties whose representatives are entitled to participate in the meetings provided for under Article IX are parties, the rules established under such agreements shall apply in Antarctica.

Article VI: The provisions of the present Treaty shall apply to the area south of 60 degrees South Latitude including all ice shelves but nothing in the present Treaty shall prejudice or in any way affect the rights, or the exercise of the rights, of any State under international law with regard to the high seas within that area. (*Note*. There is no reference to the continental shelf of the continent in terms of the 1958 Convention on the continental shelf.)

Article IX: 1. Representatives of the Contracting Parties named in the preamble of the present treaty shall meet . . . at suitable intervals and places for the purpose of exchanging information, consulting together on matters of common interest pertaining to Antarctica, and formulating and considering and recommending to their Governments measures in furtherance of the principles and objectives of the Treaty . . .

2. Each Contracting Party which has become a party to the present Treaty by accession under Article XIII shall be entitled to appoint representatives to participate in the meetings referred to in paragraph 1 of the present Article, during such time as that Contracting Party demonstrates its interest in Antarctica by conducting substantial scientific research activity there such as the establishment of a scientific station or the despatch of a scientific expedition.

Article XII: 1. (a) the present Treaty may be modified or amended at any time by unanimous agreement of the Contracting Parties whose representatives are entitled to participate in the meetings provided for under Article IX. Any such modification or amendment shall enter into force when the depositary Government has received notice from all such Contracting Parties that they have ratified it. (b) Such modification or amendment shall thereafter enter into force as to any other Contracting Party when notice of ratification by it has been received by the depositary Government. Any such Contracting Party from which no notice of ratification is received within a period of two years from the date of entry into force of the modification or amendment in accordance with subparagraph 1 (a) of this Article shall be deemed to have withdrawn from the present Treaty on the date of expiration of such period. 2. (a) If after the expiration of thirty years from the date of entry into force of the present treaty, any of the Contracting Parties whose representatives are entitled to participate in the meetings provided for under Article IX so requests . . . a Conference of all the Contracting Parties shall be held to review the operation of the Treaty. (b) Any modification or amendment to the present Treaty which is approved at such a Conference by a majority of the Contracting Parties there represented, including a majority of those whose representatives are entitled to participate in the meetings provided for under Article IX . . . and shall enter into force in accordance with the provisions of paragraph 1 of the present Article.

Article XIII: 1. The present Treaty shall be subject to ratification by the signatory States. It shall be open for accession by any State which is a Member of the United Nations or by any other State which may be invited to accede to the Treaty with the consent of all the Contracting Parties whose representatives are entitled to participate in the meetings provided for under Article IX of the Treaty.

In July 1981 a Special Consultative Meeting on Antarctic mineral resources was convened at Buenos Aires to consider the work to be done on a mining regime for the Treaty area. After seven years of negotiations the text of a Convention on the Regulation of Antarctic Mineral Resource Activities (referred to as the Wellington Convention) was adopted on 2 June 1988 at the meeting of the Consultative Parties (Wellington, 2 May to 2 June 1988) and declared open for signature at Wellington on 25 November 1988. Of great interest is the way in which the 'area of application' of the Convention is described in Chapter I, Article 5. Said Article 5 reads as follows.

1. This Convention shall, subject to paragraphs 2, 3, and 4 below, apply to the Antarctic Treaty area.
2. Without prejudice to the responsibilities of the Antarctic Treaty Consultative Parties [defined in Chapter I, Article 1 as the Contracting Parties entitled to appoint representatives to participate in the meetings referred to in Article IX of the Treaty] under the Antarctic Treaty and measures pursuant to it, the Parties agree that this Convention shall regulate Antarctic mineral resource activities which take place on the continent of Antarctica and all Antarctic islands, including all ice shelves, south of 60 degrees south latitude and in the seabed and subsoil of adjacent offshore areas up to the deep seabed.
3. For the purposes of this Convention 'deep seabed' means the seabed and subsoil beyond the geographic extent of the continental shelf as the term continental shelf is defined in accordance with international law.

Note. By including the continental shelf of Antarctica in the area of application the Draft Convention extended its reach to areas north of 60 degrees south latitude, leaving no prospective area free for non-regulated mining activities except for those directed at the deep seabed. The latter however were and are covered by the rules of Part XI of the 1982 Convention on the Law of the Sea.

Within a year it became clear that due to opposition from France and Australia the Convention would never enter into force and for the time being remain a Draft Convention. Instead, the representatives agreed at their meeting in May 1991 in Madrid to adopt a supplemental Protocol on Environmental Protection to the Treaty. The area of application of the Protocol is the Antarctic Treaty area which is the area defined as such in Article VI of that Treaty. Contrary to the Draft Convention any reference to the continental shelf of the continent is left out. Article 1 copies the definition of 'Consultative Parties' as given in Chapter I, Article 1 of the Draft Convention. Mining matters are dealt with in Article 7. Said article simply reads: 'Any activity relating to mineral resources, other than scientific research, shall be prohibited'.

Note. Since the prohibition laid down in Article 7 is only applicable with respect to the Treaty area, it means that mining activities on the continental shelf of Antarctica must be considered to be allowed provided these activities take place north of 60 degrees south latitude and from a mining base similarly situated.

The Protocol may be modified or amended as set out in Article 25 as follows:

> 1. The Protocol may be modified or amended at any time in accordance with the procedures set forth in Article XII (1) (a) and (b) of the Treaty. 2. If, after the expiration of 50 years from the date of entry into force of this Protocol, any of the Antarctic Treaty Consultative Parties so requests . . . a conference shall be held as soon as practicable to review the operation of this Protocol. 3 . . . 4 . . . 5 (a) With respect to Article 7, the prohibition on Antarctic mineral resources activities contained therein shall continue unless there is in force a binding legal regime on Antarctic mineral resource activities that includes an agreed means for determining whether, and if so, under which conditions, any such activities would be acceptable. This regime shall fully safeguard the interests of all States referred to in Article IV of the Antarctic Treaty and apply the principles thereof. Therefore, if a modification or amendment to Article 7 is proposed at a Review Conference referred to in paragraph 2, it shall include such a binding legal regime.

10.3.3 The 1962 Agreement between the Netherlands and Germany

10.3.3.1 Background

The contested area is formed by the estuary of the River Ems. In order to promote the development of the oil and natural gas reservoirs which were assumed to be present in the subsoil of the contested area, without awaiting a final determination of the boundary between the two States, the Netherlands and the Federal Republic of Germany signed on 14 May 1962 an Agreement for that purpose. In accordance with this Agreement a so-named Common Area (the Area) was created. This Area was bounded by the borders as respectively claimed by the two States. Furthermore a line (further referred to as the Line) was drawn more or less along the centre line (*thalweg*) of the river dividing the Area in two more or less equal parts. The significance of the Line is not explained, apparently because both State Parties did not want to prejudice the final fixing of a boundary.

Articles 4 through 10 of the Agreement are declared to be applicable to the oil and natural gas deposits present in the subsoil of the Area prior to the start of the production.

10.3.3.2 Main Provisions

The Agreement's main provisions are the following:

Article 4. 1. With respect to exploration and exploitation activities and installations used for these activities German law is applicable on the German side of the Line and Netherlands' law on the Netherlands' side of the Line. 2. Contracting Parties may in accordance with their own petroleum legislation grant petroleum rights which are valid for the whole Area, but the rights so granted and the rights already existing upon this Agreement entering into force may only be exercised in accordance with the provisions of this Agreement.

Note. On the Netherlands' side petroleum rights with respect to the whole Area were comprised in a single concession mainly covering land territory. On the German side two separate, independent concessions were granted which taken together were covering the whole Area.

Article 5. 1. The German concession holders on the one hand and the Netherlands' concession holder on the other are entitled to receive one half of the quantity of oil and gas that is produced (from the deposits present in the subsoil of the Area prior to the start of production). 3. Costs that can reasonably be attributed to the exploration and production of the aforesaid quantity are also equally divided.

Article 7. 1. The concession holders on the one side of the Line shall for the purpose exploration and production closely cooperate with the concession holders on the other side. To this end they shall exchange work programmes for the Area and their results. 2. For the purpose of their cooperation the concession holders shall conclude as soon as possible agreements of cooperation which will provide for the following matters: (i) the manner in which the oil and gas reserves are calculated and the outcome of such calculations; (ii) particulars about the sharing of production and costs; (iii) a procedure for the settlement of disputes. 4. The agreements of cooperation must to be submitted to the Contracting Parties for information.

10.3.3.3 Agreements of Cooperation

In accordance with the Agreement two agreements of cooperation, which are virtually identical, were concluded: each having on the one side the Netherlands' concession holder and on the other the two German concession holders. Under both agreements the Netherlands' concession holder was appointed as the operator. In accordance with the Article 7.2 of the Agreement the agreements of cooperation contain rules and procedures concerning the calculation of the quantity of natural gas originally present in the subsoil of the Area and of the quantity that actually can be recovered therefrom the latter always being a fraction of the former. Since no wells were drilled in the Area itself and the gas that was actually being produced in the wells located on the land territory of the Netherlands' concession could never be identified as being gas that originated from the reservoirs in the subsoil of the Area, the quantity of natural gas that could be recovered from said reservoirs and that had be shared in equal proportions between the concession holders had to be calculated. In this context the agreements of cooperation provide for the only solution possible which is:

(i) to estimate on the basis of the available data the size and characteristics (porosity, permeability) of the reservoirs in the subsoil of the Area and the quantity of the natural gas contained therein;
(ii) to estimate on the basis of available data the numerical value of the applicable recovery factor;
(iii) to calculate the total volume of natural gas (to be produced) to which each group is entitled on the basis of the estimates mentioned under (i) and (ii);
(iv) to agree on the period over which these estimated volumes should be produced and delivered by the operator to each group; and
(v) to agree on a time schedule for a regular (in fact annual) revision of the estimates and calculations.

It should be added that the annual revisions of the calculation of the quantity of natural gas recoverable from the Area had to be approved by the Contracting Parties.

Note. The natural gas deliveries from the operator to the two German concession holders have long since stopped. It was agreed between the parties and the two governments that a volume of 20 billion cubic metres had been over-delivered, for which the Netherlands' concession holder would have to be compensated. The matter of the amount of compensation to be paid has in the meantime been resolved by arbitration.

10.3.4 THE 1974 AGREEMENT BETWEEN JAPAN AND THE REPUBLIC OF KOREA CONCERNING JOINT DEVELOPMENT OF THE SOUTHERN PART OF THE CONTINENTAL SHELF ADJACENT TO THE TWO COUNTRIES

10.3.4.1 Background

The division of continental shelf adjacent to Japan and the Republic of Korea was an issue about Japan and the Republic of Korea disagreed, Japan wishing to apply the median line principle as described in Article 6 of the 1958 Convention on the Continental Shelf, whereas the Republic of Korea wanted to apply the natural prolongation principle as this was introduced and formulated by the International Court of Justice at The Hague in its Judgment of 20 February 1969 in the *North Sea Continental Shelf Cases* (see *Note* hereinafter). The matter came to the foreground in 1970 when Korea for the purpose of developing offshore mineral resources laid claim on parts of the continental shelf which in the view of the government of Japan, applying the median line principle, belonged to Japan. After protracted negotiations on the matter the two countries concluded two Agreements: the one Agreement establishing a boundary in the northern part of the continental shelf adjacent to the two countries (Agreement of 5 February 1974) and the other establishing a Joint Development Zone (JDZ) in the southern part of that area (Agreement of 30 January 1974).

In the Agreement of 5 February 1974 it was agreed to delimitate the northern part of continental shelf concerned by constructing a median line up to the intersection with latitude 32 degrees 57 minutes north. In the Agreement of 30 January 1974 which created the JDZ the latter was described as an area bounded by straight lines connecting the points defined in the Agreement (Article II). The northwesterly boundary of the JDZ, starting at the end point of the boundary established under the Agreement of 5 February 1974 (the northern boundary) and ending at the intersection with latitude 30 degrees 46 minutes north, is constructed in accordance with the median line principle and in fact a continuation in southern direction of the northern boundary.

10.3.4.2 Basic Provisions

Article II: This article defines and describes an area of 82,665 square km in the southern part of the continental shelf adjacent to both countries, which is designated as a Joint Development Zone (JDZ) for the purpose of the joint exploration and exploitation of petroleum resources in that area.

Article III: The JDZ may be divided into sub-zones, each of which shall be explored and exploited by concessionaires of both Parties.

Article IV: Each Party shall authorize one or more concessionaires with respect to each sub-zone. When one Party authorizes more than one concessionaire with respect to one sub-zone, all such concessionaires shall have an undivided interest and shall be represented by one concessionaire.

10.3.4.3 Main Provisions

Article V: Concessionaires of both Parties shall enter into an operating agreement to carry out jointly exploration and exploitation of petroleum resources in the JDZ. Such operating agreement shall provide inter alia for the following: (i) details relating to the sharing of petroleum resources and expenses in accordance with Article IX; (ii) designation of operator; (iii) treatment of sole risk operations; (iv) adjustment of fisheries interests; (v) settlement of disputes. The operating agreement and modifications thereof shall enter into force upon approval by the Parties.

Article VI: The operator shall be designated by agreement between concessionaires of both Parties.

Article IX: Concessionaires of both Parties shall be respectively entitled to an equal share of petroleum resources extracted in the JDZ. Expenses reasonably attributable to exploration and exploitation of such petroleum resources shall be shared in equal proportions between concessionaires of both Parties.

Article X: The concessionaires shall have the right of exploration and of exploitation. The duration of the exploration right shall be eight years from the date of entry into force of the operating agreement. The duration of the exploitation right shall be thirty years from the date of establishment of such right. Concessionaires of both Parties may apply to the respective Parties for an extension of an additional period of five years. When commercial discovery of petroleum resources is made during the period of exploration right, concessionaires of both Parties may apply to the respective Parties for the establishment of exploitation right. When the Parties recognize that commercial discovery is made, each Party may request its concessionaire concerned to present an application for the establishment of exploitation right.

Article XI: Concessionaires of both Parties shall be required to drill a certain number of wells during the period of exploration right in accordance with a separate arrangement to be made between the Parties.

Article XIII: Concessionaires of both Parties shall release twenty-five per cent of the original sub-zone concerned within three years, fifty per cent of such sub-zone within six years and seventy-five percent of such sub-zone within eight years from the date of entry into force of the operating agreement. The size, shape and location of the area to be released and the time of release shall be determined by agreement between concessionaires of both Parties. If there is no area mutually proposed for release, concessionaires of both Parties shall release fifty per cent of the areas respectively proposed for release. A concessionaire may unilaterally release the total sub-zone concerned after two years have elapsed from the date of entry into force of the operating agreement.

Article XIV: Either Party may, subject to its own laws and regulations concerning the protection of concessionaires, cancel exploration right or exploitation right of its concessionaire after consultations with the other Party, if such concessionaire fails to discharge any of its obligations under this Agreement or the operating agreement.

Article XVI: In the application of the laws and regulations of each Party to petroleum resources extracted in the JDZ, the share of such petroleum resources to which concessionaires of one Party are entitled under Article IX shall be regarded as petroleum resources extracted in the continental shelf over which that Party has sovereign rights.

Article XVII: Neither Party (including local authorities) shall impose taxes or other charges upon the concessionaires of the other party with respect to: (i) exploration or exploitation activities in the JDZ; (ii) income derived from such activities; (iii) the possession of fixed assets in the JDZ necessary for carrying out such activities; or (iv) the sub-zone with respect to which such concessionaires are authorized. Each Party (including local authorities) may impose taxes and other charges upon its concessionaires with respect to: (i) exploration or exploitation activities in the JDZ; (ii) the possession of fixed assets in the JDZ necessary for carrying out such activities; or (iii) the sub-zone with respect to which such concessionaires are authorized.

Article XIX: Except where otherwise provided in this Agreement, the laws and regulations of one Party shall apply with respect to matters relating to exploration or exploitation of petroleum resources in the sub-zone with respect that Party has authorized concessionaires designated and acting as operators.

Article XX: The Parties shall agree on measures to be taken to prevent collisions at sea and to prevent and remove pollution of the sea resulting from activities relating to exploration or exploitation of petroleum resources in the JDZ.

Article XXI: When damage resulting from exploration or exploitation of natural resources in the JDZ has been sustained by nationals of either Party or other persons who are residents in the territory of either Party, actions for compensation for such damage may be brought by such nationals or persons in the court of one Party: (a) in the territory of which such damage has occurred, (b) in the territory of which such nationals or persons are resident or (c) which has authorized the concessionaire designated and acting as the operator in the sub-zone where the incident causing such damage has occurred. The court of one Party in which actions for compensation for such damage have been brought shall apply the laws and regulations of that Party.

Article XXIII: If any single geological structure or field of petroleum resources extends across any of the lines specified in paragraph 1 of Article II (*Note*. Meant here are the straight lines bounding the JDZ) and the part of such structure or field which is situated on one side of such lines is exploitable, wholly or in part, from the other side of such lines, concessionaires and other persons authorized by either Party to exploit such structure or field (hereinafter referred to as 'concessionaires and other persons') shall through consultations seek to reach agreement as to the most effective method of exploitation of such structure or field. When agreement concerning such method is reached the agreement (including modifications thereof) shall enter into force upon approval by the Parties. In cases of exploitation under the aforesaid agreement petroleum resources extracted from such structure or field and expenses reasonably attributable to exploitation of such petroleum resources shall be shared among concessionaires and other persons in proportion to the quantities of producible reserves in the respective parts of such structure or field which are situated in the area with respect to which they have been authorized by either Party. The above provisions shall apply mutatis mutandis with respect to exploitation of a single geological structure or field of petroleum resources extending across lines bounding the sub-zones of the JDZ.

Article XXIV: The Parties shall establish and maintain the Japan – Republic of Korea Joint Commission (hereinafter referred to as 'the Commission') as a means for consultation on matters concerning the implementation of this Agreement.

Article XXV: The Commission shall inter alia: review the operation of this Agreement and, when necessary, deliberate on and recommend to the Parties measures to be taken to improve the operation of this Agreement; receive technical and financial reports of concessionaires which shall be submitted annually to the Parties; observe operations of operators and installations and other facilities for exploration or exploitation of petroleum resources in the JDZ; study problems, including those relating to the application of laws and regulations of the Parties, unexpected at the time of entry into force of this Agreement, and when necessary, recommend to the Parties appropriate measures to solve such problems.

Article XXVI: Any dispute between the Parties concerning the interpretation and implementation of this Agreement shall be settled, first of all, through diplomatic channels. Any dispute which fails to be so settled shall be referred for decision to an arbitration board composed of three arbitrators, with each party appointing one arbitrator and the third arbitrator to be agreed upon by the two arbitrators so chosen or the third arbitrator to be appointed by the government of a third country agreed upon by the two arbitrators, provided that the third arbitrator shall not be a national of either Party.

Article XXVII: Exploration and exploitation of petroleum resources in the JDZ shall be carried out in such a manner that other legitimate activities in the JDZ and its superjacent waters such as navigation and fisheries will not unduly affected.

Article XXVIII: Nothing in this Agreement shall be regarded as determining the question of sovereign rights over all or any portion of the

JDZ or as prejudicing the positions of the respective Parties with respect to the delimitation of the continental shelf.

Article XXXI: This Agreement shall be ratified. (*Note*. Both Parties have done so.) This Agreement shall remain in force for a period of fifty years and shall continue in force thereafter until terminated by either Party by giving three years written notice to the other Party. Notwithstanding the foregoing provisions when either Party recognizes that petroleum resources are no longer economically exploitable in the JDZ, the Parties shall consult with each other whether to revise or terminate this Agreement.

Note. The two Agreements of 1974 were made in the period that the 1958 Convention on the Continental Shelf was still in force. Also known to the Parties was the Judgment of 20 February 1969 of the International Court of Justice at The Hague in the *North Sea Continental Shelf Cases*. In its Judgment the Court held that any coastal State has an original right to those areas of the continental shelf which constituted the natural prolongation of its land territory into and under the sea. The Republic of Korea based its position on that Judgment maintaining that the continental shelf adjacent to the two countries should be divided taking into account the natural prolongation of its land mass. When in 1994 the 1958 Convention was replaced by the 1982 Convention on the Law of the Sea the extent of the continental shelf was described in Article 76 of the Convention in accordance with the Court's Judgment of 1969, albeit coupled to a minimum distance of 200 nautical miles from the shoreline (thus coinciding with the Exclusive Economic Zone). Since both Parties reserved their position on the delimitation of the continental shelf, the final determination of the boundaries, if it would ever come to that, would have to be guided by said Article 76. From the outset it can be said that the outcome of such final determination is far from clear since both countries may at least claim the aforementioned minimum area of the continental shelf stretching to 200 nautical miles from its shore.

10.3.5 The 1989 Agreement between Australia and Indonesia

The Agreement of 11 December 1989 made between Australia and the Republic of Indonesia created the Timor Gap Zone of Cooperation (the Zone) in an area of the continental shelf situated between the (then) Indonesian Province of East Timor and northern Australia, an area measuring 62,000 square km.

The Zone is divided in three parts, designated respectively as Area A, which is the middle and largest part, Area B and Area C. Areas B and C fall under the jurisdiction of Indonesia and Australia respectively, but each State has accepted to keep the other informed about the administration of its petroleum legislation with respect to its Area and, more importantly, to surrender to the other ten per cent of the relevant tax revenues collected by it from the producers in its Area.

In Area A, the area in the middle, there shall be joint control by the Contracting States of the exploration for and exploitation of petroleum resources, aimed at achieving optimum commercial utilization thereof and equal sharing between the Contracting States of the benefits of the exploitation of petroleum resources as provided for in this Treaty.

Operational activities in Area A are governed by the Petroleum Mining Code which is annexed to the Agreement as Annex B. Petroleum operations in Area A shall be carried out through production sharing contracts, a model of which is annexed to the Agreement as Annex C. The law applicable to a production sharing contract shall be specified in the contract.

Note 1. The model contract has to a large extent been modelled on the at the time latest Indonesian type of production sharing contract with the Joint Authority being substituted for Pertamina.

Note 2. In accordance with the objective of this Agreement as stated above the revenues accruing to the Joint Authority under the production contracts in Area A will be distributed to the two Contracting States in equal proportions.

The joint administration shall be exercised by the Ministerial Council consisting of those Ministers who may from time to time be designated for that purpose by the Contracting States provided that at any time there shall be an equal number of Ministers designated by each Contracting State, and by the Joint Authority who shall be responsible to the Ministerial Council.

The Ministerial Council shall inter alia: give directions to the joint Authority on the discharge of its functions; amend the Petroleum Mining Code and modify the Model Production Sharing Contract to facilitate petroleum operations in Area A; shall have to approve the following:

(a) production sharing contracts which the Joint Authority may propose to enter in with corporations;
(b) the termination of a production sharing contract;
(c) the variation, with the agreement of the contractor, of the following provisions of the contract:
    (i) the Joint Authority's or the contractor's production share;
    (ii) the operating cost recovery provisions;
    (iii) the term of the contract; and
    (iv) the contract area relinquishment provisions;
(d) the transfer of rights and responsibilities by contractors to other corporations that will then become contractors; and
(e) the distribution to Australia and the Republic of Indonesia of revenues derived from production sharing contracts in Area A (discussed in *Note 2* placed herein below).

The Joint Authority is required to divide Area A into contract areas, issuing prospecting approvals and commissioning environmental investigations prior to contract areas being advertised, advertising of contract areas, assessing applications and making recommendations to the Ministerial Council on applications for production sharing contracts. For the purpose of awarding contracts Area A has been subdivided into fourteen contract areas. Each contract area is subdivided in geometric blocks measuring five minutes by five minutes.

The Joint Authority shall to enter into production sharing contracts with corporations subject to Ministerial Council approval, to supervise the activities of the contractor pursuant to the requirements of the Petroleum Mining Code and the terms and conditions set out in the contract; to recommend to the Ministerial Council the termination of production sharing contracts where contractors do not meet the terms and conditions of those contracts; etc.

The Joint Authority is required to issue regulations and giving directions under the Petroleum Mining Code on all matters related to the supervision of and control of petroleum operations including on health, safety, environmental protection and assessments and work practices, pursuant to the Petroleum Mining Code and to make recommendations to the Ministerial Council to amend the Code and to modify the Model Production Sharing Contract consistent with the objectives of this Treaty.

The Contracting States shall cooperate to prevent and minimize pollution of the marine environment arising from the exploration for and exploitation of petroleum in Area A. The Joint Authority shall issue regulations to protect the marine environment in Area A and shall establish a contingency plan for combating pollution from petroleum operations in that Area.

Contractors shall be liable for damage or expenses incurred as a result of pollution of the marine environment arising out of petroleum operations in Area A in accordance with contractual arrangements with the joint Authority and the law of the state in which a claim in respect of such damage or expenses is brought.

If any single accumulation of petroleum extends across any of the boundary lines of Area A and the part of such accumulation that is situated on one side of a line is exploitable, wholly or in part, from the other side of the line, the Contracting States shall seek to reach agreement on the manner in which the accumulation shall be most effectively exploited and on the equitable sharing of the benefits arising from such exploitation.

For the purposes of the taxation law related directly or indirectly to: (a) the exploration for or the exploitation of petroleum in Area A; or (b) acts, matters, circumstances and things touching, concerning, arising out or connected with any such exploration or exploitation, Area A shall be deemed to be and be treated by each Contracting State as part of that Contracting State. In the application of the taxation law in Area A Contracting State shall grant relief from double taxation in accordance with the Taxation Code contained in Annex D. In accordance with said Code the business profits or losses of a person other than an individual (e.g., a contractor) derived from or incurred in Area A in a year shall be reduced by fifty per cent.

Note. 'Taxation law' means the federal law of Australia or the law of the Republic of Indonesia, from time to time in force, in respect of taxes to which this Treaty applies (excepting tax agreements between States).

Any dispute arising between the Contracting States concerning the interpretation or application of this treaty shall be resolved by consultation or negotiation between the Contracting States.

Each production sharing contract entered into by the Joint Authority shall contain provisions to the effect that any dispute concerning the interpretation or application of such contract shall be submitted to a specified form of binding commercial arbitration. The Contracting States shall facilitate the enforcement in their respective courts of arbitral awards made pursuant to such arbitration.

The Agreement contains the assurance that, in case the Agreement ceases to exist following the conclusion of a further agreement between the States regarding the permanent delimitation of the continental shelf in the area of the Zone, any production sharing contract then existing shall continue to apply to each State, or its competent authority nominated in place of the Joint Authority in so far as the contract is to be performed within the area of its newly established jurisdiction.

Note 1. The Agreement was made against the background of the rules and provisions of the 1982 Convention on the Law of the Sea, including the definition of 'continental shelf' contained in paragraph 1 of Article 76. This worked in favour of the Republic of Indonesia, which was faced with the geographical disadvantage of having the deep Timor trough close by. Since this trough could be considered to form the outer edge of the continental margin, Indonesia could rely on the second part of the definition reading 'or to a distance of 200 nautical miles from the baselines from which the breadth of the territorial sea is measured where the outer edge of the continental margin does not extend to that distance', and claim a continental shelf that extends to a distance of two hundred nautical miles measured from the East Timor shoreline.

Note 2. On 20 May 2002 the Indonesian Province of East Timor became an independent State named *Timor Lorosae* (East Timor). The new State shall have assumed all the rights and obligations of the Republic of Indonesia as Contracting State under this Agreement. Consequently any reference made in this Agreement to the Republic of Indonesia must be taken *mutatis mutandis* to be a reference to East Timor. Nonetheless the Contracting States (Australia and East Timor) agreed in July 2002 to modify the terms of the Agreement to the effect that all revenues accruing to the Joint Authority under production sharing contracts in Area A would be shared between the Contracting States in a 90/10 (formerly 50/50) proportion in favour of East Timor.

10.4 INTERSTATE UNITIZATION AGREEMENTS

10.4.1 Background

The area of a petroleum authorization may happen to be partially bounded by a line of demarcation separating the land or offshore territories of two adjacent or opposite States. In such a situation, a straddling petroleum reservoir falls partly under the jurisdiction of the one State and partly under the jurisdiction of the other State.

Note. Such a situation has been foreseen and provided for in the 1988/1996 UK model clauses (Schedule 4 to the Petroleum [Production] [Seaward Areas] Regulations 1988 as amended in 1995 and in 1996). In the model clause 'Directions as to Oil Fields Across

International Boundaries' it is stipulated that where the Minister is satisfied that any strata in the licensed area or any part thereof form part of an oil field, other parts whereof are in an area to which the Minister's powers to grant licences do not apply and the Minister is satisfied that it is expedient that the oil field should be worked and developed as a unit in cooperation by the licensee and all other persons having an interest in any part of the oil field, the Minister may from time to time give by notice in writing to the licensee such directions as the Minister may think fit, as to the manner in which the rights conferred by the licence shall be exercised.

The desired joint exploitation of such an 'interstate' straddling petroleum reservoir requires an agreement between the States concerned.

In interstate boundary agreements dividing and demarcating the continental shelf between two adjacent or opposite coastal States the possibility of the presence of straddling petroleum reservoirs is usually taken into account and provided for. A good example of such an interstate agreement containing a commitment regarding joint development of straddling petroleum reservoirs is the Agreement of 10 March 1965 concluded between the United Kingdom and Norway. This Agreement (Article 4) contains the following stipulations:

> If any single petroleum reservoir extends across the dividing line and the part of such reservoir, which is situated on one side of the dividing line, is exploitable, wholly or in part, from the other side of the dividing line, the two States are obliged, in consultation with the licensees, if any, to seek to reach agreement as to the manner in which the petroleum reservoir shall be most effectively exploited and the manner in which the proceeds deriving therefrom shall be apportioned.

Clearly, the fact that a reservoir extends across the dividing line is in itself not sufficient to trigger the obligation to cooperate in developing such reservoir. The requirement is that it shall be technically possible to exploit the one part of the reservoir from the other side of the dividing line. Furthermore, it is not required that the straddling reservoir should be exploited as a unit; the possibility of less stringent forms of cooperation in development operations is left open. Apparently, the only firm condition imposed upon the contracting States is the condition that provision should be made for an allocation of the proceeds of exploitation between the two sides.

Note. A similar agreement was made by the United Kingdom with the Netherlands (Agreement of 6 October 1965).

With respect to the North Sea continental shelf, there are several examples of interstate agreements regarding the exploitation (by the respective licensees) of straddling petroleum reservoirs. In fact, the reservoirs, which were the subject of the interstate agreements, extended across the median line dividing the North Sea continental shelf between the United Kingdom on the one side and the opposite coastal States on the other. A recent example is the interstate Markham Agreement between the UK and the Netherlands (officially known as: 'Agreement between the Government of the United Kingdom of Great Britain and Northern Ireland and the Government of the Kingdom of the Netherlands Relating to the Exploitation of the Markham Field and the Offtake of Petroleum therefrom'), which was signed on 26 May 1992. The Markham Agreement is based on the boundary Agreement of 6 October 1965 between the United Kingdom and the Netherlands. The Markham Agreement has much in common with the Statfjord Agreement as described hereinafter and it is clear that the latter has stood model for the former.

10.4.2 THE 1979 STATFJORD AGREEMENT

In terms of volumes of oil and associated gas the most important example of an interstate unitization agreement is the interstate *Statfjord* agreement concluded between the UK and Norway (officially known as: 'Agreement between the Government of the United Kingdom of Great Britain and Northern Ireland and the Government of the Kingdom of Norway relating to the Exploitation of the Statfjord Field Reservoirs and the Offtake of Petroleum therefrom'), which was signed on 16 October 1979.

To a great extent the Statfjord Agreement followed the pattern of the earlier Frigg Field Reservoir Agreement which was entered into by the same governments on 10 May 1976. The Statfjord Agreement contains the basic principle that the straddling reservoirs have to be exploited as a single unit, but in contrast to the earlier Frigg Agreement, no indications were given about the manner in which the governments wished to see the Statfjord reservoirs so exploited. The respective licensees were required to enter into agreements between themselves for the purpose of regulating the exploitation in accordance with the provisions of the interstate agreement. Such agreements of cooperation (in fact the unitization agreement in the proper sense of the word) and any proposed amendment thereof needed the prior approval of the two governments.

The governments agreed to ensure that for the purposes of the exploitation of the unit the licensees should appoint a unit operator, whose appointment would be subject to the approval of the two governments. The two governments had to agree on the determination of the limits of the reservoirs, of the estimated total reserves and of the apportionment of the reserves as between the Norwegian and the UK parts of the continental shelf. 'Reserves' are described as the volume of oil present in the reservoirs before the start of production, measured and determined as 'stock tank oil originally in place', in accordance with the procedures to be described in the agreements of cooperation between the licensees. The licensees are required to submit to the governments a proposal for such determinations.

Licensces are required at all times to share the production of Statfjord petroleum in such a manner that the cumulative volume of Statfjord petroleum received by the licensees of the UK Government and the cumulative volume of Statfjord petroleum received by the licensees of the Norwegian government shall each correspond to the apportionment of the reserves pursuant to the apportionment agreed between the two governments, before the production commencement date (which date is defined as the date of loading of the first tankship). This requirement however is made subject to the further requirement that licensees must secure that at the time the production from the Statfjord reservoirs ceases the share of the total volume of Statfjord petroleum received by the licensees of the UK Government, and the share of the total volume of Statfjord petroleum received by the licensees of the Norwegian government shall each correspond to the final apportionment of the reserves.

Either government may request a review of the limits of the reservoir, of the total amount of reserves and of the apportionment thereof or any of them in order to arrive at a re-determination. The time schedule for the successive re-determinations is laid down in the interstate agreement itself. The first date for a re-determination was set on the second 1 January following the aforementioned production commencement date. Whenever the apportionment of the reserves is revised the licensees shall adjust deliveries between themselves to ensure that imbalances deriving from the superseded apportionment are made good. All re-determinations shall have effect from the first day of the month following the month when the re-determination has been agreed by both governments, the intention being that such agreement will be reached on or soon after the scheduled date of the relevant re-determination.

Licensees are required to exploit the Statfjord field reservoirs in accordance with a programme which has been approved by the two governments. The governments may jointly agree to its amendment at any time. The approval, amendment and implementation of the programme shall take account of the objective that the Statfjord Field reservoirs be exploited in such manner as to prevent the waste of petroleum and minimize losses of reservoir energy.

In the event of the expiry, surrender or revocation of any production licence or any part of such a licence relating to any part of the Statfjord reservoirs, the government, which issued the licence, is obliged to ensure that the exploitation of the said reservoirs is continued in accordance with the terms of the interstate agreement and the agreements of cooperation. In particular the government concerned must take one of the following steps:

(i) issue a new licence in replacement of the licence which has expired or been surrendered or revoked; or
(ii) itself conduct such exploitation as if it were a licensee; or take such other action to continue the exploitation of the reservoir as the two governments may agree.

Each government is obliged to procure access to offshore installations and equipment for its inspectors and for the inspectors of the other government. Inspections of installations concern not only the installations themselves but also the operations carried out on such installations, and the inspectors of both governments have to satisfy themselves that the fundamental interests of their government with regard to safety, petroleum production and measurement are met.

Profits, gains and capital from the exploitation of the Statfjord Field Reservoirs will be taxed in accordance with the laws of the United Kingdom and Norway respectively. This principle is worked out in separate protocols.

The Agreement stipulates that the jurisdiction each State has under international law over the continental shelf remains unaffected by the Agreement. It is stressed in the Agreement that installations located on the UK part of the continental shelf shall be under UK jurisdiction and installations located on the Norwegian part of the continental shelf shall be under Norwegian jurisdiction.

In contrast with the Frigg Agreement, no specific means of transportation or any specific destination for the oil and associated gas production are prescribed. Instead, the Agreement provides for the right of each government and its licensees to take directly, from the Statfjord Area to its territory, that share of Statfjord oil and associated gas production to which these licensees are entitled, notwithstanding the location of the installations from which the Statfjord petroleum is produced. The offloading into tankship has been adopted as the final solution of the oil transportation problem. For the transmission of the associated gas a pipeline has been built bringing the gas to *Karstö* on the Norwegian mainland.

Note. From the given examples regarding an interstate agreement on unitization it may be concluded that the joint exploitation of a straddling reservoir is not a matter that is left to the right holders on either side of the dividing line, but is directed by the respective governments that issued the petroleum authorizations in question.

Chapter 11

Cooperative Exploration and Production Agreements

11.1 HISTORICAL BACKGROUND AND DEVELOPMENT

11.1.1 The Early Period

In the wake of the boring of the first oil well by Drake in 1859, many landowners or their leaseholders in Pennsylvania and Ontario tried their luck and became oil producers. They worked on small-sized plots of land, on which they tried to install as many drilling derricks as space allowed in order to be able to produce oil at the highest rates possible. Although not having a clear idea about subsoil oil reservoirs and the flowing behaviour of oil therein, they were anyway convinced that any oil not produced by their wells would be produced by the wells drilled by their neighbours. Nonetheless, these pioneer producers worked their wells independently from one another, without any form of cooperation being considered necessary or desirable. This non-cooperative attitude was completely in line with the way gold prospectors in those days were accustomed to operate. The gold miner worked its staked-out plots inside the gold field (riverbank, etc.), independent from anyone else and was prepared to defend his claim territory against any newcomer. In contrast with the gold miner however the pioneer oil producer considered and treated his neighbours as competitors being under the impression that he and his neighbours were draining one and the same oil pool and that the oil contained in the pool would flow into and be caught in the neighbour's wells, if not timely caught in his own wells. Remarkably, among the producers and in the courts of law there was never any doubt regarding the producer's right to take possession and become the owner of any oil that entered any well drilled (vertically) by the producer within the confines of his land property or lease area. This right was referred to as 'the rule of capture' or *profit a prendre*. On the other hand, intentional deviation of a bore hole into the neighbour's domain and draining oil by means of such deviated hole was considered stealing and non-permissible. The question of who owns the oil in the subsoil reservoir (oil in situ) received generally little attention, the producers being only interested in producing their wells at maximum rates and selling a maximum quantity at the wellhead to the buyers from the numerous refineries that meanwhile had been installed for refining the oil into kerosene. In the course of time a market for oil developed in which many small volume producers traded with many small volume kerosene refiners. On the side of the producers, situations of over-supply alternated with periods of shortage due to the wasteful way in which the oil was being produced. This instability on the supply side caused hefty price fluctuations. In contrast with the producers the refinery owners (known as oil companies) aimed at organization and concentration of commercial power whereby Rockefeller's Standard Oil Company, established in 1870, restructured as the Standard Oil Trust in 1882, took the lead. When around 1885 oil was discovered outside Pennsylvania and oil production had started to spread over the US continent Standard Oil, later followed by the other oil companies, changed its policy of restricting itself to monopolizing the refinery sector. Instead, the oil companies started to intervene in the business of the producers by acquiring production leases for their own risk and account. The pattern that now established itself on the US continent was that of an oil company following the oil prospector and waiting for him to make a discovery. As soon as this would happen, the oil company would try and buy or lease any land on which the discovery was made, together with any land in the surrounding area. In this manner the oil companies sought to forestall that any prospective acreage would fall into the hands of speculators. In the process the oil companies were collecting large-sized leases thereby reducing the chance that an oil field would be developed and produced by a multitude of persons, resulting in a waste of capital, over-production and reservoir damage.

Note. In reaction to the inroads that the oil companies were making into the production sector, independent producers in the oldest producing areas were looking for protection and tried to find such protection in cooperating with independent refiners, forming a joint venture rather than an integrated venture.

Although the number of independent, small-sized lease holdings gradually diminished as a result of bundling and integration processes, on numerous occasions the new larger-sized lease areas still proved to be small compared with the size of the underground reservoir or proved to have the wrong orientation in respect thereof. In either case the new leaseholders were confronted with a situation in which a reservoir was shared between two or more of them. Over time, as more became known about petroleum reservoirs in general, and about the forces (known as the drive mechanisms) that were responsible for driving the oil or gas to the borehole, the new leaseholders, sharing a common reservoir, started to realize that to produce a common reservoir in an efficient manner, with the objective of maximizing the ultimate recovery, mutual cooperation in matters of production would be necessary. In particular when it became clear that injection of water at the margins of a common oil accumulation would improve the ultimate recovery, the new leaseholders showed a willingness to consider and agree among themselves the joint, integral development and exploitation of such a reservoir. In this way the ground was prepared for cooperation between producers, a new attitude that hitherto had never been adopted.

11.1.2 The Middle East Joint Ventures

When moving into the Middle East, the major US and European oil companies established joint ventures for the purpose of acquiring concessions and producing oil from their areas except for the venture in Persia. In that country the search for oil followed the old pattern known from the early days in the US whereby the lonely prospector is being followed by an oil company waiting for the right moment (i.e., after a discovery was made) to take over the lease in question and some more leases next to it. In this case the prospector was William Knox D'Arcy, who was granted on 28 May 1901 a concession covering the whole country with (deliberate) deletion of the five northern provinces, the ones bordering on Tsarist Russia, and the oil company that one year after oil was struck in 1908 at Masjid-I-Suleiman took over his concession after having raised the necessary finance, was the Anglo-Persian Oil Company. Nonetheless an incorporated joint venture was created when for strategic reasons the UK government acquired in August 1914, a few days after the outbreak of the First World War, a controlling interest of 51.7 per cent in the company.

On 20 March 1951 Anglo-Persian's 1933 concession agreement was taken over by the newly established National Iranian Oil Company (NIOC). After three and a half years of negotiations a contract of work, covering the area of the former concession, was made between NIOC and Iranian Oil Participants Ltd., a UK company, better known as the Consortium. The Iranian Oil Participants Ltd. constituted an incorporated joint venture replacing Anglo-Iranian the former sole concessionaire. Shareholders and shareholding in the new joint company were fixed as follows: forty per cent for Anglo-Iranian Oil Company, the former concessionaire and on this occasion renamed British Petroleum Company (BP); fourteen per cent for Royal Dutch/Shell; six per cent for CFP; forty per cent for a group of US oil companies, distributed among them as follows: seven per cent for each of Standard of New Jersey, Standard of California, Socony Mobil, Texaco and Gulf Oil, and five per cent for Iricon Agency, Ltd., the latter owned by nine US oil companies, among which Standard of Ohio, Atlantic Refining and Getty Oil, the latter each having 0.417 per cent.

For the ventures in Iraq and the various countries of the Arabian Peninsula incorporated joint ventures were created from the outset because the participating companies were aware of the fact that they were dealing with large and largely unexplored areas that might require huge investments.

In the end, the concessions made in the 1930s in the Middle East Gulf area including the successor contract of work agreed with NIOC were all held by joint venture companies, the shares in which were held by affiliates of the participating major groups.

These joint venture companies were:

(i) Iraq Petroleum Company (IPC) and its sister-companies (companies in which the shares were held by the same companies holding shares in IPC, also known as IPC-group companies), representing a joint venture of BP, Royal Dutch/Shell, CFP, Exxon, Mobil and Partex (the IPC group). IPC group companies held concessions in respectively Iraq (three separate concessions), Qatar, Abu Dhabi and Oman;
(ii) the Iranian Oil Participants (the Consortium), a joint venture of BP, Royal Dutch/Shell, CFP, Gulf; and, as of 1954, holding a contract of work in respect of the area of Anglo-Iranian's former concession;
(iii) Kuwait Oil Company, a joint venture of BP and Gulf, covering the land territory of Kuwait; and
(iv) Aramco, a joint venture of Texaco, Exxon, Chevron and Mobil (the Aramco partners).

Each joint venture company represented an incorporated joint venture. The joint venture company held, maintained and operated the concession concerned. It sold the realized production to its shareholders or rather to the trading companies established by the shareholders for this purpose. The trading companies bought their respective shareholders' production entitlements at cost and, in their turn, sold the production to their own shareholders at the applicable posted price. The trading companies were responsible for paying income taxes with respect to their profits calculated on the basis of the posted price. In its turn, the joint venture company, as concessionaire, was responsible for paying the royalties and fulfilling the other obligations imposed by the concession. The shareholders decided on policy matters and were responsible for making capital contributions, mostly in the forms of loans to the joint venture company. These loans were paid back by the company out of the sales proceeds. The shareholders' main responsibility was to decide on the development of production capacity determining future production levels, and to approve the capital and operating costs budgets. In view of the almost unlimited reserves contained in the respective concessions the production capacity had no natural constraints and could be developed solely and entirely in accordance with the long-term marketing requirements of the individual shareholders. In accordance with nomination procedures each shareholder made known (nominated) the quantity of oil that it required over three year cycles, the first year of the cycle divided in quarters. Each shareholder was entitled to nominate its shareholding percentage interest of the estimated production capacity as announced by the company in advance of the nomination procedure, but was free to nominate a lesser quantity, if temporarily it needed less oil for its markets. If due to under-nominations the installed production capacity was not fully utilized, a second round of nominations might follow in order to give participants that nominated their full entitlement the opportunity to become over-nominators within the limits of any surplus capacity. Nominations for the quarters of the first year of the cycle had to be firm with an offtake obligation attached to it; nominations for second and third year of the cycle were allowed to be tentative, subject to a further narrowing down as the year moved forwards in the cycle. On the basis of the nominations the operator would prepare the production schedule, aiming to produce sufficient oil to satisfy the total shareholders' demand, meaning the sum of the individual nominations. Capital contributions were not fixed in accordance with shareholding interests, but based on the shareholders' individual final nominations over the nomination cycle. The flexibility in making nominations, coupled with the flexibility in capital contributions, worked quite well against the background of an unlimited production potential, for participants, whose shareholding interest in the produced volumes often did not correspond with their marketing requirements, and for governments continuously pressing to increase production. Because of these flexible arrangements production capacity could be expanded and financed and production levels could be increased (in accordance with the wishes of the government), even if some of the participants might not need any such additional capacity and would, if operating the concession on their own and not being a participant in a joint venture, find it difficult to make additional capital available for that purpose.

Around 1975 governments in the Middle East Gulf area completed their takeover of the concessions and concessionaires' other assets falling under their respective jurisdictions. The Middle East joint ventures were dissolved or completely restructured, but the experience gained with operating oil fields within the framework of a joint venture was not lost. On balance, this experience had been favourable, at least for the majors. Working together within the framework of the Middle East joint ventures had enabled the majors to control the supply to the market and to avoid price-cutting competition among them. In fact, market sharing between and among the majors actually took place when implementing the production nomination and allocation procedures envisaged by the joint venture agreement. The Middle East joint ventures were unique in the sense that production capacity could be developed in accordance with the marketing requirements of the participants, without being constrained by the size of the available reserves, and perhaps more importantly the abundant reserves could be developed in an inexpensive manner, making Middle East oil very competitive in the global oil market.

11.2 THE CONTEMPORARY JOINT VENTURE

11.2.1 Principles and Objectives

The example set by the Middle East joint ventures was followed by explorers and producers elsewhere in the world, even before the demise of said joint ventures. First of all, the major oil companies themselves showed their preference for extending their cooperation in exploration and production activities to areas outside the Middle East. Well-known examples are the joint venture formed by Royal Dutch/Shell and BP for operating the licences and leases granted to them in Nigeria, and the joint ventures between Royal Dutch/Shell and Exxon in respect of respectively the Netherlands (onshore and offshore, including Western Europe's largest onshore gas field), and the UK part of the North Sea continental shelf, where subsequently many important oil and gas fields were discovered. Soon thereafter most commercial oil enterprises including state-owned enterprises embraced the joint venture approach when contemplating to start an exploration/production venture, so much so that setting up a joint venture for the purpose of jointly applying for or acquiring petroleum rights and for jointly conducting the authorized exploration and production operations has become standard practice.

The main reasons for this practice are the wish not to say the absolute necessity to spread and minimize the various risks attached to the intended petroleum venture and to share among the participants the generally huge investments that will required.

These risks can be summed up as follows:

(i) Geological risk. This is the risk that no commercial petroleum will be discovered;
(ii) Technical risk. This is the risk inherent in having to operate under difficult and hazardous conditions, such as difficult or extreme weather and sea conditions, floating icebergs and/or having to work in deep water;
(iii) Development risk. This is the risk that the petroleum encountered in the reservoir proves difficult to produce due the negative characteristics of the reservoir and/or of the petroleum itself;
(iv) Environmental risk. This is the risk that the operations undertaken are causing damage to the environment. The environmental risk is inherent in having to work, from an environmental point of view, in a sensitive area. The ultimate environmental risk is the risk that the operations have to be abandoned when it appears that they would have to be carried out under such restricted conditions that the intended operations become too costly;
(v) Political risk. This risk is nowadays perhaps the most powerful argument for setting up a joint exploration and production venture (see Chapter 2). In particular Western private sector oil enterprises when considering an exploration and production venture in a politically risky State are keenly interested in seeking the cooperation of a commercial oil enterprise owned by the State in question.

Apart from the wish to spread and minimize the aforementioned risks and to share large-scale investments, commercial oil enterprises may have other, secondary reasons for willingly seeking to cooperate with other similar enterprises. These reasons are:

(i) to improve the chance to be successful in a bidding procedure. This is a powerful reason for a Western private sector oil enterprise to seek cooperation with a state-owned oil enterprise generally, but by preference with an enterprise that is owned by the (prospective) host State;
(ii) to obtain a participating interest in a licence or contract of work already awarded to another commercial oil enterprise; or
(iii) to benefit from the superior technology and/or geological knowledge or experience of another commercial oil enterprise. This is a powerful argument for a state-owned oil enterprise to seek cooperation with a Western private sector oil enterprise.

Commercial oil enterprises, wishing for any one of the aforementioned reasons to set up and participate in a joint exploration and production venture, must start by agreeing between and amongst themselves the very basic elements of the intended joint venture such as:

(i) the selection of the area for the intended venture;
(ii) the distribution of the individual participating interests; and
(iii) the appointment of an operator.

This will usually be the participant with the largest participating interest, and/or the participant who possesses the right type of technology, and/or technical capability for undertaking the intended joint venture.

Otherwise, if there is little to choose between the participants, it will be the participant which took the initiative with regard to the venture. The operator will also coordinate the bidding for the required petroleum rights in accordance with the terms and conditions of a bidding agreement made by the prospective participants:

(iv) the type of joint venture, i.e., an incorporated joint venture or a non-incorporated one.

In case the prospective participants opt for the incorporation of their joint venture, a joint venture company will have to be established, in which the participants become shareholders with individual shareholdings in proportion to their respective participating interests. This was the practice followed in respect of the former Middle East joint ventures. But after the demise of these joint ventures prospective participants have shown to prefer the non-incorporated form of cooperation whereby each person holds an undivided participating interest in the still to be applied for petroleum rights and in the venture that will be based thereon.

Note concerning (iv). The main attraction of a non-incorporated joint venture arrangement compared to an incorporated one is that it gives the participants a direct entitlement to production. By contrast, participants/shareholders may only claim or expect cash in the form of dividends. Such is the attraction of having direct access to production that in cases where legal requirements compel participants to form a joint venture company, the participants/shareholders will aim to construct and enter into a cooperative arrangement (a shareholders' agreement) between and among themselves that will accord them the same rights, benefits, obligations and liabilities as they would enjoy or have within the framework of an non-incorporated joint venture. Compared with the former Middle East joint ventures, the contemporary non-incorporated joint venture is basically less complicated in as much there is no flexibility in the offtake of production and no flexibility in individual capital contributions. Participants in a contemporary joint venture are obliged to offtake their undivided participating interest share of the production made available by the operator in accordance with the budgets and production schedules approved by the participants, and to contribute to capital and operating costs in the same fixed proportions. Usually, a small degree of offtake flexibility is provided by a special offtake agreement. Such an offtake agreement will provide for the possibility of uneven liftings of production (allowing under-liftings and over-liftings over a short period) and for the exchange of production volumes between the participants to bring the individual liftings back into balance within the shortest time possible (in order to forestall participants manipulating their offtake in an attempt to benefit from ever present world market price fluctuations). In other respects the contemporary joint venture arrangements have become more complicated. In particular, the attention that is paid to the protection of the economic interests of minorities has given rise to elaborate and complex sole risk and non-consent options, as will be discussed hereinafter in more detail.

If the prospective participants wish and are legally not restrained from doing so to opt for the non-incorporated joint venture, as is generally the case for reasons described above, agreement on the more detailed but still rather basic features of such type of joint venture should be reached at an early stage in order to forestall difficulties when a full-fledged joint operating agreement has to be made. Said features can be summed up as follows:

(i) to apply for, acquire and hold in an undivided participating interest the necessary petroleum right with respect to the area that has been selected for the joint venture operations;
(ii) to form a management committee for formulating policies and for approving the work programmes, development plans and corresponding budgets as prepared by the operator and in which committee all participants are represented with voting rights in accordance with their participating interests;
(iii) to oblige the operator to conduct the joint operations in accordance with work programmes, development plans and corresponding budgets as approved by the said management committee; and
(iv) to share in proportion to their participating interests any expenditure and expenses, any liabilities and risks associated with the joint operations, any production, that may be forthcoming from these operations and to which the participants are entitled under the rules of the licence or contract, and similarly share any other assets acquired in the course of the joint operations.

Participants in a non-incorporated joint venture set out the rules governing their cooperation in a joint operating agreement (JOA). In the course of time a certain standardization of these rules of cooperation has taken place.

11.2.2 Standard Conditions of the JOA

11.2.2.1 Introduction and Recitals

Identification of the participants and a description of the authorization (be it an exclusive licence or a contract of work), on which the joint venture is based.

11.2.2.2 Definitions

It is customary to precede the main body of the agreement with an article containing definitions of important terms and expressions. It allows the text of the remainder of the agreement to be concise and easy to read. As required or desired this article may be confined to a limited number of typical expressions or include any expression that has a particular meaning in the context of the operations or procedures described in the agreement.

11.2.2.3 Duration

The main body of the JOA endures for as long as the underlying authorization remains valid and is held by two or more participants. Some articles are declared to remain in effect until certain events have been properly dealt with, such as the settlement of disputes or the abandonment of wells, field installations and other facilities.

11.2.2.4 Purpose and Scope

Basically, the cooperation takes place within the framework of a jointly owned authorization, be it an exclusive licence or contract of work, and is aimed at the production of oil and natural gas to the mutual commercial benefit of the participants. Mutual cooperation starts with the search for petroleum accumulations and ends at the point where the operator delivers to each participant its share of the oil and natural gas production to which the participants collectively are entitled under the rules of the authorization granted. Not all operations falling under the authorization are carried out as joint operations for the joint account of all participants. Departures from the joint venture principle are allowed. To this end a JOA will provide for a non-consent option, permitting participants to stay out of operations adopted and approved by the other participants possessing the majority required to take binding decisions, and for a sole risk option, permitting part of the authorized operations to be carried out for the sole risk and account of one or more but less than all participants. By contrast, the scope of the joint venture and the extent of the joint operations may be extended to include the joint disposal of natural gas production. Such joint disposal and marketing of the natural gas production may be necessary in view of the requirements and peculiarities involved in the handling and transportation of natural gas and of its marketing. When the marketing is done on a joint basis, each participant dedicates its share of the natural gas production to the same long-term gas sales agreement (alternatively selling its share of the production on identical terms to the same buyer).

11.2.2.5 Participating Interests

The distribution of the participating interests among the participants belongs to the basic conditions of the joint venture agreement. In principle, the distribution does not have to be even, but an uneven distribution gives some participants a greater weight than others. The right balance between uneven interests has to be found in appropriate voting rules accompanied by procedures to protect the economic interests of frustrated participants (i.e., by incorporating non-consent and sole risk options).

11.2.2.6 Basic Rights, Obligations and Liabilities

Participants own undivided interests in the underlying authorization, the oil and gas production made available by the operator and to which the participants as licensee or contractor collectively are entitled under the terms of the said authorization, and in any other assets acquired by operator for or in the course of the operations (unless the ownership of those assets have to be or are automatically transferred to the state party in accordance with the terms of the underlying contract of work).

All obligations imposed under the authorization and all liabilities arising in connection with the conduct of the authorized operations and any and all expenditures incurred by the operator, and charged by the latter to the joint account or to any particular sole risk account, as the case may be, are shared and borne by the participants in proportion to their respective (relevant) participating interests, except to the extent that the operator is solely liable (see below). Each participant is obliged to indemnify the operator or any other participant having to make a payment with respect to any such obligation or liability, if not charged to a joint or sole risk account, in proportion to its respective participating interest. Each participant has the basic duty to pay its proportional financial contributions in advance, in cash and in the currency as requested by the operator. Any JOA will contain a cash call procedure providing for cash calls to be made well in advance of the moment the operator is required to make payments in order to avoid the occurrence of cash shortages and to forestall that the operator should have to use its own funds.

11.2.2.7 The Operator

The JOA records the appointment of the operator of the joint venture and the fact that the person so appointed agrees to act as operator. As said before the appointment of the operator will already have taken place in the initial stage of setting up the joint venture. In rare cases not the selected participant itself but one of its affiliated companies may become the operator. Such operator has no participating interest in the joint venture. Then there is of course always the possibility that the participants decide to appoint an independent third party as operator. This could be necessary if the participants have little experience with oil matters. If any of the participants (or any of its affiliated companies) act as operator, it is recorded that such participant will make no profits nor has to suffer losses as a consequence of acting as operator.

The rights and duties of the operator are listed in the JOA. The operator's rights comprise the following:

(i) the exclusive right to conduct and carry out the authorized operations;
(ii) to demand cash contributions in advance from the participants;
(iii) to represent the participants vis-à-vis the government and/or state party and vis-à-vis other third parties; in the latter context the operator shall represent the participants in disputes and the settlement thereof with third parties;
(iv) to employ and make use of sub-contractors and agents, provided the rules for awarding contracts as incorporated in the agreement (e.g., international tendering, specifications and final award subject to management committee's approval, etc.) are being observed;
(v) to employ expatriate and national personnel on the operator's conditions (level of salaries, currency of payment, special benefits, numbers, description of required qualifications, etc.);
(vi) to make use of the technical and personnel services of its parent company or of its specialized affiliated companies and to pay for such internal services at the rates provided for in the JOA.

The operator's main duties consist in:

(i) preparing and submitting for the review and approval of the management committee annual work programmes and corresponding budgets; special work programmes (i.e., work programmes meant for the evaluation and appraisal of any discovery made and for the abandonment of wells, fields and installations) and corresponding budgets and overall plans for the development of a discovery (declared by any or all participants to be commercial); and
(ii) carrying out the said annual, special work programmes and overall plans in the form as finally approved (and possibly amended) by the management committee, subject to operator's right to incur over-expenditures within certain narrow limits expressed as a percentage of total budget and as a percentage of the individual budget item being overspent. Some JOAs provide for an Authority for Expenditure (AFE) procedure. This procedure requires the operator, before entering into any commitment or incurring any expenditure under an approved work programme and budget, to submit an AFE in respect thereof. The information to be included in an AFE is set out in the accounting procedure. AFEs have to be approved by the management committee. Voting requirements with respect to decisions regarding an AFE will be as prescribed with respect to routine operational matters, because the AFE concerns expenditure and commitments under an already approved budget. To the extent that the management committee approves an AFE the operator will be authorized and obligated to proceed with such commitment or expenditure;

(iii) maintaining the authorization in good standing. This means that the operator acting as the representative of the licensee or contractor has the duty to carry out all operations in accordance with the relevant conditions of the licence or contract (e.g., the condition that licensee or contractor should carry out the operations in accordance with good oil field practices and without causing pollution or other damage to the environment). In this connection, the operator is further responsible for making the required payments (e.g., royalty) in time, submitting work programmes, giving notification of (commercial) discovery, submitting evaluation and appraisal programmes and plans of development, all as prescribed by the rules of the licence or contract of work, in order not to give the authorities cause to cancel or revoke the petroleum right concerned. If the underlying authorization is a production sharing contract the operator acts on behalf of the contractor, and as contractor the operator has similarly identical duties and tasks with regard to preparing and submitting for approval work programmes and development plans, executing approved work programmes, awarding major contracts, etc. as it has as the operator of the joint venture, the difference being that under a production sharing contract the supervisory body is the state party, whereas under the JOA supervision over the conduct of the operations is exercised by the management committee. Nevertheless, the work programmes, overall plans and corresponding budgets are the same and participants must cooperate to determine and fix the joint operations programme and sole risk programme, if any, in time so as to allow the operator as contractor to adhere to the time schedules stipulated in the production sharing contract;

(iv) procuring and maintaining all insurance required under applicable law and/or petroleum right and such further insurance as the management committee may decide, provided always that each participant may refuse to participate in such additional insurance, obtaining instead its own insurance (e.g., a participant belonging to a large oil group may obtain insurance from one of its affiliated companies and opt for self-insurance), if considered adequate and acceptable by the management committee;

(v) carrying a special liability for damages and losses arising in connection with its conduct of the operations, in case such damage or loss has been caused by the wilful misconduct or gross negligence of senior supervisory personnel of operator or any of its affiliates (if such affiliates have been engaged by operator for the purpose of the operations). In such case the operator or its affiliated companies carry the sole liability and the other participants do not have to indemnify the operator to the extent of their participating interest as they otherwise would have been obliged to do. Excluded from operator's special liability are any consequential damages, environmental damage, damage caused to a producing reservoir, loss of production and the cost of pollution control and that of environmental rehabilitation. These damages and costs, when occurring and irrespective of how caused, have to be borne by all participants in proportion to their respective (relevant) participating interests.

The provisions regarding the operator are completed with a procedure dealing with the resignation, removal and succession of the operator.

11.2.2.8 The Management Committee

The JOA stipulates the establishment of a management committee, in which all participants are represented with voting rights in proportion to their respective participating interests. The task and mandate of the management consists of formulating policy, providing guidelines for the operator, and to make binding decisions in joint venture matters, provided any such decision is supported by the voting majority required therefore as specified and stipulated in the JOA. The required voting majorities may range from unanimity to any kind of qualified majority to a simple majority, depending on the subject matter to be decided upon, on the number of participants and the distribution of participating interests among the latter, and on whether the JOA contains a non-consent option and/or a sole risk option. Where the JOA prescribed unanimity for binding decisions, then obviously a non-consent option is not needed, but a sole risk option always is. Meetings of the management committee are called by the operator. The main business at the meetings of the management committee consists of reviewing, discussing, amending, if necessary, and finally approving in accordance with the voting rules of the JOA:

(i) the annual work programmes and corresponding budgets as prepared and submitted by the operator;
(ii) a notification of a (commercial) discovery and any subsequent plan of appraisal or development, as the case may be;
(iii) production schedules;
(iv) programmes and budgets concerning the abandonment of production facilities; and
(v) conditions and specifications of the international tendering of major contracts and the award thereof.

11.2.2.9 The Non-consent Option

The non-consent option is an option available to any participant permitting the latter to opt out of participating in an operation that was adopted and approved by the management committee in accordance with the voting rules of the JOA, provided the participant in question had voted against the operation, when it was proposed in a meeting of the management committee, and provided the operation proposed was not an operation aimed at fulfilling any obligatory work programme or an operation necessary for the abandonment of a production facility (abandonment operation). After use is made of the non-consent option, the remaining, consenting participants are not obliged to proceed with the proposed operation. In turn each of them has the right to withdraw its given consent and withdraw from the proposed operation. If all remaining participants decide to withdraw their consent, the operation will be cancelled. (*Note.* As observed before, if unanimity is required for taking binding decisions within the management committee, a non-consent option is superfluous. Any participant which is opposed to a proposed operation or investment can protect its economic interest by simply voting against the proposal.)

The procedure to implement the non-consent option consists in a description of the type or category of operations in respect of which the non-consent option may be exercised. Generally, the same type and category of operations/investments which may be carried out as a sole risk operation, are made subject to the non-consent option, including exploration and appraisal drilling, the development of a discovery, and production operations involving the application of secondary or tertiary recovery methods, but excluding always, for practical reasons, seismic work.

11.2.2.10 The Sole Risk Option

In contrast with the non-consent option, a sole risk option aims to protect the economic interests of any participant, whose proposal of making investments or further investments is rejected in a meeting of the management committee, or who opposes a proposal to abandon a production facility, being itself in favour of continuing production. The sole risk option permits the participant, who wishes to make use of this option, to request the operator to carry out the rejected operation or to operate the production facility, which was earmarked for abandonment, for said participant's sole risk and account.

The implementation of the sole risk option is governed by the following requirements, rules and conditions:

(i) a description of the type and category of the operations which qualify to be executed as a sole risk operation. If such operations are specified and a listing of them is taken up in the JOA, it means that any operation not so listed cannot be carried out as a sole risk operation. If a general description is used (in the sense that qualifying operations are described as any and all operations that (also) can be carried out as a joint operation the following operations have specifically to be excluded: (a) any operation with forms part of a development plan or part of an enhanced recovery scheme, such as the drilling of individual development wells or the installation of individual enhanced recovery facilities; and (b) any operation aimed at the abandonment of a production facility. Obviously the latter operation can never be allowed to be carried out as sole risk operation. In fact, if an abandonment proposal is made and this proposal is rejected by the management committee, the exploitation of the production facility concerned will simply be continued as a joint operation as before. The participant, who made the proposal for abandonment, has then the choice between contributing its proportional share of the expenses required for the continuation of the use of the production facility concerned, or to withdraw from the joint venture. If after the rejection of a proposal for abandonment of a production facility, a further proposal is made to embark on a scheme of enhanced recovery, the participant having initially proposed abandonment may now make use of its right of non-consent with regard to the proposed enhanced recovery scheme. If so, the consenting participants or anyone of them have the right, but not the obligation, to carry out the proposed scheme as a sole risk operation;

(ii) appropriate notification (sole risk notice) and submission of complete information, including an estimate of the costs involved, about the operation proposed to be carried out as a sole risk operation. If a qualifying operation is proposed as a sole risk operation because one or more participants have exercised their non-consent option, if available with respect to that operation, the latter are not allowed to participate in the proposed sole risk operation;

(iii) joint operations take precedence, no interference or hindrance is allowed;

(iv) the operator (whether or not a sole risk participant itself) has the exclusive right and duty to carry out any sole risk operation;

(v) non-sole risk participants have to be kept fully informed about the operation, its progress and the results obtained. Upon the completion of any particular sole risk operation, other than a sole risk development operation; a sole risk continuation of exploitation or a sole risk enhanced recovery scheme, the operator must prepare and submit a complete report;

(vi) upon receiving the operator's aforesaid report non-sole risk participants have the right to re-enter and join in any follow-up operation against the payment of a reimbursement in cash of their share of the costs incurred plus a premium in cash (as far as sole risk seismic work is concerned) or in kind (i.e., in oil and gas production generated by the field discovered as a result of a sole risk drilling operation). The said premium, which is expressed as a multiple of the amount of the reimbursement, is meant to compensate the sole risk participant(s) for the risk taken with their operation. A higher risk is compensated by a higher premium;

(vii) sole risk participants have to bear all costs and liabilities associated with the sole risk operation, in which they are participating, and have to indemnify the non-sole risk participants from any and all costs and liabilities arising in connection with such operation (but the operator's special liability is maintained);

(viii) in cases where the underlying authorization is a production sharing contract special provisions regarding the allocation and distribution of cost oil and the contractor's share of profit oil to the joint operations and any particular sole risk operation respectively are required.

11.2.2.12 Disposition of Production

Any participant is entitled to receive at the point of delivery described in the JOA its participating interest share (as modified with respect to sole risk production) of the oil and gas production made available to the participants at said point by the operator in accordance with the terms of the underlying licence or production sharing contract. Delivery of oil production to which the participants are entitled takes place within the framework of an offtake agreement, allowing oil liftings deviating from the proportional shares but only for limited periods. Natural gas production will usually be disposed of on a joint and integrated basis and sold to a single buyer in accordance with the terms of a single long-term gas sales agreement (or a number of identical gas sales agreement, each of the latter covering the share of an individual participant).

11.2.2.13 Default

Any participant has the basic obligation to make its proportional cash contribution in response to monthly cash calls issued in advance by the operator. Any participant failing to do so is in default. The obligation to transfer cash to the operator in response to a cash call is an absolute obligation from which no participant is excused, even not so if a valid claim can be made to have been affected by *force majeure*. Notice of a default is given by operator to the defaulting participant and any other participant. The other, non-defaulting participants are obliged to make up the cash shortage caused by the non-contribution of the defaulting participant. A default is only allowed to exist for a specified period, usually sixty days, unless special circumstances exist which allow an extension of this period. If before the expiry of the period or any extension thereof the defaulting participant has not remedied its default, the participant has to withdraw from the venture and surrender its participating interest to the remaining participants. During the period of the default the operator may sell the oil and gas entitlement, if any, of the defaulting participant and recover the cash contribution due from the realized proceeds.

11.2.2.14 Assignment of Interest

Any participant may freely assign its interest to anyone of its affiliates, subject to giving proper notice to the other participants. If a participant wishes to assign its interest to a third party, not being an affiliate, or to another participant, it must notify the non-involved participants offering the latter the opportunity to acquire the interest on the same or equivalent terms and conditions as have been agreed in a final form with the intended assignee (subject only to the preferential rights of the other participants). If none of the non-involved participants are interested to take the whole interest the participant may complete the transaction with the intended assignee, provided the terms and conditions as notified to the non-involved participants, and on which they based their decision not to make use of their preferential right remain the same. If the intended assignee is one of the participants and if anyone of the non-involved participant expresses an interest in an acquisition on the terms offered, then the offered interest will be shared among them in proportion to their respective participating interests. If controlling shares in a participant or controlling shares in an affiliate holding the participating interest are on offer, or a third party or another participant makes an offer for a controlling interest in a participant or in an affiliate holding the participating interest, the participant or affiliate must offer its participating interest to the non-involved participants on reasonable and fair terms. Any direct assignment of a participating interest involves the assignment of a corresponding share in the underlying authorization. Since the latter assignment requires the approval of government or state party, as the case may be, assignments of a participating interest are usually made subject to the condition that such approval is obtained.

11.2.2.15 Withdrawal

In principle, any participant may withdraw from the joint venture at any time subject only to giving a few months' written notice. After expiry of the notice period the withdrawal becomes effective. However, no participant is permitted to withdraw from the joint venture where the minimum exploration work programme and the minimum exploration expenditure commitments have not been fulfilled or the signature bonus has not been paid, unless the participant wishing to withdraw guarantees to contribute its (former) participating interest of the commitments after and irrespective of its withdrawal. After the effective date of its withdrawal, a withdrawing party remains liable for its participating interest share of all costs of joint operations or sole risk operations, that had been approved by a decision taken by respectively the management committee or the sole risk participants prior to the notice of withdrawal, unless the withdrawing participant had been a non-consenting participant. Similarly, a withdrawing party remains liable for all obligations and liabilities arising from events or acts that took place or were committed prior to the effective date of withdrawal. A withdrawing participant remains liable for future costs, resulting from joint commitments made prior to the notice of withdrawal, the most important of which are the costs of abandonment involving production facilities. A withdrawing participant may be required to provide security in the form of a bank guarantee for these future liabilities. The share to be contributed or to be guaranteed by a withdrawing participant in respect of abandonment liabilities is not necessarily the withdrawing participant's original participating interest share but in principle a share that takes into account the length of time that the withdrawing participant had enjoyed the use of or has benefited from the asset (well, field or installation) about to be abandoned. The withdrawing participant's participating interest to be surrendered will be distributed over the remaining participants in proportion to their respective participating interests, unless another distribution is agreed among them. The withdrawing participant's participating interest to be surrendered will be distributed over the remaining participants in proportion to their respective participating interests, unless another distribution is agreed among them. If the remaining participants refuse to accept the withdrawing participant's interest the joint venture will be terminated. As far as the jointly owned authorization is concerned a withdrawing party must assign its participating interest therein to the remaining participants in accordance with the applicable rules of the authorization. Similarly as for external assignments, an internal assignment requires the approval of the government or the state party as the case may be. If such approval is not forthcoming, a formal internal assignment cannot take place. The withdrawing participant remains formally, i.e., from the point of view of the government or the state party, one of the participants. The withdrawing participant is compelled then to stay in the joint venture but its participating interest is held for the risk, account and benefit of the remaining participants. The latter are solely responsible and have to keep the withdrawing participant secured and indemnified against any claims or demands brought against the latter by government, state party or any other third party.

11.2.2.16 Confidentiality

Each participant is obliged to keep all information concerning the joint venture including information about the underlying authorization or information about or generated by the operations undertaken by the operator strictly confidential and may only disclose said information to its parent company and affiliated companies. Any participant wishing to disclose such information to independent third parties needs the consent in writing of the other participants. No such consent will be needed in certain special circumstances, such as when a participant wishes to disclose information to financial institutions in order to obtain a loan or guarantee; to a stock exchange, where its shares or the shares of its parent company are listed, as required in accordance with the latter's rules; to a consultant to the extent needed by the latter in order to be able to render his services; to arbitrators in the context of the settlement of a dispute; or to an independent third party being a bona fide prospective buyer for the whole or part of the disclosing participant's interest. Usually, a disclosing participant is obliged to obtain from the person, to whom information is disclosed under the aforementioned circumstances, the undertaking to keep confidential the information disclosed to it and not to use such information for any purpose other than the purpose for which the disclosure was made. Furthermore any participant is free to disclose information if disclosure is required under applicable legislation. Each participant, in particular the operator, may put its proprietary technology at the disposal of the participants for use and application in the conduct of the joint or of any sole risk operations on such conditions as the participant intending to make its technology available may formulate. If the other participants do not agree to the conditions proposed by the disposing participant they have the right to reject and prohibit the use and application by the operator of the proprietary technology offered. The usual condition proposed is that the proprietary technology is applied on a 'black box basis', meaning that participants will be informed about the results obtained by the application of the technology without being informed about the technology itself. This approach leaves the receiving participants in the dark about the technology and makes it not possible for them to verify the results.

Note. As said before a participant may have been selected as operator exactly because he is in possession of superior technology and/or technical knowledge and is therefore supposed and expected to apply his technology and knowledge to the benefit of the joint venture without too many strings attached. Therefore the issue of confidentiality concerns not so much the operator but any of the other participants which although in possession of superior technical knowledge or technology have for one reason or another not been selected as operator. Meanwhile the participant (usually the operator) who puts its or its affiliated companies' proprietary technology at the disposal of the joint venture (or any particular sole risk operation) will charge the costs incurred to the joint or sole risk account. The other participants may object to this if the proprietary technology is applied on the aforementioned 'black box basis'.

11.2.2.17 *Force Majeure*

If as a result of *force majeure* circumstances any one participant is not able to fulfil its obligations under the JOA, other than the obligation to pay any amounts required to be paid, then the participant must immediately give notice to the operator and all other participants. The participant's obligations affected by *force majeure* shall be suspended for as long as necessary for the participant to put an end to the *force majeure* situation. If the operator is affected by *force majeure* it must resign as operator and transfer the operatorship to another participant. If the *force majeure* situation endures for an unreasonably long period the affected participant should leave the joint venture and sell its participating interest to the remaining participants or an independent third party in accordance with the rules of the agreement. As noted before, a participant is never excused by *force majeure* when failing to make its cash contributions in response of a cash call from the operator. Any such failure makes the participant into a defaulting party with all the consequences thereof as set out in the JOA. Preferably, *force majeure* is defined in the same terms as used in the underlying authorization. Generally, *force majeure* circumstances are circumstances which are beyond the reasonable control of the participant concerned, including in particular strikes, lockouts and other labour disputes. A participant will not be compelled to settle any such dispute except on terms acceptable to it.

11.2.2.18 Notices

Any JOA requires that all notices required between participants (e.g., a sole risk notice, a default notice or a notice announcing a change of address) be in writing, in the language specified and be delivered in person, by registered mail, by courier service or by electronic means. A notice under the JOA shall be deemed received by the participant, to which the notice is addressed, if delivered at the address of the participant specified in the JOA. This address may be changed by notice to all other participants.

11.2.2.19 Settlement of Disputes

Generally, participants, if of different nationalities, prefer to settle disputes arising between them in connection with their joint venture through international commercial arbitration by three arbitrators. The less preferable alternative is to start a procedure before the courts of law of the host country. International arbitration is preferred firstly because the JOA is an international agreement binding parties of different nationalities and secondly because the arbitration procedure allows participants to choose the individuals who will act as arbitrators, which offers the possibility to select persons who are familiar with petroleum operations and with the problems that may arise between commercial enterprises engaged in that activity. Additionally, if the host country is a non-Western country, the law chosen by the participants to govern their JOA will almost certainly not be the law of the host country. But the latter law is the only laws which with the national courts of law are familiar. If under these circumstances a procedure before a national court of law would be started, such a court would have to apply to an unfamiliar problem an equally unfamiliar system of law. To avoid these problems, participants generally choose the arbitration procedures and rules of the International Chamber of Commerce (ICC), but those of the UN Commission on International Trade Law (UNCITRAL) are also frequently referred to in a JOA. If the choice has fallen on ICC-arbitration the JOA will stipulate that any dispute shall be finally settled by three arbitrators appointed in accordance with the ICC Rules. As a rule, it is further stipulated that the nationality of the third arbitrator or referee shall be different from the nationality of any one of the participants. Furthermore the language of the arbitration procedure and the place of arbitration are specified. Additionally it is stated that the arbitral award is binding on the participants and may be entered into any court of law having jurisdiction.

11.2.2.20 Governing Law

Participants have to make a choice regarding the law system that will govern their JOA. The choice to be made depends to a large extent on the nationalities of the participants. If the law of the home country of one of the participants is chosen it would give this particular participant an advantage because the participant has or at least is assumed to have a superior knowledge about the law of its home country and the way this law is applied in practice. It should not be overlooked that the governing law is also the law that shall have to be applied in case of arbitration, in particular that part of the law that deals with contractual arbitration procedures. Generally, participants prefer to select a well-developed and sophisticated system of law that happens not to be the national law of any of them.

11.2.2.21 Accounting Procedure

The accounting procedure is an integral part of the JOA. This part provides for the rules in accordance with which the operator should keep and maintain the books of the joint account. The same rules apply and will be applied by the operator with respect to any particular sole risk account. The accounting procedure contains the cash call procedures, the classification of costs (whereby the classification used in the contract or applicable fiscal legislation is adhered to), rules about the acquisition and disposal of materials, payment for services rendered or materials supplied by operator's parent company or affiliated companies, audit rights to be exercised by the non-operating participants, the taking of inventories, etc. By preference, the rules of keeping books of accounts, that are customary within the group of companies, to which the participant/operator belongs, are adopted. This is one of the privileges of being selected and appointed as the operator.

11.3 THE UNITIZATION AGREEMENT

11.3.1 BACKGROUND

If a single, continuous petroleum reservoir extends across the (vertical projection of) the boundary of an authorization, into the adjacent area in respect of which another authorization has been granted, the question arises in which manner the said petroleum reservoir (referred to as a straddling petroleum reservoir) should be developed and produced by the holders of the authorizations involved. It makes a principal difference whether the authorizations applicable to the adjacent areas are production licences or contract of works. Where contracts of work are involved, the state party in question is holder of the underlying general or specific licences with title to petroleum from the adjacent contract of work areas and in this capacity the only possible and legitimate participant in any development of the straddling reservoir. Hence, in the situation of adjacent areas being subject to contracts of work the development of the straddling reservoir will be completely dictated by the state party (see Chapter 5). Obviously, the operational arrangements to be made by and between the contractors operating on the adjacent areas albeit under the supervision of the state party will be similar to arrangements agreed between the holders of production licences applicable to the adjacent areas. In the following only the latter situation will be considered.

In principle, the holder(s) of a production licence on either side of the common licence boundary are free to extract oil from their respective part of the straddling reservoir in whichever way they prefer as long as their method of extraction is in accordance with the rights conferred and obligations imposed by their production licence. Within this framework they are free to decide whether or not and in which manner to cooperate with each other in the exploitation of the straddling reservoir, unless any form of cooperation is imposed upon them as one of the conditions of their respective licences.

In the early days of oil production independent producers, which found themselves sharing a petroleum reservoir (oil pool) with others operating on adjacent property, were not interested in any form of cooperation. These pioneers considered themselves to be involved in a competition for the production of oil originating from the common pool. Hence they followed the practice of drilling as many wells as could be accommodated within the boundaries of their respective plot of land and of producing these wells at the highest rate possible convinced that in this manner they would secure for themselves the largest possible share of the pool. As soon as producers gained a better insight into and understanding of the characteristics of the oil reservoirs and of the driving mechanism causing oil to flow through the pores and fissures such rock towards the borehole, they came to appreciate that by cooperating with one another they could avoid wasting their money on competitive drilling and still achieve a higher ultimate recovery. The technical necessity and economic benefit of some form of cooperation became still more apparent and compelling when the injection into the reservoir of certain flow stimulating substances, such as steam, water, natural gas or CO_2, proved to be needed in order to achieve a maximum efficient recovery. To be effective, the injection wells may have to be drilled on one side of the common licence boundary, whereas the additional production resulting from the application of the stimulating measures is produced by and through the wells located on the other side. Under these circumstances no injection scheme will be implemented if no form of cooperation is agreed beforehand. This may be different if only the water-filled part of the reservoir rock, in which the petroleum accumulation is located, extends into the adjacent area and if the required injection wells have to be drilled into that part. In such a situation, the licensees involved may need to do no more than to conclude a simple agreement permitting the drilling and operation of the injection wells against the payment of an appropriate compensation.

11.3.2 Principles and Objectives

A unitization agreement embodies and provides for cooperation between two or more persons holding separate production licences in respect of a straddling reservoir. Its objective is to develop and produce such a straddling reservoir in an integrated manner for the common benefit of all the holders of the licences involved. Its basic principle is that costs, liabilities, petroleum production and other benefits of the intended joint exploitation, are shared among and between the said holders on the basis of their respective shares in the unit. A unitization agreement may be described as a simplified form of a joint operating agreement to which some special features are added. These special features then concern the identification and demarcation of the straddling petroleum reservoir and the allocation of the percentage shares therein to each individual participating licence. The percentage shares are referred to as the tract participations. If a participating licence is held by more than one person, those co-licensees form a group jointly entitled to and holding the tract participation assigned to their licence. Hence, the tract participation is held and shared among these co-licensees in proportion to their respective percentage participating interests in the jointly held licence. A participant's unit percentage interest is then calculated by multiplying the said tract participation by the said percentage participating interest. The participants are thus brought into a position of being holders of undivided participating interests in the unit as if they were joint holders of a single licence granted in respect of said unit. On the other hand, if a participant (single participant) is the sole owner of a participating licence, its unit percentage interest is equal to the tract participation assigned to that licence.

11.3.3 THE PRE-CONTRACTUAL PERIOD

11.3.3.1 Subjects to Be Agreed

The pre-contractual period is the period lying between the moment that persons holding licences recognize and accept that there is a single continuous petroleum reservoir, which extends beyond and across the boundary of their respective licences and that this reservoir needs to be unitized, and the moment of signing of the unitization agreement. During this period the prospective unit participants have to reach agreement (to be laid down in the aforesaid unitization agreement) on the following subjects:

(i) the limits, extent, and configuration of the common petroleum reservoir to be unitized;
(ii) the mineral characteristics and mechanical condition of the reservoir rock and the characteristics of the petroleum contained therein;
(iii) a definition of 'petroleum reserves' which will have to be divided between the participating petroleum rights;
(iv) the manner in which the aforesaid petroleum reserves should be so divided (the calculation of the initial tract participations);
(v) the time schedule for any re-determinations of the limits of the unit, total petroleum reserves and tract participations;
(vi) the appointment of a unit operator;
(vii) the outline of the development plan; and
(viii) sharing or non-sharing of pre-unitization costs.

11.3.3.2 Collection of Information and Cost Sharing

At the start of the pre-contractual period the knowledge about the reservoir rock, its configuration and the petroleum contained therein is incomplete. Hence, the main activity during the pre-contractual period consists in collecting information regarding the straddling reservoir, including carrying out seismic work and the drilling of wells. In principle, any such work has to be paid for by the group, on whose side of the common boundary the work has to be carried out. Sharing of costs in proportion to unit percentage interests starts not before the unitization agreement has become effective, unless and to the extent the negotiating groups agree otherwise. Collection of information is stopped as soon as the groups feel that there is a reasonable basis on which to conclude a unitization agreement with confidence. Anyway, finding the right balance between reducing uncertainties and the costs of collecting additional information is one of the most difficult issues to be solved in the pre-contractual period. The matter becomes more complicated if at the start of this period more information about the one part of the reservoir is available than about the other, simply because the group (in this context also comprising a single participant) on the one side of the common boundary had started its exploration operations earlier than the group on the other side. In many cases drilling has taken place on the one side, while perhaps only seismic work has been done on the other. Collecting more information is then the responsibility of the group that is lacking in this respect and any make-up work has to be carried out before the negotiations on the formation of the unit can be continued.

11.3.3.3 The Division of the Petroleum Reserves

Having agreed that sufficient information is available with respect to both parts of the reservoir, the negotiating groups have to reach agreement on the methods, formulae and information to be used to assess and establish the initial tract participations (such methods and formulae to be used also for any subsequent re-determination). In principle, each individual tract should be entitled to receive the quantity of petroleum that can be extracted or recovered from it. However implementing this principle is not simple. Information is needed on the limits of the individual tracts, the quantity of petroleum contained therein and the applicable recovery factor, i.e., the fraction of the quantity in situ that can be recovered. Starting point is always trying to make an estimate of the quantity of petroleum in situ. This quantity is found by calculating the quantity of petroleum occurring per cubic metre of reservoir rock and the volume of the reservoir rock saturated with oil or natural gas. Using this method the Stock Tank Oil Initially in Place (STOIIP) or the Gas Initially in Place (GIIP) is found. As far as oil is concerned samples have to be taken and analyzed in order to determine the quantity of gas dissolved in the oil and the shrinkage factor to arrive at the stock tank oil volume. Basing the allocation of the reserves on the aforesaid recovery principle proves difficult for two reasons.

Firstly, the recovery factor is difficult to determine in as much its value depends on:

(i) the mineral characteristics of the reservoir rock (grain size distribution, clay distribution, cementation, porosity, permeability, etc.);
(ii) the mechanical condition of the reservoir rock, such as the presence of sealing or non-sealing faults;
(iii) the flow properties of the petroleum to be extracted from the reservoir rock, such as viscosity; and
(iv) on the methods of extraction that will be applied within the framework of a development plan still to be agreed, such as the tightness of the drilling pattern, the use of horizontal drilling and the application of enhanced recovery methods, e.g., the injection of either water, natural gas, steam or CO_2.

Note. All these factors and methods of development cannot be finally decided upon before the development operations have got under way and a few development wells have been drilled. Moreover, the mineral characteristics of the reservoir rock may change as a result of the compaction of the reservoir rock caused by the extraction of the petroleum. Any such change will certainly affect the numerical value of the aforesaid recovery factor.

Secondly, if the reservoir rock belonging to the unit appears not to be homogeneous, such that the one part of the unit has excellent reservoir characteristics and the other part not, then more effort should be spent on the poorly developed part (e.g., by drilling of more or more expensive wells and/or by applying expensive enhanced recovery techniques) in order to match, if at all possible, the high recoveries expected to be achieved (at much lower cost) in the well-developed part of the reservoir, or it should be accepted that an important part of the petroleum in situ will remain behind in the other, poorly developed part. It follows that, as far as heterogeneous reservoirs are concerned, a proper and fair implementation of the division of the unit on the basis of quantities of petroleum recoverable from the individual tracts must be accompanied by an accounting system under which costs are allocated between the well-developed part of the unit and the less well-developed part. Such cost allocation would be a purely theoretical accounting exercise because it is the intention that the unit will be developed in an integrated manner.

11.3.3.4 Adjustment of STOIIP/GIIP Calculation Method

It will only be acceptable solely and exclusively to calculate tract participations on the basis of either the STOIIP or the GIIP, if the reservoir rock is homogeneously developed, i.e., if the petroleum in situ can be extracted throughout the unit with the same efficiency and effect. If it appears, in the light of the information available or becoming available in the course of the information collection work, that this is not the case and that it is to be expected that in the course of the production operations in some parts of the unit more oil or natural gas will remain behind than in other parts, then a division of the petroleum reserves on the basis of either STOIIP or GIIP must be adjusted in order to take account of these differences. Within this framework negotiating groups have several options. They could agree to ignore the poorly developed parts of the unit and leave such rock volumes outside the calculation of STOIIP or GIIP. It could also be agreed to adopt a formula in accordance with which the tract participation (TP) is calculated partly, say X per cent, on the basis of STOIIP or GIIP and partly, say (1-X) per cent, on the basis of a property (P) of the reservoir rock, i.e., the property, which is mainly responsible for the excellent recovery (meaning a high value of the recovery factor). This approach yields a formula of the general form of: TP = [X per cent STOIIP] + [(1-X) per cent P]. In this manner, the parts of the unit benefiting from a high recovery factor are weighted in favour of greater tract participation Obviously, after having decided to make use of such a weighted formula, the groups have still to agree on which reservoir rock property should be used as 'P', on the value of the factor X and, not to forget, on the limits of the unit itself.

11.3.3.5 Re-determinations

During the pre-contractual period participants have also to reach agreement on the timing of the re-determinations. Re-determinations work retrospectively up to the moment the unit was formed and the initial tract participations were determined. A re-determination re-establishes with respect to that particular moment the mineral characteristics and mechanical properties of the reservoir rock, the limits of the unit, and the volume and characteristics of the petroleum contained therein and recalculates, on the basis of this new data, the apportionment of the (possibly on the same occasion re-determined) unit. The so re-calculated apportionment is assumed and considered to have been in effect from the date the initial apportionment had been in effect and applied. Re-determinations are necessary because, due to information obtained from the drilling of the development wells and from additional seismic work, more and more details about the extent and configuration of the unit become known. Chances are that the initial determinations will prove not to be correct. Furthermore, as far as the unit itself is concerned, it may appear that contrary to earlier expectations the unit proves to be in communication with other petroleum reservoirs in the vicinity causing an in-flow of petroleum from these other reservoirs. On the other hand it may also possible that the unit proves to be smaller than originally thought due to the presence of sealing faults, causing a compartmentalization of the unit. Under these circumstances participants have to re-establish the limits of the unit which may result in the inclusion of the nearby reservoirs in the unit, (unless the petroleum extracted from the other reservoirs can easily be accounted for and in one way or another be kept separate from unit-production), or may lead to the exclusion of certain compartments from the unit. If participants were not given the assurance that the initial tract participations and unit percentage interests would regularly be re-examined and, if need be, re-determined in the light of the latest information about the rock strata, much more appraisal work on either side of the dividing line and for the sole account of the right-holders concerned would have to be done during the pre-contractual period, causing unnecessary drilling and unnecessary costs.

Re-determinations have important consequences. Participants will acquire with retroactive effect new unit percentage interests. This means that participants' rights and obligations, which are expressed in terms of their respective unit percentage interests, such as the obligation to contribute cash and the entitlement to a percentage share of unit production, have to be revised and the appropriate adjustments have to be made. Obviously, the spacing in time and number of the re-determinations depend on the size and complexity of the unit, as well as on the possible presence of other petroleum reservoirs in the vicinity which in a later stage may prove to be in communication with the unit. If, in practice, it appears that only a small part of the reservoir is extending across the common boundary, and that this part can easily be exploited by wells drilled on the other side, then the licensee in possession of the smallest part may simply agree to forego any re-determination. Independent experts may have to be called in if participants are unable to agree on which data and factors to be used in the calculations carried out to estimate the initial and subsequent tract participations. Somctimes it is agreed between the groups involved to entrust from the outset to independent experts the whole matter of the identification and demarcation of the unit and the determination and re-determination of tract participations.

11.3.3.6 Appointment of a Unit Operator

In complex and time consuming cases negotiating groups may decide, before starting the negotiations, to appoint an operator, whose first task will be to organize the whole process of making the intended unitization agreement. Usually special negotiating committees are set up over which the various subjects are distributed.

11.3.4 THE CONTENTS OF A UNITIZATION AGREEMENT

11.3.4.1 Participating Interests

The agreement identifies and describes the participating oil companies and the respective licences held by them, either solely (single participants) or in undivided interests (groups).

11.3.4.2 Definitions

With respect to definitions there is no difference with a customary joint operating agreement, except for the addition of some specific expressions related to the unitization process, such as ‘the unit’, ‘tract participation’, ‘unit percentage interest’, ‘re-determinations’, ‘petroleum reserves’, ‘STOIIP’, ‘GIIP’, ‘recovery factor’, etc.

11.3.4.3 Purpose and Scope

The scope and purpose of a unitization agreement is the exploitation of the unit for the joint account of the participants on the basis of sharing costs, liabilities, oil and gas production and other benefits in proportion to the participants’ respective unit percentage interests, and in accordance with the provisions of the agreement and the respective licences.

11.3.4.4 Description and Demarcation of the Unit

The unit is described in geological terms, accompanied by a demarcation thereof. The demarcation is on the one hand depicted on a map, which also contains the boundaries of the participating licences, and on the other defined in terms of coordinates. The area so demarcated and defined is the unit area. Generally, it is stipulated that the unit as defined is subject to re-determination from time to time in accordance with the rules provided for by the agreement.

11.3.4.5 Duration

The unitization agreement will endure as long as the participating licences endure, including any extensions thereof.

11.3.4.6 Tract Participations and Unit Interests

The agreement establishes the initial tract participations and the unit percentage interests of the individual participants. These tract participations are subject to re-determinations from time to time in accordance with the provisions of the agreement. Unit percentage interests are subject to change as a result of the re-determination of the tract participations and as a result of assignments.

11.3.4.7 Re-determinations

The limits of the unit, the petroleum reserves contained therein and the division of the unit between the tracts (i.e., tract participation), are subject to re-determination from time to time. The methods, formulae and information to be used for the division of the unit will be described in an annex to the agreement. It should be made clear therein whether the division of the unit (and its own possible re-determination) will be based on reserves in situ, on recoverable reserves or on a mixture of both types of reserves.

11.3.4.8 Adjustments after Re-determination

Any re-determination of tract participations resulting in a change in percentage figures has retroactive effect. The unit percentage interests are correspondingly changed. Any change requires adjustments of the past cash contributions and offtake of unit production. Over-contributors are paid back with interest. Under-lifters receive a make-up right at the expense of the over-lifters, to be exercised within a specified period and restricted to a certain percentage of the unit percentage interest of the over-lifter. Any remaining imbalance is to be settled in cash.

11.3.4.9 Unit Operator

One of the participants is appointed as operator of the unit. Authorities and duties of the unit operator are similar to those described in a customary joint operating agreement. The unit operator is authorized to make cash calls on each individual unit participant in proportion to the latter's unit percentage interest. A unit participant not responding to the cash call within the period specified for such response is in default (see below). In complex cases prospective participants may appoint an operator prior to signing the unitization agreement. The so-appointed operator is charged with the task to organize the negotiations taking place in the pre-contractual period (see above). Any appointment of a unit operator should be made in consultation with the competent authorities and after having received their approval.

11.3.4.10 Unit Management Committee

The operations are supervised and controlled by a unit management committee in which each individual unit participant is represented with the right to cast a vote in proportion to its unit percentage interest.

The main task of the unit management committee is:

(i) to supervise the implementation of the development plan in accordance with the principles and outline thereof as laid down in an annex to the agreement;
(ii) the award of major contracts; and
(iii) the supervision of the re-determinations in accordance with the rules and procedures annexed to the agreement, which may include the supervision of the work of independent experts, if the re-determinations have been entrusted to such persons.

Decisions of the management committee are binding on the participants if taken with the required majority of votes. Under most circumstances a simple majority would be acceptable. However if a group or a single participant has a majority tract participation, it may be decided that the voting majority required for binding decisions should always include the votes of participants belonging to other groups, so that each group is represented in the majority by at least one of its members. Considering that the principles and outline of the unit development plan are laid down in an annex to the agreement, and have been arrived at in the pre-contractual period by agreement of all participants, any significant change to this plan must of necessity require unanimity.

11.3.4.11 Unit Development Plan

Development of the unit has to take place within the framework of the development plan agreed and set out in an annex to the agreement. A different plan may be adopted by the management committee, if so warranted by new information or other circumstances. Any significant change can only be adopted by unanimous vote.

11.3.4.12 Sole Risk Operations

In the context of a re determination, each group (in this context also meaning any single participant) has the right to execute for its own account and risk seismic work and/or a drilling operation within the boundaries of its tract for the purpose of obtaining information expected to have a significant bearing on the outcome of a forthcoming re-determination. The sole risk work will be carried out by the unit operator. If the unit management committee so agrees, any well drilled in this context may, if suitably located and fitting into the unit development plan, be taken over by the unit and used as a production well. In such event, the costs incurred by the sole risking group will be reimbursed to it. If a sole risk well has yielded data which appears to be of significant influence on the outcome of a re-determination, the unit is obliged to take over the well and the sole risking group shall be compensated for the costs incurred by it. The question of whether data obtained from a sole risk well have made such a significant contribution that the well must be taken over by the unit, may be submitted to an independent expert for a binding decision.

Note. Within a group and within the framework of the applicable joint operating agreement, any one participant may have made the proposal to execute an operation for obtaining additional information. Such a proposal will only be presented as a proposal from the group, if the management committee of that particular joint venture so decides. It will not be possible that the proposed operation is executed as a sole risk operation for the sole account of the proposing participant, because any such operation (being part of a development plan, i.e., the unit development plan) will not qualify as a sole risk operation in terms of the joint venture concerned.

11.3.4.13 Non-unit Operations

Any unitization agreement may provide depending on the circumstances of the case that the unit drilling and unit production installations and unit field pipelines may be used for the development and production of reservoirs located outside the unit. For such use, a tariff has to be paid to the unit owners. This is a very useful provision in cases where the unit is surrounded by small, independent, non-straddling reservoirs which cannot be exploited on a stand alone basis.

11.3.4.14 Default

The principal obligation of any unit participant is to make proportional cash contributions in response to cash calls from the unit operator, if and when made in accordance with the procedures of the agreement. A lack of response for whatever reasons to such cash calls constitutes default. The customary joint venture rules about the consequences of default will be applicable, but if the defaulting participant belongs to a group, and not all members of that group are defaulting, its default will be treated under the applicable rules of that group's joint operating agreement. This means that its default will be made good by the other members of the defaulting participant's group and that, if the default continues beyond a specified period, the defaulting participant must withdraw from the group's joint venture and assign its interest to the other members.

11.3.4.15 Withdrawal

Each participant has the right at any time to withdraw from the unit in accordance with the usual conditions as contained in a customary joint operating agreement, the most important of which is the condition that the withdrawing participant remains liable for paying its percentage unit share of the costs of settling liabilities that arose before the effective date of withdrawal, and an appropriate share of abandonment expenditure arising in the future with respect to unit facilities installed prior to the effective date. If a participant is a member of a group, its withdrawal will only affect the composition of its group and will be treated under the rules of that group's joint operating agreement. This means that its interest will be assigned to the other members of its group, unless all the other members of the group similarly decide to withdraw from the unit.

11.3.4.16 Assignment of Unit Interests

Assignments of unit interests may be voluntary or compulsory. Similar conditions apply as incorporated in a customary joint operating agreement: namely free assignment to affiliated companies and a matching right for other participants in case a transfer to an independent third party or another participant is intended. Compulsory assignments are the assignments to be made to the other participants in case of default and withdrawal. If the assignor belongs to a group the intended assignment will be treated under the rules of that group's joint operating agreement, and in principle only affects the composition of that group. In all cases the approval of the competent authorities will be required.

11.3.4.17 Other Conditions

Other conditions, such as conditions concerning confidentiality of information, giving and receiving notices, settlement of disputes, and governing law, are similar as those contained in a customary joint operating agreement.

11.3.4.18 Approval

The unitization agreement and its annexes will have to be approved by the competent authority (or authorities if international unitization is involved). Usually, any re-determination of the limits of the unit, the total reserves and the tract participations, are subject to the competent authority's approval.

11.3.4.19 Accounting Procedure

The unit operator must keep the books of the unit accounts in accordance with the rules of a customary joint venture accounting procedure. Where an interstate unitization is concerned the books must classify costs in such a manner that the rules of the applicable income tax legislations or cost recovery procedures can be met.

11.3.4.20 Annexes

The unitization agreement should be accompanied by the appropriate number of annexes covering the following subjects:

(i) the geological and topographical description of the unit;
(ii) the definition and calculation of the petroleum reserves contained in the unit;
(iii) the allocation of the reserves to the participating exploitation rights (tract participations);
(iv) the methods and information to be used in the successive re-determinations of the reserves, tract participations and the extent of the unit; and
(v) the procedures to be followed in appointing an independent expert and a description of the task to be performed by such expert.

Bibliography

Aghazadeh, G. 'The Imperative Need for Oil Producer-Consumer Cooperation in the 1990s'. *Middle East Economic Survey*, 3 June 1991.

Ait-Laoussine, N. 'OPEC Could Slash Output, Still Meet Own Needs'. *Petroleum Intelligence Weekly*, 22 September 1980.

Ait-Laoussine, N. 'The 1986 OU Price War: an Economic Fiasco'. *Middle East Economic Survey*, 6 October 1986.

Ait-Laoussine, N. 'Recent Developments in the Oil Market, 1986 Revisited?'. *Middle East Economic Survey*, 22 February 1988.

Ait-Laoussine, N. & J. Wood-Collins. 'Energy and the Environment'. *Middle East Economic Survey*, 19 March 1990.

Al-Chalabi, I. 'The Oil Market: The Shape of Things to Come'. *Middle East Economic Survey*, 23 October 1989.

Al-Farsy, F. *Modernity and Tradition, The Saudi* Equation. Kegan Paul International Ltd: London, 1990.

Anderson, R.N. 'Oil Production in the 21th Century'. *Scientific American*, March 1998.

Baker, A. 'Oudook for the Coal Industry'. In Proceedings of the Mining Students Society's Symposium on Potential Resources for the Coming Century (Delft University of Technology, Delft, November 1992).

Ballou, G.T. 'Oil Industry and Government Relations, A Study of Emerging Patterns'. *The Exploration and Economics of the Petroleum Industry*, vol. 4 (The Southwestern Legal Foundation, Gulf Publishing Company, Houston, Texas, 1966).

Bambridge, M. 'Unitisation Agreements'. In Proceedings of the Conference on Oil and Gas Agreements (Langham Oil Conferences Ltd, London, November 1996).

Banks, F.E. 'Some Economic Aspects of the World Coal Market'. *Resources Policy*, December 1989.

Barbier, E.B. 'The Global Greenhouse Effect'. *Natural Resources Forum*, February 1989.

Barents, R. 'Legal Aspects of Dutch Energy Policy'. *Journal of Energy and Natural Resources Law*, 1982.

Baum, V. 'World Bank-Aid for Third World Energy Projects'. *Petroleum Economist*, March 1986.

Beal, A. 'North Sea Joint Bidding and Joint Operating Agreements'. In Proceedings of the Conference on Oil and Gas Agreements (Langham Oil Conferences Ltd, London, November 1996).

Bley, L. 'Environment-Greenhouse Gas Emission Divisions'. *Petroleum Economist*, October 1990.

Boehmer-Christiansen, S. 'Marine Pollution Control: UNCLOS III as the Partial Codification of International Practice'. 7 *Environmental Policy and Law*, 1981.

Bohm, J.T.C. 'Waste Products Generated by Petroleum Operations: Dumping, Discharging, Prevention or Re-use?'. In Proceedings of the Mining Students Society's Symposium on Dealing with the Environmental Problem in the Mining and Petroleum Industry (Delft University of Technology, Delft, April 1990).

Bourne, L. 'United States – The Importance of Imported Oil'. *Petroleum Economist*, August 1990.

Bourne, L. 'Whither OPEC Production Now?'. *Petroleum Economist*, September 1990.

Breij, A.C.M. *The Coal Mines Went Open, the Coal Mines Closed Down* (ICOB cv, Alphen aan de Rijn, the Netherlands, 1991).

Broches, A. 'The Convention on the Settlement of Investment Disputes: Some Observations on Jurisdiction'. 5 *Columbia Journal of Transnational Law*, 1966.

Broecker, W.S. & G.H. Denton. 'What Drives Glacial Cycles?'. *Scientific American*, January 1990.

Brondel, G. 'Co-operation Between the European Community and the Developing Countries in the Energy Field'. In Proceedings of the United Nations Symposium on Financing of Petroleum Exploration and Development in Developing Countries (Athens, 22–27 April 1985).

Brothwood, M. 'Legal Aspects of Gas Carriage'. *4 Journal of Energy and Natural Resources Law*, vol. 6, 1988.

Brown, E.D. 'The UN Convention on the Law of the Sea 1982'. *Journal of Energy and Natural Resources Law*, 1984.

Brown, R. 'Choice of Law Provisions in Concession and Related Contracts'. 39 *The Modern Law Review*, November 1976.

Brown, R. 'The Relationship between the State and the Multinational Corporation in the Exploitation of Resources'. *International and Comparative Law Quarterly*, January 1984.

Brown, R. 'Contract Stability in International Petroleum Operations'. 29 *The UNCTC Reporter*, 1990.

Brtjmfield, J.B. 'State and Federal Laws and Regulations on Offshore Leases, Drilling and Production'. *Ralph Slovenko's Oil and Gas Operations: Legal Considerations in the Tidelands and on Land* (Tulane University School of Law, Claitor's Law Books, Claitor's Book Store, 1963).

Bums, N. 'The Oil Industry in Transition'. *Oil & Gas Journal*, 8 November 1982.

Cameron, P. The Regulation of Energy Supply: The Netherlands'. *Journal of Energy & Natural Resources Law*, Supplement, December 1989.

Campbell, C.J. & J.H. Laherrere. 'The End of Cheap Oil'. *Scientific American*, March 1998.

Campbell Jr, N.J. 'International Law Developments concerning National Claims to and in Offshore Areas'. *Ralph Slovenko's Oil and Gas Operations: Legal Considerations in the Tidelands and on Land* (Tulane University School of Law, Claitor's Law Books, Claitor's Book Store, 1963).

Carpio, D.T. 'The World Bank's Approach to Financing Petroleum Projects'. In Proceedings of the United Nations Symposium on Financing of Petroleum Exploration and Development in Developing Countries (Athens, 22-27 April 1985).

Champion, D., C. Girard, & T. Daubignard. 'Oil and Gas Reserves of the Middle East and North Africa'. *Natural Resources Forum*, August 1991.

Clark, W.C. 'Managing Planet Earth'. *Scientific American*, September 1989.

Daintith, T. (ed.). *The Legal Character of Petroleum Licences: A Comparative Study* (University of Dundee, Centre for Petroleum and Mineral Law Studies, 1981).

Dalgaard-Knudsen, E. 'Exploitation Concessions: Contracts or Permits? Contributions from the Norwegian Phillips/Ekofisk Case'. *3 Journal of Energy and Natural Resources Law*, vol. 5, 1987.

Date-Bah, S.K. 'The Contract as a Mechanism for Taxation'. *Journal of Energy & Natural Resources Law* 1988.

Davidson, A.G.H. 'Tax Issues'. In Proceedings of the Conference on Unitisation of Oil and Gas Fields (Langham Oil Conferences Ltd, London, May 1991).

Davis, G.R. 'Energy for Planet Earth'. *Scientific American*, September 1990.

Davis, M.J. 'The Challenge of a Changing World'. *Ralph Slovenko's Oil and Gas Operations: Legal Considerations in the Tidelands and on Land* (Tulane University School of Law, Claitor's Law Books, Claitor's Book Store, 1963).

Davis, W. 'Global Energy Issues from the Standpoint of the Private Corporation'. *Houston Law Review*, vol. 17:1005, 1980.

De Sitter, L.U. 'Petroleum Geology', *Servire's Encyclopaedic, Servire* (The Hague, 1950).

Decker, H. 'Oil's Future – Its Challenges and Opportunities'. *The Economics of the Petroleum Industry*, vol. 2 (The Southwestern Legal Foundation, Gulf Publishing Company, Houston, Texas, 1964).

Delaume, G.R. 'Convention on the Settlement of Investment Disputes between States and Nationals of Other States'. 1 *International Lawyer*, vol. 1, 1966.

Denekamp, E.J. 'The European Energy Charter'. In Proceedings of the Mining Students Society's Symposium on Potential Resources for the Coming Century (Delft University of Technology, Delft, November 1992).

Despraires, P. 'Oil Demand Prospects'. 19 *Middle East Economic Survey*, May 1986.

Desprairies, P. 'The World Energy Situation'. *Natural Resources Forum*, February 1989.

Driessen, C.F. 'Oil Concessions: Governments and Oil Companies & Oil Laws in Malta and Elsewhere'. Paper given at the Malta Oil Seminar (October 1971).

Driver, J.M. *The Templars, Holy Warrior Monks of the Ancient Lands*. London: Kandour Limited, 2007.

Dryland, N.H.H. 'Government View on Promoting Investments'. In Proceedings of the Mining Students Society's Symposium on the Current Political and Economic Outlook for the Mining and Petroleum Industry (Delft University of Technology, Delft, November 1987).

Ely, N. 'Summary of the Mining and Petroleum Legislation of the United States of America'. In Proceedings of the ECAFE Seminar on Mining Legislation and Administration (Manila, October 1969).

Ely, N. 'Changing Concepts of the World's Mineral Development Laws'. In Proceedings of the IBA Seminar on World Energy Laws (Stavanger, Norway, May 1975).

English, W. 'Unitisation Agreements'. In Proceedings of the Conference on Oil and Gas Agreements (Langham Oil Conferences Ltd, London, December 1994).

European Commission. 'Energy for a New Century: The European Perspective'. *Energy in Europe*, Special Issue, July 1990.

Fee, D. *Petroleum Exploitation Strategy*. Belhaven Press: London, 1988. Fouda, S.A. 'Liquid Fuels from Natural Gas'. *Scientific American*, March 1998.

Frihagen, A. 'The Legal Protection of a Production Licence Holder against New Regulations and Other Government Interference'. In Proceedings of the 10th annual Bergen Oil and Gas Conference (Solstrand, Norway, December 1987).

Fulkerson, W., R.R. Judkins, & M.K. Sanghvi. 'Energy from Fossil Fuels'. *Scientific American*, September 1990.

Gallins, G. 'Bilateral Investment Protection Treaties'. *Journal of Energy and Natural Resources Law* 1983.

Gault, I.T. 'Joint Development of Offshore Mineral Resources – Progress and Prospects for the Future'. 3 *Natural Resources Forum*, vol. 12, 1988.

George, R.L. 'Mining for Oil'. *Scientific American*, March 1998.

Glraud, P.N. 'International Coal Prices'. *Natural Resources Forum*, November 1989.

Graedel, T.E. & P.J. Crutzen. 'The Changing Atmosphere'. *Scientific American*, September 1989.

Greenwald, G.B. 'Natural Gas Contracts Under Stress: Price, Quantity and Take or Pay'. *Journal of Energy and Natural Resources Law* 1987.

Greenwald, G.B. 'Encouraging Natural Gas Exploration in Developing Countries'. 3 *Natural Resources Forum*, vol. 12, 1988.

Gribben, R. 'Kuwait's Treasure Chest'. *Petroleum Economist*, September 1990.

Grotens, A.H.P. 'The Dutch Gas Supply System in the 1990s'. In Proceedings of the Workshop 'The Eve of the 7th Round' (organized by the Firms Loyens & Volkmaars and Nauta Van Haersolte. The Hague, November 1989).

Hammer, A. *Hammer: Witness to History*. Simon & Schuster Limited: 1987.

Hangher, L. & T. Daintith. 'The Regulation of Energy Supply: The United Kingdom'. *Journal of Energy and Natural Resources Law*, Supplement, December 1989.

Hancher, L. 'A Single European Market for Oil and Gas – The Legal Obstacles'. 2 *Journal of Energy and Natural Resources Law*, vol. 8, 1990.

Harders, J.E. 'The EEC's Continental Shelf: A "Terra Incognita"'. 4 *Journal of Energy and Natural Resources Law*, vol. 8, 1990. HEARD, Wilbur W. 'Pending Legislation to Amend the Natural Gas Act'. *Ralph Slovenko's Oil and Gas Operations: Legal Considerations in the Tidelands and on Land* (Tulane University School of Law, Claitor's Law Books, Claitor's Book Store, 1963).

Heren, P. 'Natural Gas – Green and Growing Markets'. *Petroleum Economist*, August 1990.

Hlgginson, A.J. 'Comparison of Terms to Attract Petroleum Development in Four Countries: The Arab Republic of Egypt'. In Proceedings of the IBA Energy Law Seminar, vol. 2, 1979.

Hlllyer Jr, H.H. 'A Primer on Producer Price Regulation'. *Ralph Slovenko's Oil and Gas Operations: Legal Considerations in the Tidelands and on Land* (Tulane University School of Law, Claitor's Law Books, Claitor's Book Store, 1963).

Hodgshon, S. & B. Popescu. 'Survey of World Petroleum Activity'. *Petroleum Economist*, October 1989.

Horgan, J. 'Fusion's Future: Will fusion-energy reactors be too complex and cosdy?'. *Scientific American*, February 1989.

Horn, N., et al. 'Legal Problems of Codes of Conduct for Multinational Enterprises'. In *Studies in Transnational Economic Law*, vol. 1 (Kluwer: Deventer, 1980).

Houghton, R.A. & G.M. Woodwell. 'Global Climatic Change'. *Scientific American*, April 1989.

Htun, N. 'Environmental Perspectives and Response Options'. In Proceedings of the Mining Students Society's Symposium on Dealing with the Environmental Problem in the Mining and Petroleum Industry (Delft University of Technology, Delft, April 1990).

Hughes, I.W.G. 'The UK Department of Energy's View on Unitisation'. In Proceedings of the Conference on Unitisation of Oil and Gas Fields (Langham Oil Conferences Ltd, London, May 1991).

Hurst, C. *The Continental Shelf*. Transactions of the Grotius Society, vol. 34, reproduced in the Proceedings of the 1950 ILA Copenhagen Conference on the Rights to the Seabed and its Subsoil (1950).
Huxtable, A. 'Joint Bidding Agreements'. In Proceedings of the Conference on Oil and Gas Agreements (Langham Oil Conferences Ltd, London, December 1994).

ILA. 'The Exploration and Exploitation of Minerals on the Ocean Bed and its Subsoil'. In Proceedings of the ILA Conference (Buenos Aires, 1968).
ILA. *Removal of Installations in the Exclusive Economic %one*. Report of the Netherlands Branch (October 1983).
ILO. *Social and Economic Effects of Petroleum Development Programmes*. Consultant's Report (International Labour Office, Geneva, September 1984).

Jaidah, A.M. 'OPEC, Non-OPEC and Cooperation in Pursuit of Price and Market Stability'. *Middle East Economic Survey*, 11 March 1985.
Jaworek, W.G. 'Forecasting the Future of the International Petroleum Industry'. *The Exploration and Economics of the Petroleum Industry*, vol. 5 (The Southwestern Legal Foundation, Gulf Publishing Company, Houston, TX, 1967).
Johnson, C.J. 'Establishing an Effective Production Sharing Type Regime for Petroleum'. *Resources Policy* 1981.
Johnson, C.J. 'Ranking Countries for Minerals Exploration'. *Natural Resources Forum*, August 1990.
Johnston, F.M. 'Political Risk Market Expansion Broadens OPIC's Role'. *Risk Management*, February 1984.
Jones, A. 'Russia'. *Petroleum Economist*, Special Supplement 1992.
Jonk, G.J. *The Search for, the Demand, the Supply and the Price of Oil in the Future*. (Delft University of Technology, January 1990).
Jonker, J., J.L. van Zanden, S. Howarth & K. Sluyterman. *A History of Royal Dutch Shell*, Vols 1, 2 & 3. Oxford University Press, 2007.
Jonkman, R.M. *Modern Production Sharing Contracts*. Shell Internationale Petroleum Maatschappij B.V: The Hague, October 1990.

Kapteyn, P.J.G. 'The United Nations and the International Economic Order'. *Studies of International Economic Law, Part LI*. T.M.C Asser Institute: The Hague, 1977.
Karl, T.R., N. Nicholls, & J. Gregory. 'The Coming Climate'. *Scientific American*, May 1997.
Kellogg, W.W. 'Modelling Future Climate'. *Ambio, a Journal of the Human Environment*. The Royal Swedisch Academy of Sciences: Stockhom, 1981.
Kemp, A.G. 'Petroleum Exploitation and Contract Terms in Developing Countries after the Oil Price Collapse'. *Natural Resources Forum*, May 1989.
Kemp, A.G. & D. Rose. 'Investment in Oil Exploration and Development: A Comparative Study of the Effects of Taxation'. In Proceedings of the International Conference on Risks and Returns in Large-Scale Natural Resources Projects (Bellagio, Italy, 17–19 November 1982).

Klnna, J.C. 'Investing in Developing Countries: Minimisation of Political Risk'. *Journal of Energy and Natural Resources Law* 1982.
Konoplyanik, A. 'Time to Harmonise Russian Oil Law'. *Petroleum Economist*, April 1992.
Krapels, E.N. 'Evolution and Reform: The Art of Transition in Energy Pricing Policy'. 3 *Natural Resources Forum*, vol. 12, 1988.
Kuehne, G. 'Oil and Gas Licensing: Some Comparative United Kingdom-German Aspects'. 3 *Journal of Energy and Natural Resources Law*, vol. 4, 1986.
Kuehne, G. 'The Regulation of Energy Supply: Federal Republic of Germany'. *Journal of Energy and Natural Resources Law*, Supplement, December 1989.

Lauterpacht, E. 'Issues of Compensation and Nationality in the Taking of Energy Investments'. 4 *Journal of Energy and Natural Resources Law*, vol. 8, 1990.
Lavenant, R.P. 'Changing Concepts of the World's Mineral Development Laws with Special Reference to West Africa'. In Proceedings of the IBA Seminar on World Energy Laws (Stavanger: Norway, May 1975).
Leckow, R.B. & L.A. Mallory. 'The Relaxation of Foreign Investment Restrictions in Canada'. *ICSID Review-Foreign Investment Law Journal* 1991.
Lenczowski, G. *Oil and State in the Middle East*. Cornell University Press: New York, 1960.
Lichtblau, J.H. 'OPEC's Challenge: Surviving Until the Mid-1990s'. *Petroleum Intelligence Weekly*, 13 May 1985.
Lukman, R. 'OPEC's Efforts to Stabilise the Oil Price Structure'. In Proceedings of the Mining Students Society's Symposium on the Current Political and Economic Outlook for the Mining and Petroleum Industry (Delft University of Technology, Delft, November 1987).

Mabro, R. 'Oil Prices in the 1990s: It all Depends on Capacity'. *Middle East Economic Survey*. 6 November 1989.
MacDonald, D.S. 'Current Developments in Oil and Gas Law, Canada'. In Proceedings of the IBA Energy Law Seminar, 1981.
Mahaffie, C.D. 'The International Energy Agency in Operation'. *Houston Law Review*, vol. 17:961, 1980.
Manly Horton, Jr, C. 'Effect of Conservation Laws and Regulations on Contracts and Mineral Leases'. *Ralph Slovenko's Oil and Gas Operations: Legal Considerations in the Tidelands and on Land* (Tulane University School of Law, Claitor's Law Books, Claitor's Book Store, 1963).
Martin, J.M. 'From the Erosion of Crude Oil Prices to the Re-organisation of the Oil Industry'. *Natural Resources Forum*, May 1989.
Mcclendonj, S. 'Alternate Methods of Dispute Settlement'. *Houston Law Review*, vol. 17:979, 1980.
McHugh,J. 'Farmout Agreements'. In Proceedings of the Conference on Oil and Gas Agreements (Langham Oil Conferences Ltd, London, December 1994).
Mestad, O. *The Ekojisk Royalty Case: Construction of Regulations to Avoid Retroactivity*. (Arbeidsdokument til Solstrandseminaret, 2/3 December 1987).

Mikdashi, Z. 'Global Oil: Some Key Issues and Challenges'. *Natural Resources Forum*, August 1989.
Miller, P.P. 'The Statfjord and Murchison Oilfield Units'. In Proceedings of the IBA Energy Law Seminar. vol. I, 1979.
Miller, P.P. 'Comparison of Terms to Attract Petroleum Development in Four Countries: Indonesia'. In Proceedings of the IBA Energy Law Seminar, vol. II, 1979.
Mitchell, J.G. 'Oil on Ice'. (1997) 4 *National Geographic*, vol. 191, April 1997.
Morrison, M.R. 'Current Developments in Petroleum Exploration Offshore Malaysia'. In Proceedings of the IBA Energy Law Seminar, vol. I, 1979.
Mouton, H.W. & others. 'Deep-Sea Mining', A Report of the Deep-Sea Mining Committee of the International Law Association. Proceedings of the ILA Buenos Aires Conference, 1968.

Nazer, H.M. 'Market Stability and Energy Security'. *Middle East Economic Survey*, 6 November 1989.
Nazer, H.M. 'The Interdependence of Oil Producers and Consumers in the Oil Market'. *Middle East Economic Survey*, 3 June 1991.
New Scientist. 'Greenhouse Wars'. *New Scientist*, 19 July 1997.

Oddie. 'The Technical Basis of Unitisation'. In Proceedings of the Conference on Unitisation of Oil and Gas Fields (Langham Oil Conferences Ltd, London, May 1991).
OECD. Committee on Fiscal Affairs. *Transfer Pricing and Multinational Enterprises*. OECD (Paris, 1979).
Omorogbe, Y. 'The Legal Framework for the Production of Petroleum in Nigeria'. 4 *Journal of Energy and Natural Resources Law*, vol. 5, 1987.
Outhit, P.A. 'The Frigg Gas Field Unit'. In Proceedings of the IBA Energy Law Seminar, vol. I, 1979.

Parra, A.A. 'OPEC and Market Share: A Change of Direction'. *Middle East Economic Survey*, 26 May 1986.
Patterson, W. 'Fifty Years of Hopes and Fears'. *Nature*, vol. 449. 11 October 2007.
Pletrowski, R.E. 'International Law Applicable to Deep Sea Mining'. In Proceedings of the IBA Energy Law Seminar, vol. I, 1979.
Plnot, L. & J-M. Le. 'Petroleum Exploration and Production in France'.
Petroleum Economist, Special Supplement, September 1991.
Prahalad, C. & Y. Doz. *Responding to Host Government Policies*, *The Multinational Mission* (The Free Press: New York, 1987).

Raschke, M.G. 'World Coal Outlook: 1990s'. *Natural Resources Forum*, August 1989.
Read, A.D. *Protection of the Marine Environment – An Industry View* (E & P Forum: London, September 1989).
Record, R.H. 'Current Developments in Oil and Gas Law in the United States of America'. In Proceedings of the IBA Energy Law Seminar, 1981.

Reddy, A.K.N. & J. Goldenberg. 'Energy for the Developing World'. *Scientific American*, September 1990.

Robinson, M.S. 'Anatomy of Downstream Integration by Oil-Producing Countries'. *Natural Resources Forum*, February 1989.

Roggenkamp, M.M. *The Legal Framework of Pipelines in the Oil and Gas Industry*. Doctorate dissertation (Leyden University, Intersentia Law Publishers: Leyden, 3 February 1999).

Roland, A. 'Secrecy, Technology, and War: Greek Fire and the Defense of Byzantium, 678-1204'. *Society for the History of Technology*, 1992.

Ronne, A. 'The Regulation of Energy Supply: Denmark'. *Journal of Energy and Natural Resources Law*, Supplement, December 1989.

Rugman, A.M. et al. *International Business, Firm and Environment* (McGraw Hill International Editions: Singapore, 1986).

Salden, K. *Legal Aspects of State Jurisdiction over Offshore Oil and Gas Pipelines with Particular Reference to the ^eepipe Project* (Leyden University: December 1988).

Schneider, S.H. 'The Changing Climate'. *Scientific American*, September 1989.

Shatalov, A.T. 'The Gas Industry of Russia-Modern Conditions and Perspectives'. In Proceedings of the Mining Students Society's Symposium on Potential Resources for the Coming Century (Delft University of Technology: Delft, November 1992).

Schenck, PA. *The Past, the Present, the Future; Fossil & Contemporary*. (Delft University of Technology: Delft, 1993).

Scherer, H.J. 'Strict Liability for Environmental Damage in Germany'. *International Business Lawyer*, June 1991.

Schierbeek, P. *Mining Law and Policy* (Uitgeverij Waltman: Delft, February 1965).

Schierbeek, P. *Mine and Tours* (Delft University Press: Delft, May 1980).

Schneider, D. 'The Rising Seas'. *Scientific American*, March 1997.

Schneider, D. 'Burying the Problem'. *Scientific American*, January 1998.

Seymour, A. & N. Antill. 'OPEC Revenues and Foreign Investments'. *Middle East Economic Survey*, 14 December 1998.

Seymour, I. 'OPEC Extends Economic Olive Branch to Western World'. *Middle East Economic Survey*, 31 January 1975.

Shaw, S. Joint Operating Agreements'. In Proceedings of the Conference on Oil and Gas Agreements (Langham Oil Conferences Ltd, London, December 1994).

Shihata, I.F.I. 'The Multilateral Investment Guarantee Agency (MIGA)'. In Proceedings of the Mining Students Society's Symposium on the Current Political and Economic Outlook for the Mining and Petroleum Industry (Delft University of Technology, Delft, November 1987).

Sle, S.T. *Black Gold, White Gold and the Philosophers' Stone* (Delft University of Technology: Delft, February 1992).

Sle, S.T. *Catalysis and Reactor Design* (Delft University of Technology: Delft, 1997).

Slater, J. 'Recent Developments in United Kingdom North Sea Law'. vol. 4 (iii) *International Business Lawyer*, 1976.
Sliepcevich, C.M. 'Marine Transportation of Natural Gas'. *The Exploration and Economics of the Petroleum Industry*, vol. 5 (The Southwestern Legal Foundation, Gulf Publishing Company, Houston, TX, 1967).
Stone, R. 'The Long Shadow of Chernobyl'. *National Geographic*, April 2006.
Strongman, J.E. 'The Evolving Role of the World Bank Strategies to Attract Private Mining Investments'. In Proceedings of the Mining Students Society's Symposium on Potential Resources for the Coming Century (Delft University of Technology: Delft, November 1992).
Suplee, C. 'Unlocking the Climate Puzzle'. 5 *National Geographic*, vol. 193, May 1998.
Symonds, E. 'Investment without Risk?'. *Petroleum Economist*, May 1990.

Taverne, B.G. 'Methods of Participation of Host Countries in Crude Oil Exploration and Production Ventures in the Middle East and Northern Africa'. In Proceedings of the IBA Seminar on World Energy Laws, 1975.
Taverne, B.G. 'Concessions and New Types of Exploration/Production Contracts'. In Proceedings of the IBA Energy Law Seminar, vol. I, 1979.
Taverne, B.G. *Government, Public Enterprise and EC Law* (Leyden University, August 1987).
Taverne, B.G. 'Mineral Resources in Contested Boundary Areas'. In Proceedings of the Conference on International Boundaries (The Royal Institute of International Affairs, London, January 1992).
Taverne, B.G. *Principles of Petroleum Regulation and Government Policy* (Delft University Press: Delft, 1993).
Taverne, B.G. *An Introduction to the Regulation of the Petroleum Industry (Laws, Contracts and Conventions)* (Kluwer Law International: London/The Hague/Boston, December 1994).
Taverne, B.G. 'Production Sharing Agreements'. In Proceedings of the Conference on Oil and Gas Agreements (Langham Oil Conferences Ltd, London, December 1994/November 1996).
Taverne, B.G. *Co-operative Agreements in the Extractive Petroleum Industry* (Kluwer Law International: London/The Hague/Boston, December 1996).
Taverne, B.G. 'Production Sharing Agreements in Principle and in Practice'. Chapter 4 of *Upstream Oil and Gas Agreements* (Sweet & Maxwell: London, 1996).
Tayler, M.P.G. 'The Legal Background and Structure of Unitisation Agreements'. In Proceedings of the Conference on Unitisation of Oil and Gas Fields (Langham Oil Conferences Ltd, London, May 1991).
Thornley, N. 'Gas Transportation Agreements'. In Proceedings of the Conference on Oil and Gas Agreements (Langham Oil Conferences Ltd, London, November 1996).

Trimble, N. 'Gas Sales Agreements'. In Proceedings of the Conference on Oil and Gas Agreements (Langham Oil Conferences Ltd, London, November 1996).

Tucker, F.S. 'Bright Future for Coal'. *Petroleum Economist*, October 1989.

United Nations, Department of Technical Co-operation for Development. 'Petroleum Exploration Strategies in Developing Countries'. In Proceedings of a United Nations Meeting held in The Hague, March 1981 (Graham & Trotman Ltd, London, 1982).

Van De Vljver, J.J. et al. *Abandonment Obligations (Removal of Offshore Structures).* Sub-committee D of the E & P Forum, London, August 1985.

Van Der Llnde, C. *Dynamic International Oil Markets* (Kluwer Academic Publishers: Dordrecht, 1991.

Vareberg, T.V.A. 'Current Prospects for the European Oil and Gas Market'. In Proceedings of the Mining Students Society's Symposium on Potential Resources for the Coming Century (Delft University of Technology, Delft, November 1992).

Verloren Van Themaat, P. *The Legal Fundament of the New International Economic Order*, Studies of International Economic Law, Part II (T.M.C. Asser Institute: The Hague, 1979).

Vintner, G. 'EC Aspects'. *Petroleum Economist*, Special Supplement, September 1991.

Vock, R.D. 'Petroleum Development Agreements: Form and Drafting'. 4 *Journal of Energy and Natural Resources Law*, vol. 8 1990.

Walde, T. 'Third World Mineral Development: Recent Issues and Literature'. *Journal of Energy and Natural Resources Law*, 1984.

Walde, T. 'Mineral Development Legislation: Result and Instrument of Mineral Development Planning'. 2 *Natural Resources Forum*, vol. 12, 1988.

Waldock, C.H.M. 'The Legal Basis of Claims to the Continental Shelf'. Transactions of the Grotius Society, vol. 36, reproduced in the Proceedings of the ILA Copenhagen Conference on the Rights to the Seabed and its Subsoil (1950).

Walser, C.H. 'Multilateral Development Assistance for Petroleum Exploration and Production'. In Proceedings of the IBA Energy Law Seminar, vol. I, 1979.

Weisenborn, C.M.M. *The Obligatory Guarantee Agreement* (Law Faculty, Leyden University, December 1992).

Willheim, E. 'Australia-Indonesia Sea-Bed Boundary Negotiations – Proposals for a Joint Development Zone in the "Timor Gap"'. In Proceedings of the Seminar on Australia and the Law of the Sea (Sydney February 1987).

Wocol. *Coal – Bridge to the Future* (Ballinger Publishing Company: Cambridge, MA, 1980).

Wood, W.A. 'Legal Aspects of Foreign Investment in Oil and Gas Exploration and Development in Brazil'. *Journal of Energy and Natural Resources Law* 1989.

Woodliffej. C. 'International Unitisation of an Offshore Gas Field'. 2 *International and Comparative Law Quarterly*, April 1977.

Wright, C.J. The Role of the Industry in the IEA'. In Proceedings of the IBA Energy Law Seminar, vol. I, 1979.

Wright, M. 'Recent Decisions on the Law of Oil and Gas'. *Ralph Slovenko's Oil and Gas Operations: Legal Considerations in the Tidelands and on Land* (Tulane University School of Law, Claitor's Law Books, Claitor's Book Store, 1963).

Wyatt, D. & A. Dashwood. *The Substantive Law of the EEC* (Sweet & Maxwell: London, 1980).

Yamani, A.Z. 'The Influence of Political Decisions on Oil Economics'. *Middle East Economic Survey*, 4 July 1988.

Yamani, A.Z. 'American Dependence on Arab Oil in the Years to Come'. *Middle East Economic Survey*, 27 November 1989.

Yergin, D. *The Prize; the Epic Quest for Oil, Money and Power*. Simon & Schuster: New York, 1991.

Zakariya, H.S. 'Political Risks of Transnational Petroleum Investment: The Mitigating Role of National and International Insurance Programmes'. 2 *Natural Resources Forum*, vol. II, 1987.

Zillman, D.N. & W. Fox. 'The Regulation of Energy Supply: The United States'. *Journal of Energy and Natural Resources Law*, Supplement, December 1989.

Zillman, D.N., A.R. Lucas & G. Pring. *Human Rights in Natural Resource Development, Public Participation in the Sustainable Development of Mining and Energy Resources* (Oxford University Press, 2002).

Sources

In this book the following sources have been used:

(a) In respect of the contents of legal documents such as conventions, protocols, charters, laws, regulations, (model) contracts and petroleum agreements, as mentioned, cited and discussed throughout the text:
   (i) legal texts contained in official publications of national governments or international organizations (State Gazette, Official Gazette, Official Journal, etc.); or
   (ii) original or English translations of legal texts placed on websites of the internet by government Ministries or Departments, United Nations Organizations, OPEC or by interested non-governmental organizations; or
   (iii) legal texts found reproduced (and translated into English) in non-governmental, commercial publications, such as specialized magazines (Petroleum Times) or Barrows' Basic Oil Laws and Concession Contracts.
(b) In respect of statistical information presented in the Tables and throughout the text:
   (i) BP Statistical Review of World Energy 2007;
   (ii) *World Energy Outlook 2006* and *World Energy Outlook 2007*, publications of the International Energy Agency (OECD/IEA), Information Office, Paris, France;
   (iii) Statistical Pocketbooks 2001–2006, publications of the European Commission;
   (iv) *Middle East Economic Survey* (MEES);
   (v) *PEPS*, IHS Inc. – Energy, country analyses;
   (vi) US Department of Energy's Energy Information Administration (EIA), OPEC Brief; and
   (vii) home page (internet) of the Organization of Petroleum Exporting Countries (OPEC).

(c) In respect of information on scientific, technical, historical and government policy facts and matters:
   (i) IPCC WG I Assessment Reports 1990-2007;
   (ii) UN-Energy Report: *Sustainable Bioenergy. A Framework for Decision Makers*, New York, 8 May 2007;
   (iii) information placed by institutions, government organizations, interested parties, etc. on the website of Wikipedia, the free encyclopedia;
   (iv) policy papers and annual reports of the European Commission;
   (v) Robert Peele's Mining Engineers' Handbook, volume I;
   (vi) general and specialized trade journals of international reputation: the *Petroleum Economist, The Economist, Russian Petroleum Investor, Petroleum Intelligence Weekly, Oil and Gas Journal, Energy Law, Platt's Oilgram*, etc.;
   (vii) science publications, such as *Nature and Science*; and
   (viii) international magazines of general interest, such as *Scientific American* and the *National Geographic.*

Index

INTERNATIONAL ENERGY AND RESOURCES LAW AND POLICY SERIES

Other titles in this series:

1. David S. Macdougall (ed.), *European Community Energy Law*, 2001 (ISBN 1-85333-962-8).
2. David Dezhi Peng (ed.), *Insurance and Legal Issues in the Oil Industry*, 2001 (ISBN 1-85333-913-X).
3. Ernst J. Mestmäcker (ed.), *Natural Gas in the Internal Market*, 2001 (ISBN 1-85333-795-1).
4. Walied M.H. El-Malik, *Minerals Investment under the Shari'a Law*, 2001 (ISBN 1-85333-907-5).
5. Thomas W. Wälde & George K. Ndi (eds), *International Oil and Gas Investment*, 2001 (ISBN 1-85333-963-6).
6. Bernard G. Taverne, *An Introduction to the Regulation of the Petroleum Industry*, 2001 (ISBN 1-85966-081-9).
7. James Otto, *Taxation of Mineral Enterprises*, 2001 (ISBN 1-85966-105-X).
8. Zhiguo Gao (ed.), *International Petroleum Contracts*, 1994 (ISBN 1-85966-103-3).
9. Bernard G. Taverne, *Co-operative Agreements in the Extractive Petroleum Industry*, 1996 (ISBN 90-411-0926-9).
10. Thomas W. Wälde (ed.), *The Energy Charter Treaty: An East-West Gateway for Investment An*, 1996 (ISBN 90-411-0913-7).
11. Zhiguo Gao (ed.), *Environmental Regulation of Oil and Gas*, 1998 (ISBN 90-411-0726-6).
12. Gerald B. Greenwald (ed.), *Liquefied Natural Gas: Developing and Financing International Energy Projects*, 1998 (ISBN 90-411-9664-1).
13. Danièle Barberis, *Negotiating Mining Agreements: Past Present & Future Trends*, 1998 (ISBN 90-411-9673-0).
14. Nathalie L.T.J. Horbach, *Contemporary Developments in Nuclear Energy Law*, 1999 (ISBN 90-411-9719-2).
15. Bernard G. Taverne, *Petroleum, Industry and Governments: A Study of the Involvement of Industry and Governments in the Production and Use of Petroleum*, 2nd edition, 2008 (ISBN 978-90-411-2663-4).
16. Michael A.G. Bunter, *The Promotion and Licensing of Petroleum Prospective Acreage*, 2002 (ISBN 90-411-1712-1).
17. Morakinyo Adedayo Ayoade, *Disused Offshore Installations and Pipelines*, 2002 (ISBN 90-411-1739-3).
18. Malgosia Fitzmaurice & Milena Szuniewicz (eds), *Exploitation of Natural Resources in the 21st Century*, 2003 (ISBN 90-411-2063-7).
19. Philip Andrews-Speed, *Energy Policy and Regulation in the People's Republic of China*, 2004 (ISBN 90-411-2233-8).
20. Ilias Bantekas, John Paterson & Maidan K. Suleimenov (eds), *Oil and Gas Law in Kazakhstan: National and International Perspectives*, 2004 (ISBN 90-411-2250-8).

21. Elizabeth Bastida, Thomas Wälde & Janeth Warden (eds), *International and Comparative Mineral Law and Policy*, 2005 (ISBN 90-411-2116-1).
22. Janusz Bielecki & Melaku Geboye Desta (eds), *Electricity Trade in Europe: Review of Economic and Regulatory Challenges*, 2004 (ISBN 90-411-2279-6).